Emerging Technologies
for In Situ Processing

NATO ASI Series

Advanced Science Institutes Series

A Series presenting the results of activities sponsored by the NATO Science Committee, which aims at the dissemination of advanced scientific and technological knowledge, with a view to strengthening links between scientific communities.

The Series is published by an international board of publishers in conjunction with the NATO Scientific Affairs Division

A	Life Sciences	Plenum Publishing Corporation
B	Physics	London and New York
C	Mathematical and Physical Sciences	D. Reidel Publishing Company Dordrecht, Boston, Lancaster and Tokyo
D	Behavioural and Social Sciences	Martinus Nijhoff Publishers
E	Applied Sciences	Dordrecht, Boston and Lancaster
F	Computer and Systems Sciences	Springer-Verlag
G	Ecological Sciences	Berlin, Heidelberg, New York, London,
H	Cell Biology	Paris and Tokyo

Series E: Applied Sciences – No. 139

Emerging Technologies for In Situ Processing

edited by

Daniel J. Ehrlich

Lincoln Laboratory,
Massachusetts Institute of Technology,
Lexington, Massachusetts, U.S.A.

and

Van Tran Nguyen

Director/Physics and Technologies Research Division,
CNET- Grenoble, France

1988 **Martinus Nijhoff Publishers**
Dordrecht / Boston / Lancaster
Published in cooperation with NATO Scientific Affairs Division

Proceedings of the NATO Advanced Research Workshop on
Emerging Technologies for In Situ Processing
Cargèse, France
May 4–8, 1987

Library of Congress Cataloging in Publication Data

Nato Advanced Research Workshop on "Emerging Technologies for In Situ
  Processing" (1987 : Cargèse, Corsica)
   Emerging technologies for in situ processing / edited by Daniel J.
Ehrlich and Van Tran Nguyen.
       p.   cm. -- (NATO ASI series. Series E, Applied sciences ; no.
139)
   "Proceedings of the NATO Advanced Research Workshop on 'Emerging
Technologies for In Situ Processing,' Cargèse, France, May 4-8,
1987"--T.p. verso.
   "Published in cooperation with NATO Scientific Affairs Division.
   Includes bibliographies and index.

   1. Integrated circuits--Very large scale integration--Design and
construction--Congresses.   I. Ehrlich, Daniel J. II. Nguyen, Van
Tran. III. North Atlantic Treaty Organization. Scientific Affairs
Division. IV. Title. V. Series.
TK7874.N334 1987
621.395--dc19   88-15503
                                                          CIP

ISBN-13: 978-94-010-7130-7   e-ISBN-13: 978-94-009-1409-4
DOI:10.1007/978-94-009-1409-4

Distributors for the United States and Canada: Kluwer Academic Publishers, 101
Philip Drive, Norwell, MA 02061, USA

Distributors for the UK and Ireland: Kluwer Academic Publishers, MTP Press Ltd,
Falcon House, Queen Square, Lancaster LA1 1RN, UK

Distributors for all other countries: Kluwer Academic Publishers Group, Distribution
Center, P.O. Box 322, 3300 AH Dordrecht, The Netherlands

TABLE OF CONTENTS

PREFACE   ix

1. H. Ahmed
    In-Situ Processing Combining MBE, Lithography and
    Ion-Implantation   1

2. I. Hayashi
    Motivations and Early Demonstrations for In-Situ
    Processings For III-V Semiconductor Devices   13

3. John J. Ritsko
    Laser Etching and Microelectronic Applications   23

4. D. Bauerle, T. Szorenyi, G.Q. Zhang, K. Piglmayer,
    M. Eyett, R.Kullmer
    Laser-Induced Chemical Processing of Materials   33

5. E.L. Hu
    High Technology Manufacturing : Critical Issues for the
    Future   45

6. J.P. Harbison, P.F. Liao, D.M. Hwang, E. Kapon,
    M.C. Tamargo, G.E. Derkits Jr. and J. Levkoff
    Ultra High Vacuum Processing : MBE   55

7. P.N. Favennec, H. L'Haridon, M. Salvi, L. Henry,
    A. Le Corre, D. Lecrosnier, M.A. Di Forte Poisson,
    J.P. Duchemin
    Epitaxial Growth of III-V Materials on Implanted III-V
    Substrates   61

8. R.L. Jackson, T.T. Kodas, G.W. Tyndall, T.H. Baum and
    P.B. Comita
    Mechanisms of Laser-Induced Deposition from the Gas Phase   71

9. D. Braichotte and H. Van Den Bergh
    Time Resolved Measurements in the Thermally Assisted
    Photolytic Laser Chemical Vapor Deposition of Platinum   83

10. M. Rothschild and D.J. Ehrlich
    Excimer Laser Projection Patterning   93

11. P.H. Key, P.E. Dyer, R.D. Greenough
    Excimer Laser Patterning and Etching of Metals   105

12. G. Stengl, H. Loschner, E. Hammel, E.D. Wolf, J.J. Muray
    Ion Projection Lithography   113

13. J.E. Bouree and J. Flicstein
    UV Light-Assisted Deposition of Al on Si from TMA   121

14. M. Green, C. Aidinis and O. Fakolujo
    E-Beam Induced Decomposition of Inorganic Solids   131

vi

15. T. Dupeux, P. Deroux-Dauphin, G. Nicolas
    Electronic Connection Through Silicon Wafers   137

16. C.J. Brierley, F.W. Ainger, C. Trondle
    The Development and Use of Novel Precursors for
    Photolytic Deposition of Dielectric Films   145

17. J. Melngailis, A.D. Dubner, J.S. Ro, G.M. Shedd, H. Lezec
    and C.V. Thompson
    Focused Ion Beam Induced Deposition   153

18. R. Putzar, H.C. Petzold, U. Weigmann
    Laser-Induced Metal Deposition for Clear Defect Repair
    Work on X-Ray Masks   163

19. I.W. Boyd and F. Micheli
    Confirmation of the Wavelength Dependence of Silicon
    Oxidation Induced By Visible Radiation   171

20. A.J. Steckl, J.C. Corelli, J.F. McDonald
    Focused Ion Beam Technology and Applications   179

21. G. Auvert, Y. Pauleau and D. Tonneau
    Laser Direct Writing for Device Applications   201

22. J.L. Peyre, D. Riviere, Ch. Vannier and G. Villela
    Laser-Induced Photoetching of Semiconductors with
    Chlorine   213

23. S.J.C. Irvine, M.C. Ward and J.B. Mullin
    Recent Advances in Photo-Epitaxy for Infrared Detector
    Fabrication   221

24. E. Borsella, L. Caneve, R. Fantoni
    Synthesis and Characterization of Laser Driven Powders   233

25. C. Arnone and C. Zizzo
    Physical Properties of Laser Written Chromium Oxide Thin
    Films   241

26. E. Fogarassy, A. Slaoui, C. Fuchs
    UV Laser Induced Oxidation of Silicon in Solid and Liquid
    Phase Regime   249

27. O. Chevrier and A. Boudou
    Laser Assisted Plasma Etching of Silicon Dioxide   257

28. S. Leppavuori, J. Lenkkeri and J. Levoska
    Nd : Yag Laser Processing for Circuit modification and
    Direct Writing of Silicon Conductors   265

29. T. Baller, G.N.A. Van Veen and J. Dieleman
    Study of Excimer laser Enhanced Etching of Copper and
    Silicon With (SUB) Monolayer Coverages of Chlorine   273

30. R.B. Jackman and J.S. Foord
Surface Chemical Probes and Their Application to the
Study of In Situ Semiconductor Processing   279

Subject Index   289

PREFACE

The Workshop on Emerging Technologies for in situ Processing was held in Cargese, France on the dates 4-8 May 1987. The site was the Institute d'Etudes Scientifiques de Cargese of the University of Nice. More than seventy scientists and engineers working on microelectronics technology and semiconductor physics were in attendance. Principal support was provided by the Advanced Research Workshop Program of the North Atlantic Treaty Organization (NATO) under the administration of Drs. Di Lullo and Sinclair. Additional technical stimulus was through the NATO RSG II Group.

The ideas of in situ processing of microelectronics are based on the new and potentially revolutionary nonlithographic technologies which may permit the fabrication of semiconductor devices by dramatically simpler and more precise methods than those of current technology. A heavy reliance is on partially yet-to-be-invented beam-controlled techniques for microfabrication. Although fragments of the necessary understanding and technology had begun to appear in the scientific literature at the time of the Workshop, no previous international meeting had been organized on the topic. The workshop was organized to focus thinking on the key problems and potential implications of such a technology.

The underlying principle of in situ processing is the replacement of current indirect photoresist microfabrication sequences with direct photon-electron, and ion-beam-controlled techniques based on microscopic reactions on the semiconductor. This approach, in principle, permits an elimination of the current solvent and air exposure, known to degrade critical electronic surfaces in resist processing. One possibility that has been raised is the complete processing of electronic devices within a single advanced vacuum chamber, in something of an extension of current molecular beam epitaxy technology. A second possibility, already partly realized, is that of "real-time" fabrication (again without photoresists) for the purpose of highly precise "closed-loop" construction of microdevices. An incentive here is to streamline conception, design and fabrication sequences to allow more efficient realization of new device ideas. Current fabrication technology is ponderously slow, inordinately expensive for phototyping, and known to be a limitation in new device and circuit design.

The operating thesis of in situ processing is to combine various existing and future beam technologies with compatible current non-beam methods to achieve totally vacuum and/or real-time fabrication. The workshop was organized to discuss the current state of affairs of the relevant component technologies and identify the needs for basic and applied research. Speakers were asked to specifically discuss the potential "revolutionary" implications of in situ processing and to point to critical obstacles to such technology. In all, 34 papers were presented.

The first day of the workshop was composed of a series of widely ranging invited papers designed to define the scope of the meeting. Prof. Haroon Ahmed (Cambridge University) discussed "In Situ Processing Systems Combining MBE, Lithography, and Ion Implantation". Dr. Izuo Hayashi (The Optoelectronics Joint Research Project) then spoke on "Motivation and Early Demonstrations For In Situ processing of III-V Compounds". Dr. John Ritsko (IBM) then presented a summary of "Laser Etching and Microelectronic Applications", and Prof. Dieter Bauerle (Johannes-Kepler University) spoke on "Chemical Processing With Lasers". A review of "High Technology Manufacturing : Critical Issues for the Future" was presented by Prof. Evelyn Hu (University of California - Santa Barbara).

On the second day overview papers were given by Drs. Robert Jackson (IBM) on "Laser Deposition" and Mordechai Rothschild (MIT Lincoln Laboratory) on "Eximer Laser Projection Technology".

On the third day, Prof. Andrew Steckl summarized "Focused Ion Beam Technology and Applications" and Dr. Geoffroy Auvert (CNET Grenoble) discussed "Laser Direct Writing for Device Applications".

On the fourth day, invited papers were presented by Dr. James Harbison (Bellcore) on "Ultrahigh Vacuum Processing MBE" and by Dr. Stuart Irvine (RSRE-Malvern) on "Recent Advances in Photo-Epitaxy for Infrared Detector Fabrication".

The final day of the workshop was highlighted by an invited paper delivered by Prof. Richard Jackman (Univ. of Oxford) on "Surface Chemical Probes and their Applications to the Study of In-Situ Semiconductor Processing".

The Conference Chairmen wish to express their deeply felt thanks to a few who made critical contributions in the organization of the workshop : Foremost of these is Dr. Josselyne de Montlaur of the DRET, who provided extensive advice and pragmatic help. Critical help was also received from Dr. John Melngailis of MIT. We thank the session chairmen H. Ahmed, M. Green, J. Haigh, K. Malloy and R. Reynolds for their expertise in orchestrating the discussion and Mmes Annie Bernard, Annick Raharijaona, Brigitte Sohy, Michèle Thedrel and Mme Marie France Hanseler for their assistance. We also thank Drs. Brian Holeman and Jack Kennedy of NATO for providing some of the initial stimulus. Finally we wish to express our deep appreciation for the encouragement and support of Dr. Di Lullo who died during the time of the early planning of the workshop.

Daniel J. Ehrlich   Van Tran Nguyen

Workshop Chairmen

IN-SITU PROCESSING COMBINING MBE, LITHOGRAPHY AND ION IMPLANTATION

H. AHMED
Microelectronics Research Group, Department of Physics, Cambridge University,
Cambridge Science Park, Milton Road, Cambridge CB4 4FW.

1. INTRODUCTION

The current interest in the physics of low dimensionality has required a re-examination of the techniques for microfabrication developed originally for the fabrication of semiconductor integrated circuits. The new requirements place great emphasis on the growth of layers of high purity, crystalline perfection within the layer, a very small degree of crystalline mismatch between layers and controlled and predictable surface properties. Additionally, stringent requirements have been placed on layer thickness, layer composition and electrical, optical and mechanical properties. At present the only viable means for producing such layers is molecular beam epitaxy (MBE).

At the same time it has become necessary to fabricate in these layers structures with lateral dimensions that are extremely small, similar to such physical quantities as mean-free paths of electrons in solids, scattering distances of electrons in semiconductors, coherence lengths in superconductors, optical wavelengths, and typical electron-electron separations in 2-dimensional electron gases (2DEG). In order to make such structures it is necessary to fabricate artefacts with dimensions in the range 1nm to 100nm. Conventional techniques of optical and electron lithography are not viable and new lithographic techniques have to be used. The most promising methods are based on high voltage electron lithography and focused ion beam lithography.

In certain cases there is also a need for the etching of features, thermal treatment of layers and contact formation which again demand higher specifications in processing equipment than are needed for conventional semiconductor device and circuit processing. Finally, such practical considerations as levels of cleanliness and toxicity of materials must be taken into account. These requirements have led to the concept of fully integrated fabrication systems in which layer growth is combined with lithography, implantation, annealing, etching and overgrowth of surfaces before the substrate is removed from the vacuum environment.

A typical specification for such a concept is based on what has been achieved in different aspects of microelectronics. For example, minimum superlattice layer thicknesses less than 1nm [1], implanted layers of B and As confined to 20 nm [2], free-standing wires of 50 nm diameter [3], nanolithography with line widths down to 10nm [4,5], GaAs FET gate lengths less than 100nm and Aharonov-Bohm loops within 80nm have been produced. The next step is to carry out the processing *in situ* so that the properties of both layers and structures do not deteriorate by exposure to air or other gases between fabrication steps. Reports indicate that such exposure can degrade the surface properties of semiconductors so that there is a decrease in the carrier density at the surface, and the surface resistivity becomes higher than in the bulk [6]. Deep level electron traps and reversal in semiconductor polarity can also take place at the surface. After overgrowth such defective layers are buried and carrier transport properties changed. Evidence from optical measurements of photoluminescence from layers which have not been exposed after growth indicate that the material has superior properties. In addition to the physical properties of materials and devices there are practical advantages with *in-situ* processing systems in that they do not require clean rooms with high specifications and operators are less likely to be exposed to risks.

With *in-situ* systems which meet the specifications that have been proposed here it should be possible to explore such phenomena as transport in one-dimensional conductors, properties of 2-D electron gases with lateral confinement, short-channel ballistic effects, band-gap tuning, lateral size quantization effects and many other areas of low-dimensionality physics.

1

D. J. Ehrlich and V. T. Nguyen (eds.), Emerging Technologies for In Situ Processing, 1–11.
© *1988 by Martinus Nijhoff Publishers.*

2. COMBINED LAYER GROWTH AND MICROSTRUCTURE FABRICATION SYSTEMS

2.1 Layer growth

Combined in-situ processing systems for microelectronics have to provide the following facilities:

a. means for layer growth with compositional changes and uniform doping
b. means for forming latent images in layers to delineate patterns
c. means for developing the latent images to form microstructures
d. means for overgrowth on layers once microstructures have been formed
e. repetition of the above processes as required to complete a multi-layer, three-dimensional device or structure.

The methods available for layer growth are MBE systems, evaporation systems and more esoteric systems such as ion cluster beams. For investigation of new physics phenomena MBE is the most appropriate system. Layers of bulk GaAs may be grown with very low background impurity levels and doped in stages to give high mobility layers of n-type material with precisely controlled carrier concentration. Compositional changes can be abruptly incorporated by the production of layers of $Al_xGa_{1-x}As$ with x varied over a wide range. Careful control enables AlGaAs/GaAs heterostructures to be formed giving 2DEG layers of precise width, carrier concentration and mobility. The properties of MBE-grown material are now well established in many research centres.

2.2 The unique features of focused ion beams for *in-situ* processing

Focused ion beams can be used for a number of tasks:

a. lithography with ion-sensitive coatings on substrates
b. implantation in localized regions
c. sputtering in selected areas
d. isolation by crystalline damage around active regions
e. selective deposition of material from gases
f. inspection of surface topography or surface distribution
g. micro-sectioning of structures and devices
h. micro-analysis of surface layers

Probes may be focused to diameters less than 0.05μm, with sufficient current to carry out processes in time-scales which are practicable for special device structures. The processes are not suitable for mass production but are ideally compatible with *in-situ* processing in ultra-clean environments.

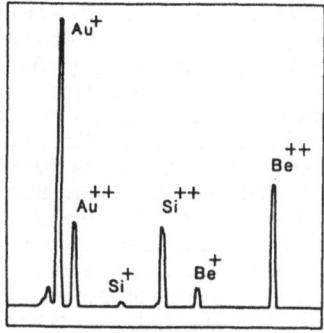

Fig. 1 Spectrum of ion emission from gold-silicon-beryllium (70-15-15 atomic %) liquid-metal field-ion source.

2.2.1 <u>Sources and optics for focused ion beams</u>   Sources for focused ion beams have been available for some time and single element sources with restricted applications have now been supplemented by alloy sources from which a large number of useful dopant species for microelectronics are available.

The principal single element source is gallium which is a p-type dopant in Si but which has limited solubility and low penetration (0.06μm at 100keV). It is, however, compatible with GaAs substrates and also useful for several other applications. Multiple-element sources from alloys are used in combination with mass filters which enable the desired species to be selected. For applications in GaAs processing the Au-Si-Be source is important [7]. It gives both the p-type dopant Be and the n-type dopant Si conveniently out of the same source. The source characteristic in Fig. 1 shows that a high output of doubly charged ions is available which enables the penetration depth to be increased (into GaAs, the ranges at 100kV for Be^{++} and for 100kV Si^{++} are approximately 0.2μm and 0.6μm respectively). For applications in silicon technology the B-Ni-Pt source has been developed [8] and other work has reported As sources as well as a number of other species which have useful applications in processing technology for devices.

The most significant requirements from sources, in addition to the appropriate species, are low energy-spread, stable emission characteristics and reasonable lifetime so that the UHV environment is not unduly disturbed. Currently, energy spreads of 5 to 10eV are obtainable, the emission stability is within a few percent over several tens of hours of operation, and lifetimes of hundreds of hours are available in good vacuum conditions from Au-Be-Si sources and Ga sources but results are generally poorer for the liquid-metal sources that provide dopants for silicon.

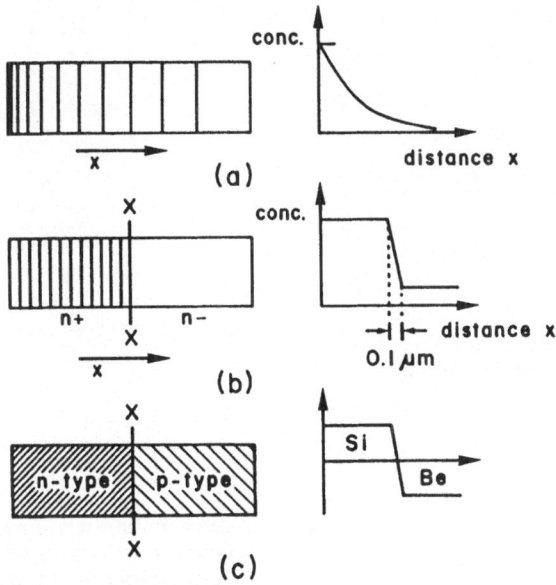

Fig. 2 Focused ion beam implantation in localized regions:
      (a) continuously graded doping distribution
      (b) step change in doping concentration
      (c) change from donor to acceptor implant.

Following the source, the electrostatic optics must have some means of ion extraction and beam focusing, with lenses, a mass filter to separate the ion species emitted by the source, beam alinement methods, a beam blanking system, apertures and beam deflexion to provide a suitable scanned field on the substrate. In some systems the full acceleration voltage is provided at the source end and in other systems the beam is formed and mass-filtered prior to acceleration to the full beam energy. For small beam sizes in the sub-micron regime, the beam size *versus* beam current relation in focused ion beam systems is determined almost entirely by the chromatic aberration coefficients of the lenses and the large energy spread from the source.

At present a current density of around $3A/cm^2$ in a beam of $0.1\mu m$ diameter represents acceptable system performance at 100kV acceleration voltage. These figures determine the time required for implantation, etching or resist exposure with focused ion beams, and confirm that the technique is limited to applications where high throughput and large areal exposure are not required. One of the most critical and novel features of focused ion beam systems is the question of mass separation. Both $\mathbf{E} \times \mathbf{B}$ filters and multiple-element magnetic filters [9] have been designed to achieve the required performance.

2.2.2. Localized implantation with focused ion beams With a focused ion beam diameter variable in the range $0.05\mu m$ to $1\mu m$ and a current density of around $3A/cm^2$ implantation into localized regions is a practical proposition. The probe may be positioned accurately to within 10nm or better and scanned in any pre-determined pattern with the aid of a hardware pattern generator with computer generated co-ordinate positions and pattern sizes. The probe may be blanked on or off rapidly as in electron beam lithography, to which the concept of focused ion beam writing owes most of its origins.

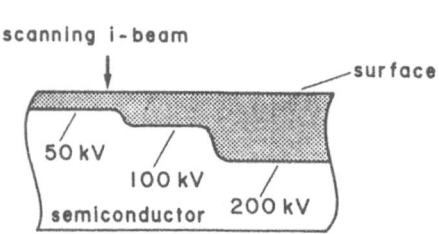

Fig. 3 Localized variations in implantation depth using focused ion beams.

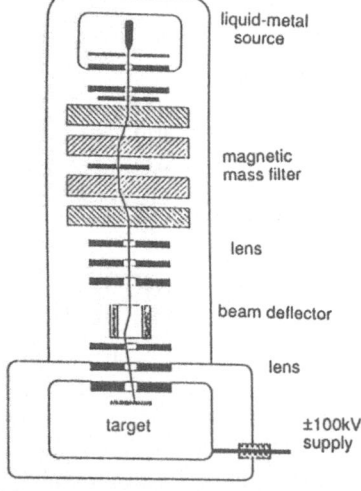

Fig. 4 Accelerating-retarding focused ion beam concept for varying beam landing energy from near zero to about 200 keV (for singly-charged ions).

Changes in doping concentration may be achieved with this probe most conveniently by changing the scan rate so that the dwell time per pixel is varied. The concentration gradient is limited by the beam diameter and the scattering within the substrate, which depends on the dopant species, the substrate and the beam energy. As shown in Fig. 2(a) the doping may be graded continuously as a function of distance and a change in concentration by a factor of 1:10 may be achieved in normal dimensions of devices and structures. The doping may also be varied in a step change at a pre-determined position. The resolution of the change is determined by the same scattering considerations as in conventional broad beam ion implantation. This is illustrated in Fig. 2(b). With alloy sources of the Au-Si-Be type the doping may be changed from n to p-type in a lateral direction, Fig. 2(c) In order to achieve this type of change some sort of alinement scheme is necessary. The ion species is selected by adjustment to the mass filter and the consequential mis-alinement of the beam is compensated by registering to marks prepared on the substrate in advance, as shown in Fig. 2 where marks along the x-x direction can be used. Thus lateral p-n junctions as well as junctions into the depth of the material may be created. A further extension of the approach can create small volumes of doped material of one polarity completely enclosed in another polarity of material by varying both the beam species and beam energy.

Another feature of the versatility of this technique is illustrated in Fig. 3. Here the energy of the implant is varied with distance so that the implant depth changes. This requires the beam voltage to be adjusted and again we are faced with the problem of beam mis-alinement which is overcome by registering to marks prepared by an earlier processing step.

The application of focused ion beam implantation to III-V compound heterostructures and superlattices requires that a wide range of implantation energies should be available since in these structures layer thicknesses range from 1nm to several hundred nanometres. Flexibility in beam penetration has been achieved by splitting the acceleration and retardation of the ion beam between the source end and the target end of the column. With this arrangement the ion gun can accelerate the beam to 100kV and the beam can either be accelerated to nearly 200kV at the target or retarded to near zero landing energy. The scheme is illustrated in Fig. 4.

The focused Ga ion beam can be used to isolate devices. The active layer on a semi-insulating substrate can be patterned under computer control to make the region around each structure electrically a high resistance. The area may be made amorphous, sputtered away completely, or made into polycrystalline regions of high resistivity, as illustrated in Fig. 5. Each approach has some advantages and disadvantages. Re-deposition can cause problems, and the resolution of the technique is limited by beam size and by the mobility of defects from their point of origin.

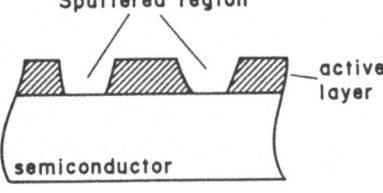

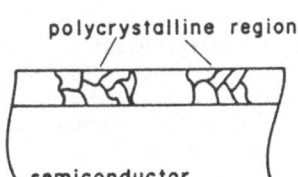

Fig. 5 Electrical and physical isolation of layers using a focused ion beam to form amorphous or polycrystalline regions or to sputter material.

3. REGISTRATION OF PATTERNED LAYERS IN SYSTEMS WITH *IN-SITU* LAYER GROWTH FACILITIES

A major problem with *in-situ* processing in which particle beams are combined with layer growth is the provision of an adequate means for registration of patterns. The problem is particularly serious in the case of focused ion beams. Registration marks may be made in substrates by etching or deposition prior to initial loading into combined systems. These are detected with an ion beam by collecting secondary electrons liberated by ion impact. Strong signals have been obtained from etched marks in GaAs substrates [10]. However, any significant overgrowth of layers on such a substrate masks this registration mark and the secondary electron signal is strongly attenuated. Similar conditions arise in systems where a resist coating is used on substrates for ion beam lithography. In experiments where the substrate may be removed from the vacuum system the area around the registration marks may be exposed separately by electron or ion lithography and the substrate re-inserted in the vacuum system. For a combined system with complete *in-situ* processing, alternative schemes have to be devised.

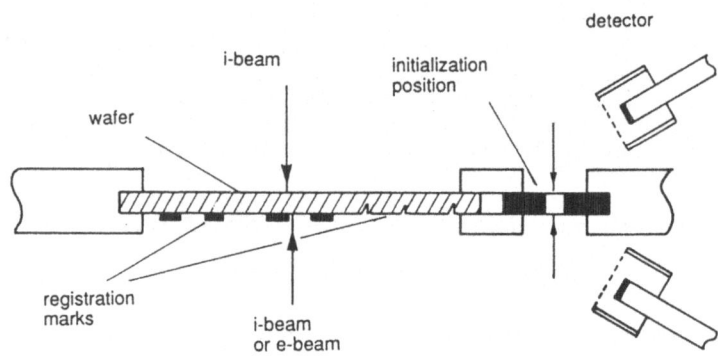

Fig. 6 Registration technique using a second particle beam to recognise registration marks on the back of a wafer.

3.1. Back-wafer registration

A successful technique which solves almost all difficulties is to use 'back-wafer' registration marks. This requires a dual-beam system in which a second electron (or ion) beam is incident on the back face of the wafer. The registration procedure is illustrated with reference to Fig. 6. A special stage is used in which both faces of the wafer may be accessed by the particle beams. Two beams from opposite directions are incident on the wafer and the stage has an initialization target area in which apertures are placed at the same planes as the wafer surfaces. The two beams are moved to the initialization regions and detect the apertures, and with reference to this common feature the positions, field size and co-ordinates of the two beams are determined, and they are adjusted with respect to each other. They are then both moved to the wafer surfaces. The upper ion beam implants while the lower ion beam etches a registration pattern. In any subsequent step the lower beam detects the registration marks and the upper beam writes in registration. In experiments with dual electron-beam systems an overlay accuracy of better than 0.5μm has been obtained [11]. In principle this accuracy could be improved to better than 0.1μm with a smaller beam size and more careful arrangement of initialization areas.

In an alternative scheme the lower ion beam may be replaced with an electron beam. In this case the registration marks in the back of the wafer must be etched before loading the substrate into the *in-situ* processing system. In this process it is necessary to use wafers polished on both faces and it is important to ensure that any layer growth processes are confined to the active surface of the wafer and that the back is not covered with successive layers. This condition is usually met in MBE and most other film deposition systems.

Back-wafer registration has many advantages for *in-situ* processing and overcomes the problems in multi-level layer processing with microstructures. This is obviously achieved at the expense of system complexity but it is a highly accurate method capable of the resolution required for sub-micron structures.

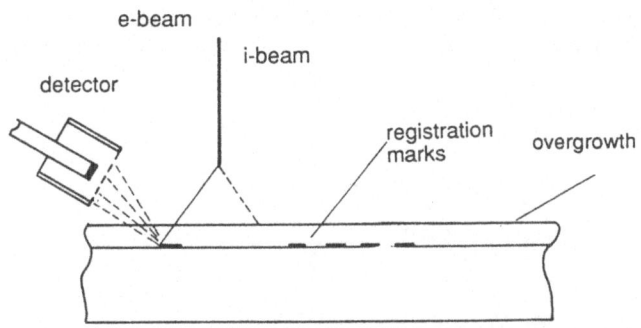

Fig. 7 Co-incident particle beam technique. Implantation is carried out with ions, and buried registration marks are recognised with the electron beam.

3.2 Co-incident electron and ion beams for *in-situ* processing

In-situ processing requires fabrication, registration and inspection without breaking the vacuum or gaseous environment. These needs can also be fulfilled by attaching a layer growth system such as MBE to a particle beam column in which both an electron and an ion beam are available so that they are able to access exactly the same region of the substrate. Such an optical design is possible using a variety of arrangements [12]. Either all-electrostatic optics or a combined electrostatic and electromagnetic arrangement may be used so that beam energy may be independently adjusted.

An ion source on-axis and an electron gun off-axis are combined so that the electron beam is bent from the off-axis position into the same optical axis as a mass-filtered ion beam. Subsequent optics requires that both beams are brought to a focus at the same position. It is also necessary to arrange that the beams may be blanked on and off and they must be scanned over the same area of the substrate. This may be achieved by preliminary adjustments in which both beams are used to form an image of the same area of a target. Adjustments to focusing and to the scanned field can ensure that identical images are formed.

The main feature of a co-incident particle beam system is that it enables registration marks to be detected with the electron beam while the ion beam is used to implant or to carry out other processes. The electron beam may also be used to inspect without damaging the surface that has been fabricated with ions. In certain applications it may be possible to expose resist with the deeper penetrating power of the electrons before implanting with ions, Fig. 7.

4. THERMAL PROCESSING TECHNIQUES SUITABLE FOR *IN-SITU* PROCESSING

Thermal processes are needed which are compatible with the ultra-clean MBE system environment. These have applications such as the annealing of ion implantation damage, the annealing or sintering of ohmic contacts to substrates, certain thermally-assisted layer growth processes, and localized metal deposition from gaseous sources. Energy sources such as lasers, lamps and electron beams are suitable for these applications and schemes can be devised for in-situ installation of these sources. The electron beam system is particularly suitable because of its greater flexibility in time - temperature cycles and its ability to deposit energy into the surface regardless of the physical properties such as reflectivity.

The suitable system is an isothermal annealing arrangement which heats up the whole substrate uniformly. Attempts at localized annealing over selected and restricted areas have met with only partial success [13]. A typical time - temperature cycle is shown in Fig. 8. The heating cycle may be controlled with a feedback arrangement using temperature measurement with a pyrometer or other means. Experiments on silicon and GaAs substrates have shown that implantation damage may be restored without causing significant diffusion. In the case of GaAs certain implants may be annealed without requiring capping layers and ohmic contacts may also be processed thermally [14]. The uniform temperature distribution in this system ensures that substrate flatness is retained so that subsequent lithography is not affected. The time - temperature cycle has been shortened into the millisecond regime in some experiments so that diffusion is restricted to less than $0.01\mu m$ while ion implantation damage is removed from the substrate.

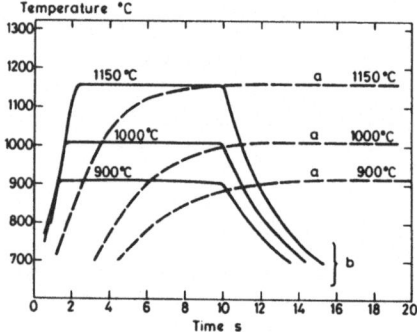

Fig. 8 Time - temperature cycles obtainable with an electron-beam rapid thermal annealing system.

5. SUB-10nm-SCALE LITHOGRAPHY

The performance of layer growth techniques with MBE has been developed to enable layers less than 1nm thick to be grown with abrupt compositional changes. It has become necessary to enhance lateral patterning techniques to achieve similar dimensions. With such techniques a

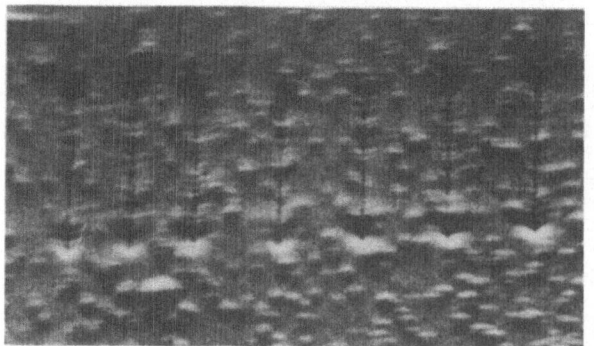

4 µm

400 nm

Fig. 9 A sub-10nm line-width pattern formed in a fluoride layer on a solid substrate (with acknowledgements to Dr S.D. Berger, Cavendish Laboratory, Cambridge University).

whole new range of low-dimensionality physics becomes possible. Several attempts have been made to achieve sub-10nm-scale structures using high resolution polymer resists and focused electron beams down to 0.5nm diameter. These have been unsuccessful because the scattering of the primary electrons and the range of the secondary electrons delocalizes the electron exposure. Furthermore, the polymer molecular chain is generally longer than 10nm and these resists cannot provide the required resolution.

Alternative techniques which avoid the need for resist and work directly on the substrate have been explored. In one approach the electron beam is accelerated to voltages in excess of 250kV so that the semiconductor substrate lattice is damaged sufficiently for differential etching to provide discrimination between damaged and crystalline areas [15]. However, the resolution obtained so far has been greater than 10nm.

The most promising approach has been to use new resists compatible with UHV *in-situ* processing systems and very high current density electron beams. Remarkable results have been achieved with this approach [16]. The beam current density is only available at present from field-emission sources and the range of resists is limited to certain compounds such as aluminium and calcium fluoride. Fig. 9 shows an example of a pattern formed with this technique on a substrate of silicon, demonstrating that sub-1.0nm patterns may be formed on useful substrates.

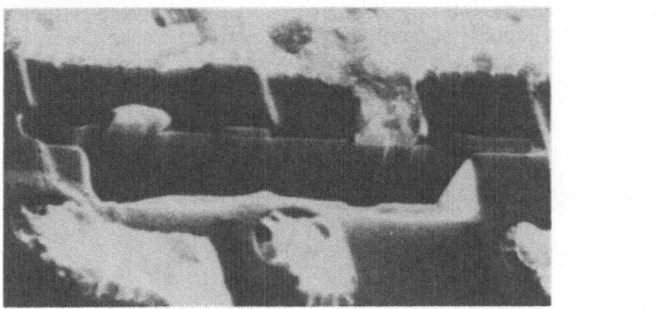

10 µm

Fig. 10   Microsectioned electronic device showing buried layers, which may be examined or electrically connected.

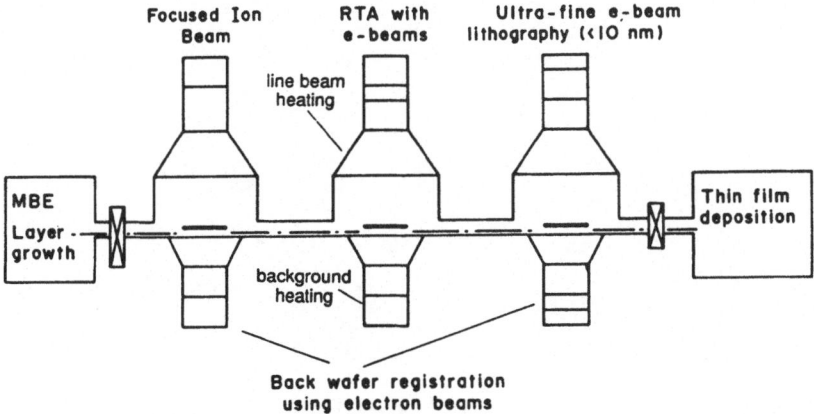

Fig. 11   Combined system for microstructure fabrication including layer growth, lithography, implantation and thermal annealing.

6. INSPECTION OF BURIED STRUCTURES AND LAYERS FABRICATED BY *IN-SITU* METHODS

Any process control in microelectronic fabrication requires inspection at various stages in order to assess and to optimize the process. The use of electron and ion beams enables in-situ inspection to be carried out. The electron beam provides high resolution images in the SEM mode of operation. It can also provide crystalline information by reflexion-diffraction from surfaces and electron channelling. Ion beams provide topographical contrast and analysis using SIMS attachments. In addition the ion beam provides a unique micro-sectioning facility [17]. This is illustrated in Fig. 10, where a section has been cut across a gate region covered with polysilicon. The section may be cut in a precisely pre-selected position and once it is cut this inspection can be carried out at high resolution with the electron beam. This technique can be adjusted to cut vias to buried layers and to buried patterns. It may also be used to form mesas for electrical isolation of layers grown on semi-insulating material.

7. NOVEL STRUCTURES

The requirements for basic physics research into low dimensionality and the need to make microstructures for this work has led to the concept of combining all the critical techniques into a single system, as illustrated in Fig. 11. Here we have a wafer-track system in ultra-high vacuum which links several processing machines, capable of very high quality in patterning and layer growth.

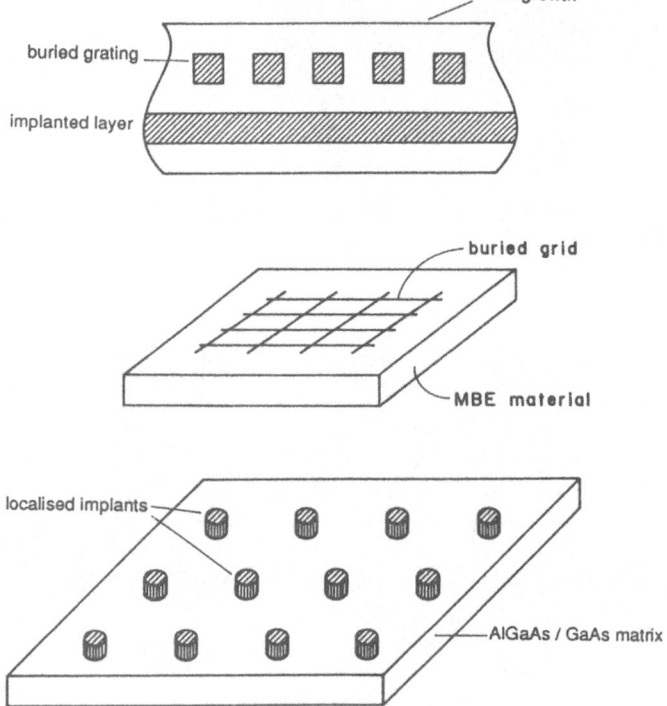

Fig. 12   Some novel structures that may be fabricated in a combined system:
      (a) buried gratings
      (b) buried grids
      (c) implanted arrays buried in a superlattice matrix.

Many novel structures may be fabricated in very high resolution combined systems. It is possible to make quantum dots or pillars in heterostructures by doping layers and patterning in a grid with a Ga ion beam. Such structures have been described as lateral superlattices [18].

Surface gratings and buried gratings may be formed by ion sputtering in multiple layer materials to make, for example, distributed-feedback structures for optical devices. The gratings gap may be infilled and overgrown as required. Phased and chirped structures are also possible. More complex structures such as loops and ring structures for investigating quantum interference effects may also be made in 2DEG. Several types of grids may also be formed by implantation, sputtering, and disordered layers, and high gallium implant doses can modify composition. Grids with uniform or non-uniform pitch surrounded by a matrix of pure crystalline material can be formed and overgrown, Fig. 12. With specially prepared material free-standing structures may be formed for investigation into electron and phonon transport.

This wide range of physics experiments is complemented by more practical and immediate applications such as the fabrication of HEMTs with superior performance to conventional devices.

ACKNOWLEDGEMENTS

The contributions of members of the Microelectronics Group and of the Semiconductor Physics Group at the Cavendish Laboratory are gratefully acknowledged.

REFERENCES

1. Couch, H., Kelly, M.J., Kerr, T, Stobbs, M.J. & Britton, E.G. (1987) Semiconductor Sci. Technol. **2**, 244.
2. McMillan, G.B., Shannon, J.M., Clegg, J.B. & Ahmed, H. (1984) J. Appl. Phys. **59**, 1081.
3. Smith, C.G., Ahmed, H. & Wybourne, M.N. (1987) J. Vac. Sci. Technol. **B3**, 314.
4. Lee, K.L. & Ahmed, H. (1981) J. Vac. Sci. Technol. **19**, 946.
5. Beaumont, S.P., Bower, P.G., Tamamura, T. & Wilkinson, C.D.W. (1981) Appl. Phys. Lett. **38**, 436.
6. Takamori, A., Miyauchi, E., Arimoto, H., Bamba, Y. & Hashimoto, H. (1984) Jpn. J. Appl. Phys. **23**, 8.
7. Reich, D.F., Fray, D.J., Evason, A.F., Cleaver, J.R.A. & Ahmed, H. (1986) Microelectron. Eng. **5**, 171.
8. Gamo, K., Inomoto, Y., Ochiai, Y. & Namba, S. (1981) *Microcircuit Engineering 81*, 359.
9. Heard, P.J., Cleaver, J.R.A. & Ahmed, H. (1983) *Microcircuit Engineering 83*, ed. H. Ahmed, J.R.A. Cleaver & G.A.C. Jones (London: Academic Press) 135.
10. Evason, A.F., Cleaver, J.R.A., Heard, P.J. & Ahmed, H. (1985) Electron. Lett. **21**, 629.
11. Jones, G.A.C. & Ahmed, H. (1979) J. Vac. Sci. Technol. **16**, 1776.
12. Cleaver, J.R.A. & Ahmed, H. (1985) J. Vac. Sci. Technol. **B3**, 144.
13. Sun, H.T., McMahon, R.A. & Ahmed, H. (1983) J. Vac. Sci. Technol. **B1**, 827.
14. Shah, N.J., Ahmed, H., Sanders, I.R. & Singleton, J.F. (1980) Electron. Lett. **16**, 433.
15. Jones, G.A.C., Blythe, S. & Ahmed, H. (1986) Microelectron. Eng. **5**, 265.
16. Kratschmer, E. & Isaacson (1987) J. Vac. Sci. Technol. **B5**, 369.
17. Kirk, E.C.G., Cleaver, J.R.A. & Ahmed, H. (1987) in press
18. Iafrate, G.J. (1986) *Physics of sub-micron structures*, ed. H.L. Grubin, K. Hess, G.J. Iafrate & D.K. Ferry (New York: Plenum Press) 301.

MOTIVATIONS AND EARLY DEMONSTRATIONS FOR IN-SITU PROCESSINGS FOR III - V
SEMICONDUCTOR DEVICES

IZUO HAYASHI
Optoelectronics Technology Research Laboratory
5-5 Tohkodai, Toyosato, Tsukuba, 300-26, Japan

1. INTRODUCTION

Silicon transistors have evolved into integrating circuits starting
from discrete devices over twenty years ago. The similar evolution will
occur in compound semiconductor optoelectronic devices, like lasers or
photodetectors. The same benefits, high performance and low cost will be
also expected by such integration of optoelectronic devices.

There are different situations in optoelectronics device integration
compared with silicon, however. They are mainly made of III -V
semiconductors instead of Silicon. In addition, optoelectronic devices are
always connected with electronic devices, thus one must deal with
Opto-Electronic Integrated Circuits, the OEIC (1), (2). The III-V
semiconductors are more fragile materials than Si, moreover simultaneous
integration of optoelectronic and electronic devices will create
difficulties. Both devices having different structures require generally
incompatible fabrication processes (2).

Process technologies for optoelectronic devices today are entirely
those for discrete devices. These may correspond to those of silicon
before the age of integration, "alloy transistor" technology (2).

In this talk, detailed description of new technologies for
optoelectronic integration, in particular, motivations for in-situ
processings and some early demonstrations will be presented.

2. MOTIVATIONS FOR NEW TECHNOLOGIES OF OPTOELECTRONIC DEVICES

Semiconductor lasers require elaborate and difficult processes among
optoelectronic devices because lasers have both an optical element a
resonant cavity, and an electronic element, a diode with carrier injection
through (p-n) junction. They are connected by recombination luminescence.
Therefore a laser structure must satisfy requirements from both sides,
thus not only high geometrical preciseness but also high quality in
materials are mandatory, defectless interfaces in particular. Therefore
new technologies for semiconductor laser integration will be described in
this talk.

It is obvious that one of the common methods for laser fabrication
today, cleaving crystals to produce laser facets, must be replaced. An
etching technique with sufficient geometrical preciseness and low damage,
which is compatible with in-situ passivative coating, is needed. The
freshly etched surface atoms must be connected perfectly with those of the
passivating layer.

In a laser structure, if an interface is a active part of the device,
atoms on one side of the interface must be precisely bonded with those on
the other side, without leaving any dangling bond defect. A fresh
atomically clean surface will disappear instantly, when it is exposed to
the normal air environment (3).

13

D. J. Ehrlich and V. T. Nguyen (eds.), Emerging Technologies for In Situ Processing, 13–21.
© 1988 by Martinus Nijhoff Publishers.

Therefore the etching must be done in clean controlled gaseous environment. A "dry" etching technique, compatible with other in-situ processing for the passivating layer, is needed.

Lasers integrated with other components on semiconductor substrate can be produced by such a dry process with sufficient reproducibility and reliability for integration. Process technology for integration must have a high degree of perfection. Yields for each device must be over 99 % when one hundred of such devices are to be integrated on a single chip, which is far beyond those needed for discrete devices.

Another example of today's laser process technologies to be replaced for integration is those for "stripe" fabrication.

The stripe structures with best performances are buried hetero structure, BH, which satisfy both optical and carrier confinement in lateral direction. However two step processes, epitaxy of the first double hetero structure DH layers, etching of mesa setructure and the second epitaxy of the confining layer is needed. "One step" DH layer epitaxy into a groove is also used.

Both of these processes leave air exposed interfaces on sides of active layers, which tend to decrese yields and reliability. If two processes, the etching and the deposition or epitaxy can be done in-situ in a clean environment, the problem will be solved. Another approach is a "planer" process, in which BH structure can be produced into the DH wafer in situ without etching by ion implantation. An ultra high vacuum process system, which contains focused maskless ion implantation, MBE and dry etching will be explained later.

A combined process system, in which a series of processes can be performed in situ in a clean controlled environment, is needed to produce complex device structures like lasers, without creating defected interfaces made by process interruption.

The conventional photolithography processes are not suitable for these devices, in which surface damage will be produced by air exposure. Lack of high quality oxide like in silicon in compound semiconductors is the basic reason for the defect creation at the surface.

3. EARLY DEMONSTRATIONS FOR IN SITU PROCESSINGS

A dry etching technique has been developed in the Optoelectronics Joint Research Laboratory (OJL), to replace the cleaving process in the OEIC fabrication (3). A vertical etching into DH layers has been obtained with enough geometrical precision and small enough damage. Lasers of equal threshold and efficiency have been demonstrated compared with those made by cleaving technique (4).

Enhanced chemical reaction with radicals during a broad beam Cl ion sputtering, is the essence of this technique (3). The similar Cl etching technique using focused ion beam assistance has also been demonstrated, which has a capability to be a part of a clean environment process system (5).

A light etching (cleaning) technique has also been developed, with which an oxidized and contaminated surface can be cleaned by radicals at elevated temperature (6). With this technique the clean environment process can be connected with a conventional air exposing process, thus the more pratical process sequence can be performed.

Details of these etching techniques will be described by Asakawa in this workshop.

The focused ion beam doping system (7) is the heart of the clean

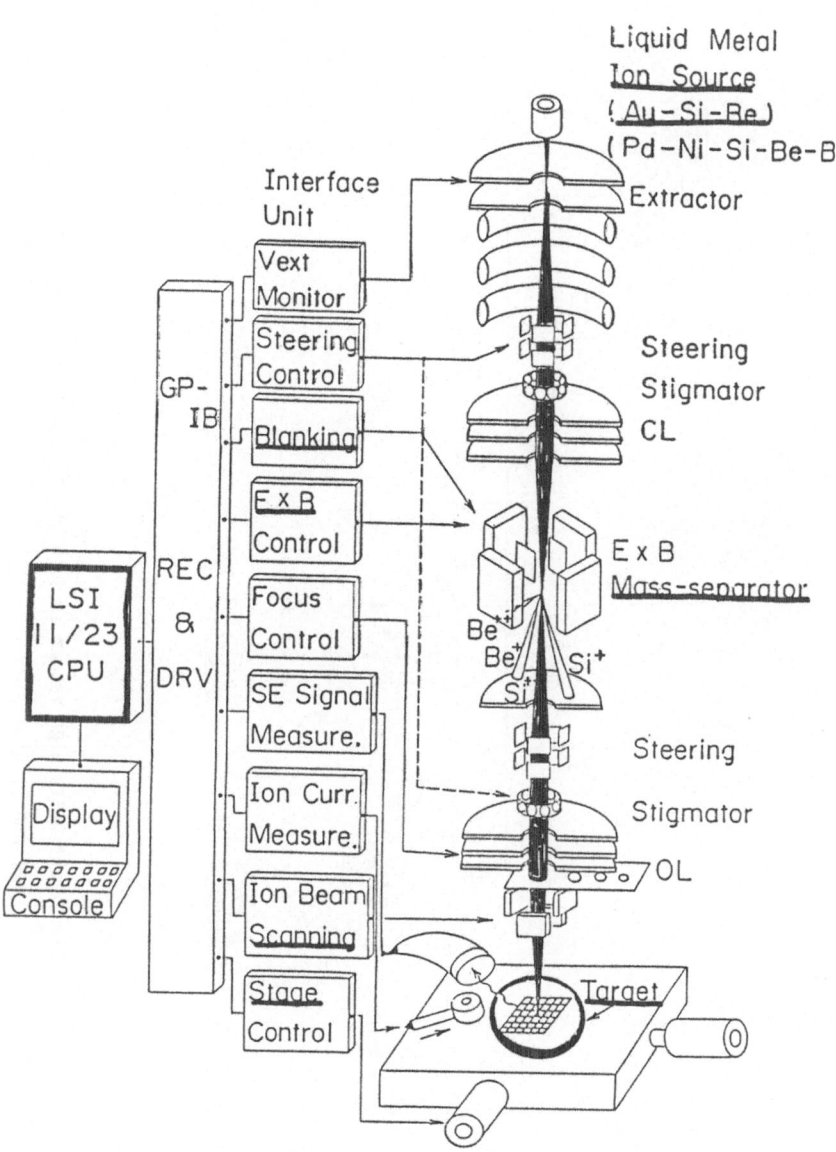

Fig.1 FIB system with computer control

16

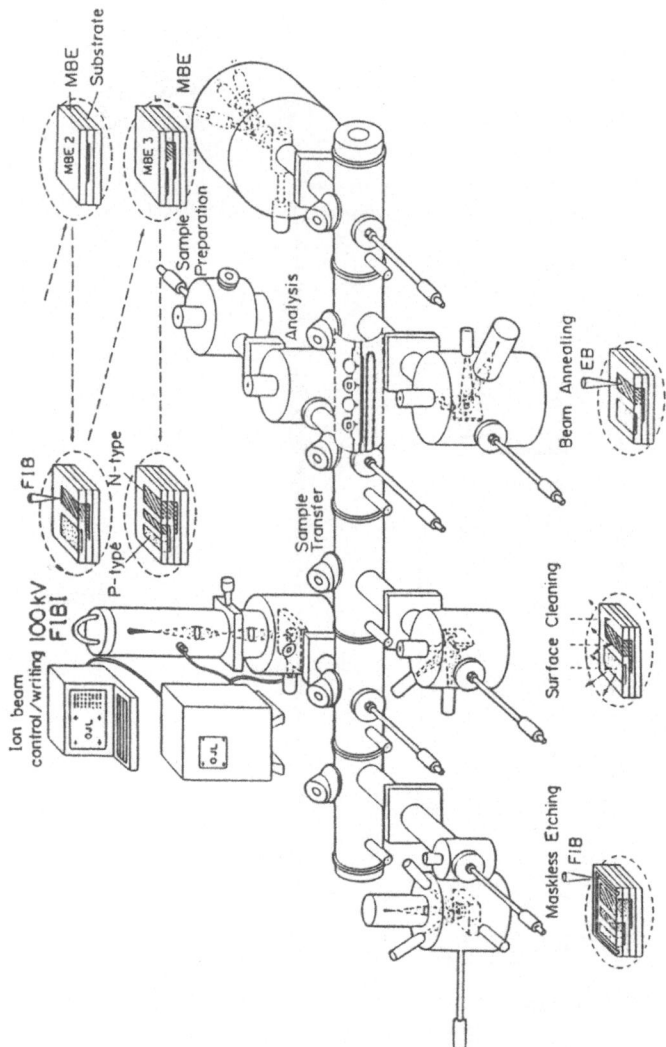

Fig. 2 The ultra high vacuum in situ processing system composed of FIB doping MBE and etching.

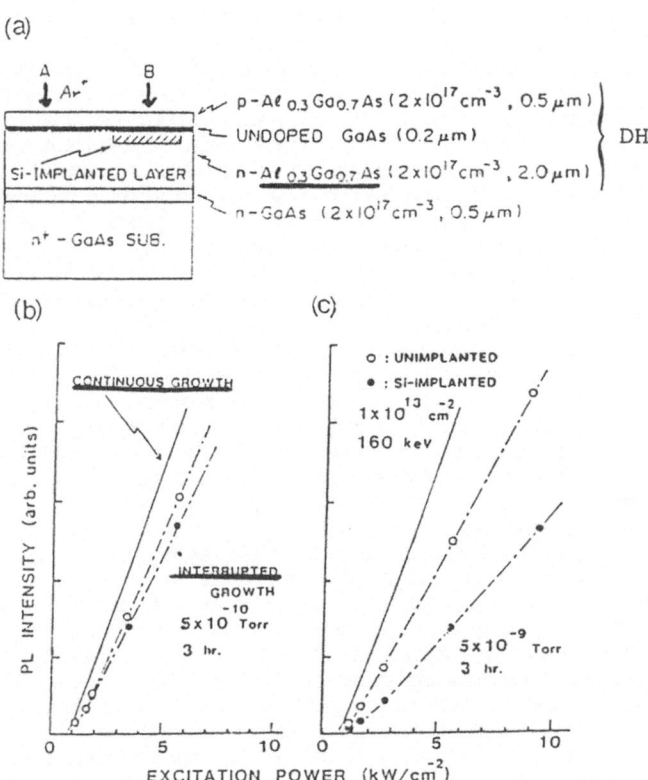

(a)

(b)

(c)

Fig. 3 A semiconductor laser structure made by MBE láyer growth
and FIB doping. The growth is interrupted a few hours at
the first AlGaAs layer. (a) FIB Si-implanted AlGaAs/GaAs
DH structure (b) Photoluminescence intensities from GaAs
layer vs input power of Ar laser for 5×10^{-10} Torr, and
(c) for 5×10^{-9} Torr.

environment (ultra high vacuum) in situ process system, developed in the Optoelectronics Joint Research Laboratory. An accelerated an focused ion beam of either n (Si) or p (Be) type with 0.1 m in diameter, can be implanted into a semiconductor crystal. A fine doping pattern can be made with arbitrary amount of dose into any portion of the pattern. Energy of the implantation can also be varied without distorting other parameters of the beam (8). A computer system with which any of such variations can be controlled by programming has been developed (8) (Fig.1). A liquid metal ion source (Au-Si-Be) showed sufficient stability for automatic operation of the FIB system. Overnight operation of fine doping pattern has been done by the computer control without human supervision.

A high vacuum of 10^{-10} Torr range, in the sample preparation room is a key feature of this FIB system, in addition to the high accuracy of the ion-optics, including a mass separator.

The ultra high vacuum in-situ processing system composed of FIB doping, MBE and etching has been studied (7) (Fig.2). It is demonstrated that a freshly grown GaAs surface in MBE can be preserved during FIB implantation. Analysis by luminescence, doping profile, secondary ion mass analyzer has shown that such interfaces showed little or no indication of defect creation during a few hours exposure of the MBE grown surface in the implantation machine (9).

A most critical testing was done with an interrupted DH layer growth for lasers. After the first semi-insulating AlGaAs layer growth in MBE, the wafer was moved into the FIB and implanted with Si ions for n type conducting stripe, before the GaAs active layer and the second AlGaAs confining layer growth. The degradation of the implanted and exposed surface of the Al 0.3 Ga 0.7 As layer, which is far more sensitive than GaAs, was small enough to keep lasing characteristics (10) (Fig. 3). If the vacuum is worse in one order of magnitude (10^{-9} Torr range) the degradation have been large enough to prevent lasing.

An experiment to test the effect of radical cleaning process combined with the MBE, has been performed. It was shown that the air exposed surface degradation of a GaAs wafer was decreased considerably with hydrogen radical cleaning. Specially designed radical gun was used to produce H, Cl or O radicals. Major part of oxygen and carbon has been removed from the GaAs surface (11).

Very recently a focused Ga ion beam etching assisted with Cl gas, demonstrated to cut a groove with fat side walls, with which laser facet of cleave equivalent can be produced (5).

A fine diffusion induced disordering (DID) process has been done with a focused Si beam implantation into an MQW layer structure (12), with which lateral hetero-structures of submicron dimension can be produced. A BH stripe laser structure has been demosntrated (12).

4. SUMMARY AND FUTURE PROSPECT

Effectiveness of the in-situ clean environment process has been demonstrated by the UHV process system in the OJL. This kind of process system can be applied to not only laser structure fabrication but also any critical fabrication steps for other optoelectronic as well as electronic devices. With cleaning or etching procedure as described, such in-situ processes can be combined with conventional processes. An OEIC will be made by using such in-situ clean environement process for optical devices, after electronic integration has been done by a conventional method.

Addition of other maskless processes for deposition or epitaxy will expand capability of the in-situ process system. A selective area deposition or epitaxy will be very attractive as a part of such in-situ processes.

In order to establish process systems for large scale OEIC fabrication, progress of a variety process technologies will be required. Low temperature processes for optoelectronic devices are needed not to degrade electronic circuits, which already have been completed on the same substrate. This is an important step to minimize the process interference.

It will take years to develop actual fabrication processes for OEIC, as was the case for Si integration. Weakness of compound semiconductor materials, the very sensitive surface in particular, requires much intense care in device fabrication.

However feasibilities of the in-situ process combination in clean environment described above, have demonstrated the possibility of pratical IC fabrication including such fragile devices like semiconductor lasers.

A variety of OEICs, starting from small scale ones for optical fiber communication or optical light sources for optical memories, then larger scale OEICs for optical interconnection of high speed electronic systems, finally OEICs for a wide range of "electronic" or "optic" systems will be developped in the next century.

REFERENCES

1. YARIV A.
    "The beginning of integrated optoelectronics circuits"
    IEEE Trans., 1984 ED-31 (11), pp 1656-1661

2. HAYASHI I.
    "Research aiming for future optoelectronic integration"
    IEE Proceedings, Vol. 133, pt. J. N° 3, June 1986, pp. 237-244

3. ASAKAWA K. and SUGATA S.
    "GaAs and AlGaAs anisotropic fine pattern etching using a new reactive
    beam etching system"
    J. Vac. Sci. and Technol. 1985, B3, pp. 402-405

4. YUASA T., ASAKAWA K., HANNOH M., SHINOZAKI K. and ISHII H.
    "Reactive-ion-beam-etched cavity GaAs - AlGaAs multi quantum well
    lasers"
    IOOC-ECOC Venice, Italy, Oct. 1985, Tech. Digest, pp. 505-508

5. TAKADO N., ASAKAWA K., YUASA T., SUGATA S., MIYAUCHI E., HASHIMOTO H.
    and ISHII M.
    "Chemically enhanced Forcused Ion Beam Etching of Deep Groove and
    Laser Mirror Facets in GaAs. under C12 Gas Irradiation Using a Fine
    Nozzle"
    Appl. Phys. Lett. to be published

6. TAKAMORI A., SUGATA S., ASAKAWA K., MIYAUCHI E., HASHIMOTO H.
    "Cleaning of MBE GaAs Substrates by Hydrogen radical beam Irradiation"
    Jpn. J. of Appl. Phys. 26 N° 2, L142, 1987

7. MIYAUCHI E. and HASHIMOTO H.
    "Application of FIB technology to maskless ion implantation in an MBE-
    Grown GaAs or AlGaAs epitaxal layer for three-dimensional pattern
    doping crystal growth"
    Vac. Science and Technol. A4 (1986) pp. 933-938

8. MIYAUCHI E. and HASHIMOTO H.
    "Maskless Ion Implantation System For 3-dimensional Fine Doping
    Structures in III-V compound Semiconductors"
    Nucl. Instrum. Methods, Feb. 1987

9. TAKAMORI A., MIYAUCH E., ARIMOTO H., BAMBA Y., MORITA T. and
    HASHIMOTO H.
    "Growth-interrupted interfaces in multilayer MBE growth of gallium
    arsenide"
    Jpn. J. Appl. Phys. 24 (1985) L414-L416

10. TAKAMORI A., MIYAUCH E., ARIMOTO H., BAMBA Y., MORITA T. and
    HASHIMOTO H.
    "Optical properties of pattern-doped GaAs/AlGaAs grown by an MBE
    system equipped with a focused ion beam implanter"
    Inst. Phys. Conf. Ser. N° 79 (1986) 247-252

11. SUGATA S. Private Communication

12. ISHIDA K., TAKAMORI T., MATSUI K., FUKUNAGA T., MORITA T., MIYAUCHI E. HASHIMOTO H. and NAKASHIMA H.
"Fabrication of Index Guided AlGaAs/GaAs MQW Lasers by Selective Disordering using Be Focused Ion Beam Implantation"
Jap. J. of Appl. Phys. Vol. 25, 1986, L728

LASER ETCHING AND MICROELECTRONIC APPLICATIONS

JOHN J. RITSKO

IBM T. J. WATSON RESEARCH CENTER, YORKTOWN HEIGHTS, NY, USA

Laser etching involves use of a laser to directly pattern a semiconductor, insulator, or metal. By either scanning a focused beam or illuminating a mask which is optically imaged onto a sample, the laser both defines and etches the desired pattern. Conventional lithographic techniques involve applying a photoresist which is exposed and developed and then separately etching the substrate and finally removing the resist. While there are more steps in conventional lithography, the processes are relatively well understood, widely practiced, and many tools are commercially available. Some laser etching techniques require further process development and only a few industrial grade tools are commercially available at this time.

The laser processes which are effective in etching different materials vary considerably depending on the material, its thickness, the substrate on which it lies, the resolution of the features required, and the size and complexity of the desired pattern. This paper will briefly review some of the more promising processes and applications of laser etching of semiconductors, insulators and metals.

Semiconductors can be etched with relatively low intensity radiation by photochemical processes involving photocreation of electron-hole pairs in the near surface region of the material. The electron energy bands near the surface are bent such that holes are swept to the surface by the fields in an n-type semiconductor and electrons are swept into the bulk of the sample. In p-type semiconductors the opposite is true. There can also be surface traps which keep both electrons and holes near enough to the surface to participate in surface chemical reactions. The reaction which etches the semiconductor involves gas phase or liquid phase molecules which combine with the material to form volatile or soluble products. A typical laser assisted chemical etching process was described by Houle[1]. Silicon is spontaneously etched by fluorine when subjected to a low pressure of XeF_2. The etch rate is enhanced by a factor of four when an unfocused Ar^+ ion laser (all lines) operating at 4 watts is incident on the surface. One of the photochemical reactions which occur involves the reaction of existing SiF_3 groups on the surface with photogenerated holes to create SiF_3^+. A fluorine atom picks up a photogenerated electron to form F^-. And finally SiF_3^+ and F^- combine to form the volatile gas SiF_4. A mass analysis of the reaction products indicates that the light significantly alters the reaction products from those observed in spontaneous etching[1].

D. J. Ehrlich and V. T. Nguyen (eds.), Emerging Technologies for In Situ Processing, 23–31.
© *1988 by Martinus Nijhoff Publishers.*

When Si is exposed to Cl_2 gas in the absence of light, spontaneous etching does not occur. However if chlorine atoms are generated in the gas above the sample by UV photodissociation, the atoms will etch Si[2,3]. The reaction involves formation of a volatile halide ($SiCl_4$) by first forming Cl^- ions by electron tunneling from the semiconductor. In heavily doped n-type Si there is a large population of electrons available for tunnelling and the etching caused by reaction with Cl radicals is reasonably rapid. Using an unfocused excimer laser at 308 nm in 100 Torr of Cl_2, single Si crystal was etched to a depth of about 20 Å for a total exposure of 1 J/cm². However, the etched depth for similar exposure conditions falls off very rapidly as the doping concentration decreases. In p-type Si no etching in the absence of light occurs even in the presence of gas phase Cl radicals. However, when the laser beam hits the sample at normal incidence, it not only creates Cl atoms in the gas but generates electron-hole pairs in the semiconductor. In p-type Si the electrons are swept to the surface by the space charge fields and can form Cl^- ions and ultimately volatile products. Under these conditions p-type Si can be etched at the rate of about 1 Å per J/cm² of exposure.

The dependence of the etch rate of Si in Cl_2 under 308 nm excimer laser irradiation parallel or perpendicular to the surface is consistent with etching requiring both Cl radicals and abundant electrons[3]. For n-type Si direct irradiation of the surface does not greatly increase the electron concentration near the surface if the sample is heavily doped. Hence there is little difference in the etch rate of parallel or perpendicular radiation. However, in lightly doped n-type Si photoelectrons can be a major portion of the total electrons in the conduction band and direct irradiation greatly enhances the etch rate[3].

Using these mechanisms Arikado et al. have created very fine VLSI patterns (64K DRAM) in Si by direct irradiation with an excimer laser in a Cl_2 gas atmosphere. The pattern was created by placing the sample in near proximity to an Al pattern on a quartz mask plate.

Gallium Arsenide and other III-V semiconductors are important for a variety of applications. Controlled anisotropic etching is necessary for the fabrication of integrated optical components and microwave devices for example. Etched features in GaAs may also be useful in coupling optical fibers to optoelectronic transmitters and receivers. Using visible laser wavelengths and specific liquid etching solutions, Osgood et al[4] have demonstrated the fabrication of high-resolution diffraction gratings and through wafer vias in Cr doped (semi-insulating) GaAs, n-type GaAs and InP. for semi-insulating, SI, GaAs low laser powers produced gratings with periods less than 1 μm. An acidic solution involving H_2SO_4, H_2O_2 and water which etches the SI GaAs at 0.7 μm/min in the dark was used. The laser enhances the etch rate locally by producing electron-hole pairs in the semiconductor. The spatial resolution is limited by the lateral spreading of the photogenerated carriers and is estimated to be on the order of 100 nm. The etching mechanism is clearly photochemical since the estimated temperature rise for the laser powers used was less than 0.1°C and wavelengths below the GaAs band gap produced negligible etching.

Low temperature photochemical etching is useful for creating high resolution patterns. However, some applications may require rapid etching of relatively coarse features. Osgood et al. showed that by focusing the laser to approximately an 8 μm spot, through wafer vias could be etched in roughly 4 minutes (30 μm/min). Under the conditions of these high intensities, 0.2 MW/cm^2, a variety of thermal and optical processes influence etching and the detailed mechanism is not easy to determine.

Galium Arsenide, n-type generally etches more rapidly than p-type. In n-type material, holes are swept to the surface and reactions such as

$$GaAs + 6h^+ \rightarrow Ga^{+3} + As^{+3}$$

lead to soluble compounds being formed. On the other hand in p-type compounds, electrons come to the surface and

$$GaAs + 3e^- \rightarrow Ga + As^{-3}$$

leads to Ga metal on the surface which is more difficult to remove.

The fabrication of high quality distributed feedback laser gratings by maskless laser etching techniques has been demonstrated by Aoyagi et al.[5]. Samples used were n-GaAs and n-InP (100) single crystals. The GaAs was etched in an aqueous solution of I_2 and KI while the InP was etched in an AZ-303 developer. Etching was accomplished with an Argon ion laser run at 457.9 nm using a conventional holographic technique.

Photoelectrochemical etching of InP and GaInAsP has been demonstrated to be important for creating several structures needed for semiconductor lasers. Mirrors of a semiconductor laser are made by a cleaving procedure normally initiated by manual scribing. This is difficult where precise alignment is necessary. Bowers et al.[6] have shown that photoelectrochemical etching can be used to make deep narrow grooves into the laser substrate to weaken the crystal along specific lines and induce cleaves along these lines when the wafer is flexed[6]. The process is carried out in an electrochemical cell. A sample of InP is immersed in a conducting solution with a low dark etch rate such as 20:1 H_2O:HCl. A bias voltage is applied and the current measured. An n-type substrate is positively biased (typically 0.4 - 0.6V) and, using a potentiostat, applied voltages can be measured with respect to a saturated calomel reference electrode. Using a laser with a wavelength such that bandgap excitations are created (for example an Ar$^+$ laser), a pattern can be etched at rates of 50 μm/hr over a large area. After one calibration run with a given mask pattern, the etch rate could be accurately predicted from measurements of the etching current and sample area. Using this method narrow, high aspect ratio grooves were etched which successfully induced cleaves for laser facets resulting in a high yield of low threshold lasers[6].

Using similar apparatus, Lum et al.[7] etched InP and GaInAsP photoelectrochemically in a 2-M HF/0.5-M KOH solution with the sample controlled between 0.2 and 0.5 volts relative to a saturated calomel electrode. A laser interferometer was used to create a sinusoidal spatial intensity variation across the sample surface. Several lasers were used, including HeNe (632.8 nm), Ar+ (488 nm) and HeCd (441.6 nm). Etching occurs by oxidation induced by photogenerated holes and the etch rate ratio between light and dark areas is greater than 100:1. The quantum efficiency for decomposition is 100%. A two dimensional hole transport model was used to fit the measured sensitivity of these holographic gratings as a function of spatial frequency. There was good agreement between the quantitative predictions of this model and the experimental data. At high spatial frequency the etching is limited by surface hole diffusion. The measured first order diffraction efficiency is 20 - 25% for 8.5 μm gratings and can approach 15 - 20% for 1.5 μm gratings. Gratings with periods as short as 0.5 μm were produced. Some gratings were of such quality that diffracted beams out to seventh-order could be observed.

In the microelectronics industry at the present time silicon dioxide is perhaps the most useful insulator. Its relative inertness makes it difficult to etch rapidly but since often only thin layers need to be patterned, slow etch rates may be acceptable. Steinfeld et al. have demonstrated a laser generated radical etching process which etches SiO_2 without the ion bombardment damage associated with the more commonly used dry etching process namely reactive ion etching[8]. They used a TEA CO_2 laser at 1 Hz with 0.2 - 1.5 J/pulse incident parallel to the sample surface about 1 mm from the sample. The sample was placed in a $CF_3 Br$ atmosphere at 5.5 Torr. Each laser pulse dissociated about 10% of the gas and the radicals produced attacked the SiO_2 resulting in products which are volatile at room temperature. The etch rate with the low power laser used was only 17 Å/min but with more powerful lasers Steinfeld et al. estimate that rates of 3500 Å/min are achievable. These rates would be comparable to those in present plasma processes.

The radical etching mechanism was also exploited by Ambartsumyan et al.[9]. They used a 25 watt pulsed TEA CO_2 laser at normal incidence to dissociate SF_6 by multiphoton dissociation. While dissociated SF_6 was sufficient to etch Si, the dissociation products are not able to etch SiO_2. Rather it was found that etching of SiO_2 occurs when hydrogen is present as well as SF_6 and there is adsorbed water on the SiO_2 surface. The etch rate was monitored by the production of the volatile product, SiF_4, which is actually formed after the laser pulse. The etch rate was estimated as $\approx$ 100 Å/min for laser intensity of 3.5 J/cm^2 using gas pressures of 2 Torr of SF_6 and 3 Torr of H_2.

Polymers can be cleanly and sharply patterned much more easily than SiO_2. This may lead to their use as self developing resists. Moreover, as the general use of polymers in semiconductor devices and interconnection packages expands, the opportunities for laser etching of polymers to become more important will also grow. The fact that polymers can be very cleanly patterned was described by Srinivasan et al.[10,11]. They used pulsed excimer radiation at 193 nm to etch polyethylene terephthalate (PET) and polymethylmethacrylate (PMMA). PMMA is a commonly used resist polymer. The high absorption coefficient of these polymers in the near UV is such that 95% of the radiation is absorbed in the first few thousand angstroms at the surface. At this wavelength the radiation is highly efficient at breaking bonds. The threshold for the etching process is only 10 mJ/cm^2 for PMMA. At lower fluences many of the bonds recombine although some react in air and oxidative photoetching occurs. Above threshold the bond breaking per unit time exceeds a critical value where ablative decomposition occurs. The driving forces for this phenomenon are the large increase in specific volume of the fragments compared to the polymer chain they replace, and the excess energy absorbed which excites the fragments vibrationally, rotationally and translationally. Since this is primarily a photochemical process at 193 nm, there is little dependence of the etch rate on sample temperature. Typical etch rates of PET in air were 1200 Å/pulse at 0.37 J/cm^2; and, for PMMA, the etch rate was 0.3 μm/pulse at 0.25 J/cm^2.

Brannon et al.[12] studied the etching of a polyimide film made by thermal curing of polyamic acid which was made from pyromellitic dianhydride (PMDA) and oxydianiline (ODA). Brannon et al. used different excimer wavelengths 248, 308 and 351 nm to quantify the etch rates and further understand the etching mechanism. For each wavelength the etch rate data fell on a curve which was well described by the equation:

$$x = k^{-1} \ln (E/E_T)$$

where x is the etch depth per pulse, k the absorption coefficient, E the laser fluence, and E_T the threshold fluence. This formula follows from the well known Beer-Lambert law of optical absorption with the assumption of a threshold fluence. From the slope of the etch rate per pulse versus the log of the fluence, an absorption coefficient was derived which was 1.6 x 10^5 cm^{-1} at 248 nm, 0.9 x 10^5 cm^{-1} at 308 nm, and 0.47 x 10^5 cm^{-1} at 351 nm. These values are all in quite reasonable agreement with the absorption coefficient as a function of wavelength measured at low intensity. The threshold fluence varied from 0.027 J/cm^2 at 248 nm to 0.07 J/cm^2 at 308 nm to 0.12 J/cm^2 at 351 nm. Brannon et al. point out that at this threshold the product of the threshold energy and the absorption coefficient is nearly constant independent of wavelength. This product is equal to the absorbed energy density at threshold ($\approx$ 5 x 10^3 J/cm^3). That it is the same at all wavelengths studied suggests that it is a material property which is required for rapid ablation. Thus Brannon et al. argue that, at the wavelengths they studied, the mechanism of polyimide etching is photothermal. The process is one in which the energy, initially absorbed as electronic excitation, is rapidly converted to vibrational heating of the solid. The intense local heating is confined by the low thermal diffusivity of the polymer and results in an explosive pyrolysis. This results in well defined clean etch features.

The etching of thin metal films can find microelectronic applications for circuit and interconnection package repair and customization. Thin film metal masks can have certain defects deleted and some devices may require trimming and shaping of conductors. Commercial laser based tools exist for many of these applications. As an example of the laser processes which are employed, Andrew et al.[13] showed that thin films of aluminum on substrates with poor thermal conductivity can be patterned with a single pulse of a UV excimer laser. Andrew et al. showed that aluminum films, 0.12 μm thick on glass substrates, could be removed by a single 308 nm excimer laser pulse at 0.2 J/cm^2. With somewhat higher pulse energies, $\lesssim$ 1 J/cm^2, films up to 1 μm could be removed. The energies needed to remove these films are far below the energy calculated to cause vaporization of the metal. The energy used corresponded more closely to the melting threshold. Hence, the mechanism of ablation was considered to be melting followed by an explosive process due to pressure buildup at the film/substrate interface as a result of decomposition or degassing of the substrate. In addition to aluminum thin films, Andrew et al. obtained qualitatively similar results for films of Ag, Au, Ni, Cu, and Cr on insulating substrates such as Mylar, Kapton, glass and fused silica.

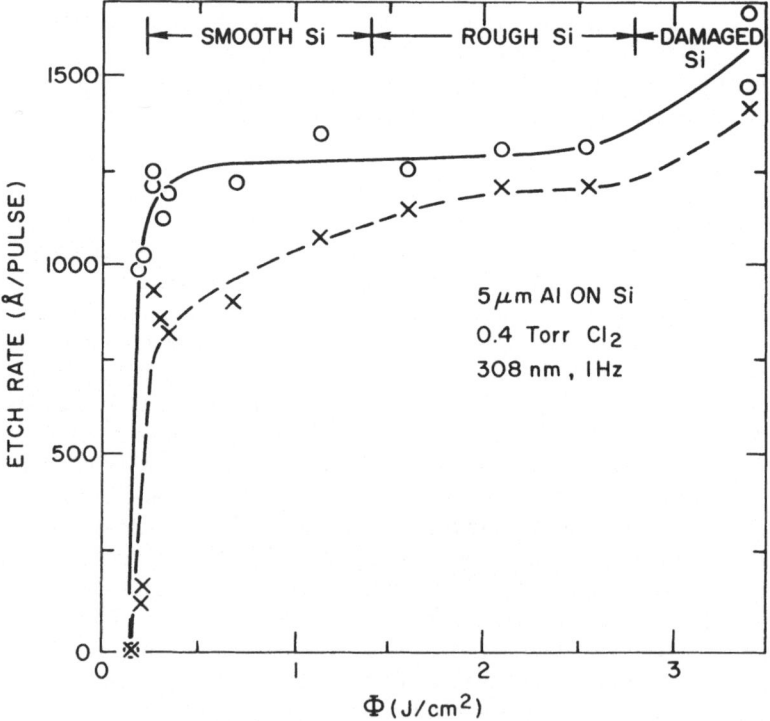

Figure 1   Etch rates on the sides of the etched area (O) and in the center when clean etching through the of Al film was obtained (X), as a function of the laser fluence on the sample. The solid and dashed curves are only guides to the eye.

A laser assisted chemical process which can be used to pattern much thicker metal films was described by Koren et al.[14,15] who etched aluminum in a Cl_2 atmosphere using an excimer laser at 308 nm. Once the protective aluminum oxide has been removed, aluminum metal films react with chlorine to form aluminum chlorides. These have low boiling points and actually sublime in vacuum leading to spontaneous etching of aluminum by chlorine. The rate for this process is very low however and is greatly enhanced by an excimer laser. Figure 1 shows the etch rate of aluminum in 0.4 Torr Cl_2 as a function of pulse fluence. In these studies a small rectangular region of the sample was etched and the depth profile was always deeper at the edges

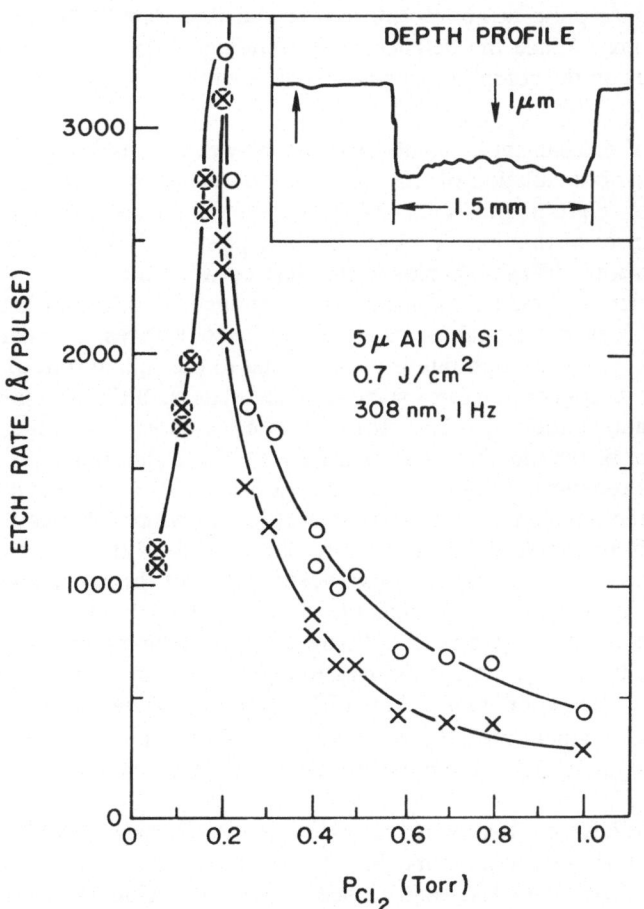

Figure 2   Sides (O) and center(X) etch rates vs the Cl_2 pressure in the cell. The curves are only guides to the eye. In the inset a depth profile is shown of a partially etched film at 0.4 Torr Cl_2 after 25 laser pulses of 0.7 J/cm² at 1 Hz.

of the rectangular region than in the center. Since the etch rates were determined by measuring the time to penetrate a film of a given thickness (usually 5 μm of Al on Si) two separate etch rates, for the center and edge regions, were recorded. These are indicated in figures 1 and 2 by crosses and circles respectively. Figure 1 shows that the etch rate is 1250 Å/pulse, essentially independent of the laser fluence between 0.25 and 3.0 J/cm^2 at a Cl_2 pressure of 0.4 Torr. At lower fluences the etch rate drops sharply with no detectable etching below 0.15 J/cm^2. At fluences above 3 J/cm^2 a very rough surface occurs along with damage (etching) of the Si substrate. The effect of the chlorine pressure on etch rate is quite dramatic as shown in figure 2. Beginning at low pressure, the etch rate at room temperature increases rapidly reaching 3300 Å/pulse at about 180 mTorr and then rapidly falling off with increasing pressure. While the detailed understanding of this effect is not known it clearly depends on the complex mechanisms of reaction and ablation.

The etching mechanism is mainly reaction between Al and Cl_2 in the dark followed by laser photoablation of the product. Photochemical effects are negligible. At constant low chlorine pressures ($\lesssim 0.2$ Torr) the etch rate is nearly proportional to the square root of the time between the laser pulses. This is consistent with the notion that reactive diffusion occurs in the time between the laser pulses and that a reaction product is formed with thickness determined by diffusion limitations. At low pressures (below the peak in figure 2) the laser removes essentially all of this reaction product and the etch rate is proportional to the square root of the time between pulses. At higher pressures something happens such that this is not the case. At pressures above about 0.2 Torr, the etch rate increases much more slowly than the square root of the time between laser pulses. The higher pressures of Cl_2 in the cell may cause redeposition of the ablated material so that each laser pulse does not fully remove the reaction product as it does at low pressure. A second speculation as to the origin of the peak in the pressure dependence of the etch rate is that the higher Cl_2 pressure inhibits the explosive volatilization of the reaction product by, in effect, inhibiting boiling of the aluminum chloride. This latter concept is partially supported by the fact that the peak in figure 2 is quite temperature dependent[14]. As temperature is increased, the peak moves to higher pressure and becomes stronger, reaching 1.4 μm/sec at 60°C in 2 Torr of Cl_2 when the illumination is 1.0 J/cm^2 at 1 Hz. At higher temperatures, the vapor pressure of the aluminum chloride is higher and greater pressures are required to suppress rapid volatilization.

In summary, it is clear that there are a wide variety of processes which can be used to directly pattern semiconductors, insulators, and metals. Laser etching shows promise in semiconductor laser and device fabrication including patterning, trench isolation, optoelectronic coupling and grating formation. Interconnection packages may be fabricated using metal and polymer etching techniques and a variety of repair, customization and trimming applications are feasible. Many of these offer considerable advantages over conventional lithographic techniques and their widespread use in the microelectronics industry is only a matter of time.

References

1. F. A. Houle, Mat. Res. Soc. Symp. Proc. vol. 29, p 203 (1984).

2. D. J. Ehrlich, R. M. Osgood Jr. and T. F. Deutsch, Appl. Phys. Lett. 38, 1018 (1981).

3. T. Arikado, M. Sekine, H. Okano, Y. Horiike, Mat. Res. Soc. Symp. Proc. 29, 167 (1984).

4. R. M. Osgood, A. Sanchez-Rubio, D. J. Ehrlich, V. Daneu, Appl. Phys. Lett. 40, 391 (1982).

5. Y. Aoyagi, S. Masuda, A. Doi, S. Namba, Japan J. Appl. Phys. 24, L294 (1985).

6. J. E. Bowers, B. R. Hemenway, D. P. Wilt, Appl. Phys. Lett. 46, 453 (1985).

7. R. M. Lum, A. M. Glass, F. W. Ostermayer, P. A. Kohl, A. A. Ballman, R. Logan, J. Appl. Phys. 57, 39 (1985).

8. J. I. Steinfield, T. G. Anderson, C. Reiser, D. R. Denison, L. D. Hartsough, J. R. Hollahan, J. Electrochem. Soc. 127, 514 (1980).

9. R. V. Ambartsumyan, Y. A. Gorokhov, A. L. Gritsenko, V. N. Lokhman, Sov. Tech. Phys. Lett. 8, 276 (1982).

10. R. Srinivasan, V. Mayne-Banton, Appl. Phys. Lett. 41, 576 (1982).

11. R. Srinivasan, J. Vac. Sci. Tech. B1, 923 (1983).

12. J. H. Brannon, J. R. Lankard, A. I. Baise, F. Burns, J. Kaufman, J. Appl. Phys. 58, 2036 (1985).

13. J. E. Andrew, P. E. Dyer, R. D. Greenough, P. H. Key, Appl. Phys. Lett. 43, 1076 (1983).

14. G. Koren, F. Ho, J. J. Ritsko, Appl. Phys. Lett. 46, 1006 (1985).

15. G. Koren, F. Ho, J. J. Ritsko, Appl. Phys. A40, 13 (1986).

LASER-INDUCED CHEMICAL PROCESSING OF MATERIALS

D. BAUERLE, T. SZORENYI, G.Q. ZHANG, K. PIGLMAYER, M. EYETT, R. KULLMER

Angewandte Physik. Johannes-Kepler-Universität Linz, A-4040 Linz, Austria

1. INTRODUCTION

Materials processing with lasers in a chemically reactive atmosphere (LCP) is a new and interdisciplinary field. The technique permits maskless single-step patterning of materials by deposition, surface modification, synthesis and etching [1,2]. The lateral dimensions of the processed regions range from several centimeters to less than one micrometer. Contrary to conventional techniques such as chemical vapor deposition (CVD), plasma deposition (PCVD) or plasma etching (PE), reactive ion sputtering (RIE) etc., LCP is not limited to planar substrates, but allows three dimensional fabrication as well.

In this paper we review recent results obtained in our group. Special emphasis is put on laser-induced gas-phase deposition (Sect. 1). Recent results on surface modification, such as the reduction and metallization of piezoelectric oxides (Sect. 2) and on laser-induced etching (Sect. 3) are briefly mentioned.

2. LASER-INDUCED CHEMICAL VAPOR DEPOSITION (LPCVD)

Laser-induced chemical vapor deposition has been demonstrated for metals, semiconductors and insulators. The deposition mechanisms may be based on pyrolytic (photothermal), on photolytic (photochemical), or on combined pyrolytic-photolytic (hybrid) activation of the chemical reaction The fundamental excitation mechanisms and the systems investigated have been reviewed in [1,2]. In the following we shall briefly discuss pyrolytic LCVD of microstructures in the form of spots and stripes. The deposition of W by hydrogen reduction of WF_6 will serve as a model system.

2.1. Deposition of Spots

Figure 1 shows scanning electron micrographs of a series of spots which have been deposited from an admixture of 5 mbar WF_6 and 500 mbar H_2 [3]. A Gaussian laser beam (647 nm Kr^+ laser; $2w^o = 7$ µm) at perpendicular incidence to the substrate was used. The substrate material was fused SiO_2 (a-SiO_2) covered with a 700 Å thick layer of sputtered W. In these experiments all parameters were held constant, except for the laser beam illumination time t_i, which was increased for each successive spot. The growth of spots is characterized by changes in both morphology and microstructure : with increasing laser beam illumination time t_i, the diameter d of spots increases, while the ratio of the diameter d and the maximum height h, i.e. d/h, decreases. The microstructure is polycrystalline with fine grains when deposition commences, and becomes

33

D. J. Ehrlich and V. T. Nguyen (eds.), Emerging Technologies for In Situ Processing, 33–43.
© 1988 by Martinus Nijhoff Publishers.

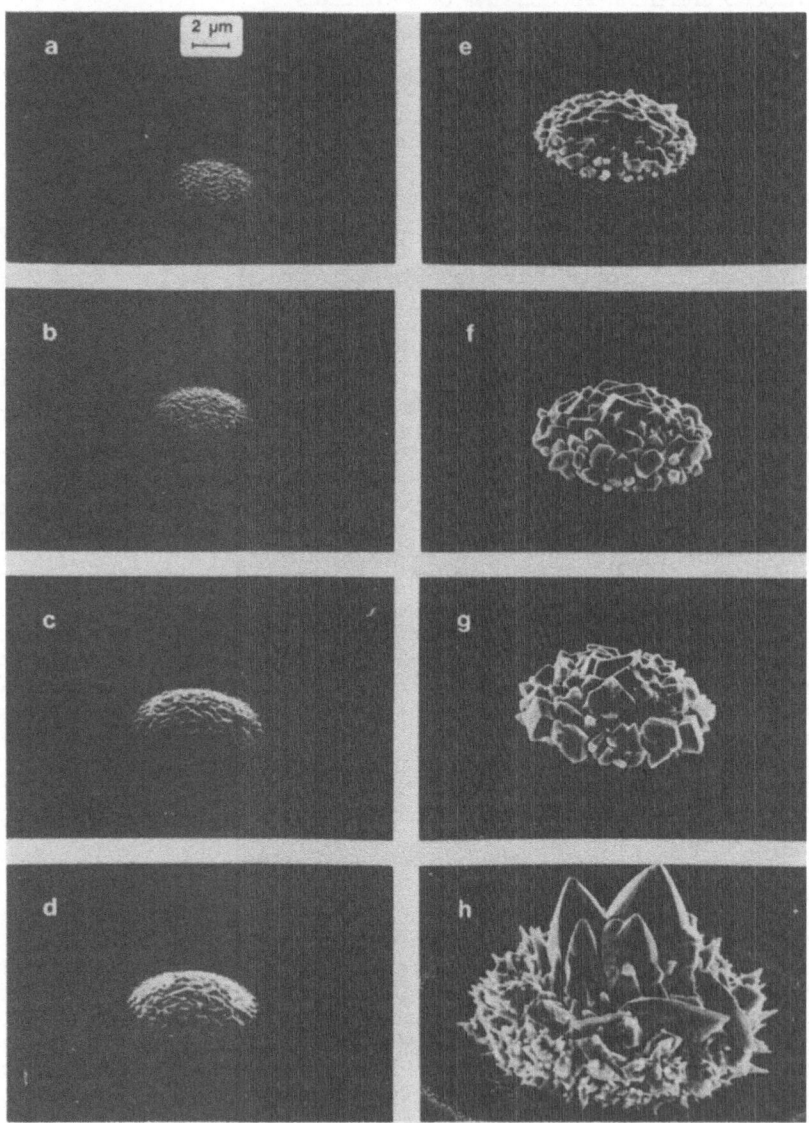

FIGURE 1. W spots deposited from an admixture of 5 mbar WF_6 and 500 mbar H_2 by means of 647.1 nm Kr^+ laser radiation (P = 155 mW, $2w_0 = 7\,\mu$m). The substrate was fused quartz (a-SiO_2) covered with a 700 Å thick layer of sputtered W. The laser beam illumination time t_i was increased from a) to h) with: a) 0.01 s. b) 0.02 s. c) 0.05 s. d) 0.07 s. e) 0.1 s. f) 0.2 s. g) 0.3 s. h) 0.5 s.

coarse grain with increasing t_i. For longer illumination times, the growth of one or several single crystals is observed (Fig. 1h). When increasing t_i even further, only one of the single crystals survives and the growth of a rod along the axis of the laser beam is observed. Similar experiments were performed with substrates of a-SiO$_2$ covered with 1000 Å amorphous Si (a-Si).

The diameter of W spots as a function of laser beam illumination time t_i is shown in Fig. 2 for various laser powers. Within the accuracy of the measurements, the spot diameters proved to be independent of the H$_2$ partial pressure which was varied between 50 and 1000 mbar. The spot diameter first increases very rapidly with t_i and then saturates at a certain value which depends on the laser power. This saturation in lateral growth is accompanied by an enhanced axial growth, i.e. the formation of a rod. For the a-Si/SiO$_2$ substrate, saturation occurs within somewhat shorter times (in about 0.5 s) and at somewhat larger diameters.

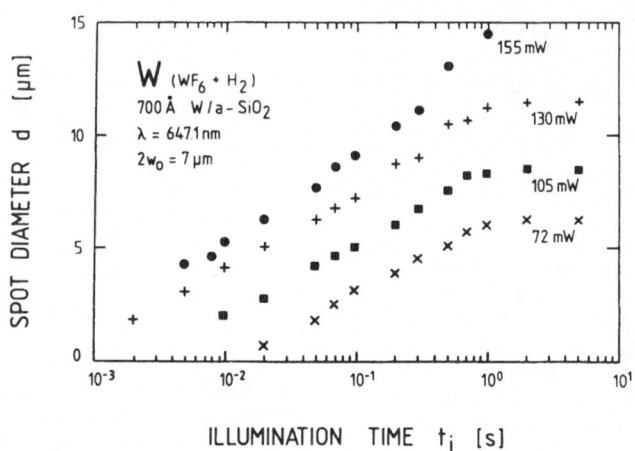

FIGURE 2. Diameter of W spots as a function of 647.1 nm Kr$^+$ laser beam illumination time. The laser focus was $2w_0$ = 7 μm. The substrate was a-SiO$_2$ covered with 700 Å sputtered W.

2.2. Direct Writing

Laser direct writing is performed by moving the substrate perpendicular to the laser beam. Some recent results are shown in Fig. 3 for the example of W stripes deposited on fused quartz (a-SiO$_2$) covered with a 1200 Å thick layer of amorphous Si (a-Si) [4]. The microstructure of the stripes is polycrystalline. The grain size depends on the laser power and the partial pressures of gases.

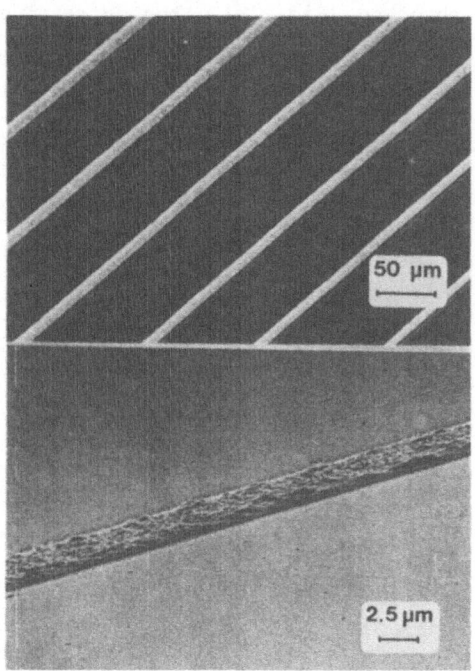

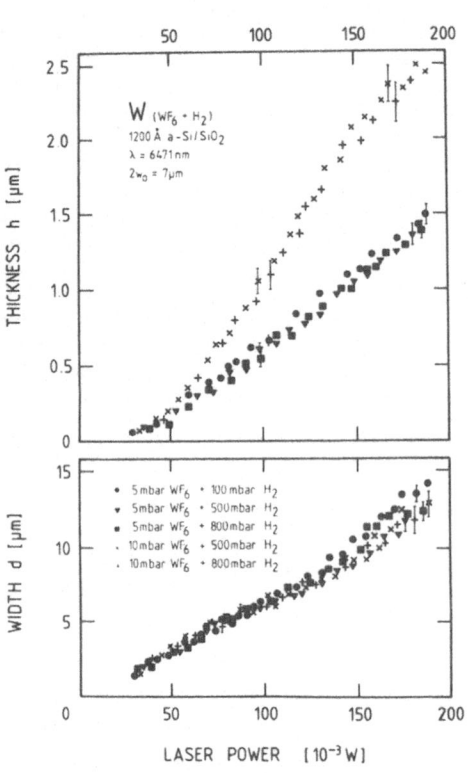

FIGURE 3. Scanning electron micrographs of W stripes deposited from mixtures of WF_6 and H_2 with 647.1 nm Kr^+ laser light ($2w_0 = 7$ μm). and a scanning velocity of $v_s = 100$ μm/s. a) $p(WF_6) = 5$ mbar, $p(H_2) = 400$ mbar, $P = 146, 132, 120, 107, 97$ and 83 mW (left to right). b) $p(WF_6) = 5$ mbar, $p(H_2) = 100$ mbar, and $P = 135$ mW. The substrate was $a-SiO_2$ covered with 1200 Å a-Si.

FIGURE 4. Thickness and width of W stripes as a function of laser power. The focus of the laser beam was $2w_0 = 7$ μm and the scanning velocity $v_s = 100$ μm/s. The substrate was $a-SiO_2$ covered with 1200 Å a-Si.

Figure 4 shows the thickness h and width d of stripes as a function of laser power for various partial pressures of WF_6 and H_2 and a scanning speed of $v_s = 100$ μm/s. Both d and h increase approximately linearly with increasing laser power. The width is, within the accuracy of the measurements, independent of both the WF_6 and the H_2 partial pressures. The thickness, on the other hand, increases with increasing WF_6 pressure but is independent of the amount of H_2. The ratios d/h are approximately 10 and 5 for WF_6 pressures of 5 and 10 mbar respectively, and they are nearly independent of laser power. The characteristic cross-section of stripes is a well-defined mesa type (nearly perfectly rectangular) with 5 mbar WF_6, and it changes to convex type (semi-circular) with 10 mbar WF_6. As already investigated for other systems, direct writing is characterized by a threshold power below which no deposition is observed [1]. For the parameters indicated in Fig. 4, this threshold value is at about $P \approx 30$ mW and, interestingly, it is independent of the partial pressures within the ranges investigated. Another feature is the increase in resolution which is observed for the lowest laser powers. Here, the width of stripes is only about 1/5 of the diffraction-limited diameter of the laser focus. This effect which originates mainly from the exponential dependence of the deposition rate on the laser-induced temperature distribution, has been observed before for other systems and is discussed in detail in [1]. Experimental results similar to those presented in Figs. 3 and 4 have been obtained also with substrates of a-SiO_2 covered with a 700 Å thick layer of sputtered W and of crystalline Si (c-Si) [5].

ELECTRICAL PROPERTIES

Figure 5 shows the resistance of stripes per unit length and their resistivity normalized to the bulk value of W ($\rho_B = 5.33 \cdot 10^{-6}$ Ωcm) as a function of laser power. Each data point in the resistivity curve represents an average of about 11 measurements on various stripes deposited on different but identically prepared substrate samples. The strong decrease in the resistance with increasing laser power is mainly due to the increase in the cross sections of stripes (see Fig. 4). The increase in resistivity with increasing laser power may be accounted for by stripe texture which becomes more pronounced the higher the laser power. It is remarkable to note that these resistivity values are the best reported for LCVD of W and they are only a factor of 1.3 to 2.3 larger than those of the bulk material.

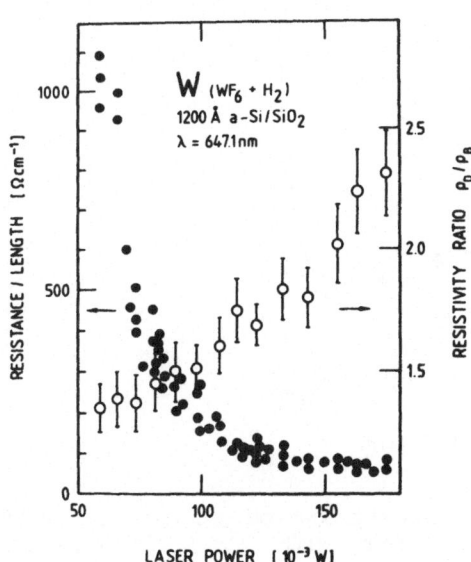

FIGURE 5. Electrical resistance (left scale) and resistivity ratio (right scale) of W stripes as a function of laser power (647.1 nm Kr^+, $2w_0 = 7$ μm). The partial pressures of WF_6 and H_2 were $p(WF_6) = 5$ mbar and 100 mbar $< p(H_2) < 800$ mbar. A value of $\rho_B = 5.33 \cdot 10^{-6}$ Ωcm has been used for bulk W. The substrate was a-SiO_2 covered with 1200 Å a-Si.

3. REDUCTION AND METALLIZATION OF PIEZOELECTRIC OXIDES

Laser-induced surface modification has been demonstrated for piezoelectric oxides such as BaTiO$_3$, SrTiO$_3$, PbTi$_{1-x}$Zr$_x$O$_3$ (PZT), Pb$_{1-3y/2}$La$_y$Ti$_{1-x}$Zr$_x$O$_3$ (PLZT) and LiNbO$_3$. These materials are insulators with a band-gap of, typically, Eg 3 eV. When scanning a cw-Kr$^+$ or Ar$^+$ laser beam across the sample surface in a chemically reducing atmosphere such as hydrogen, semiconducting or metallic stripes are produced, depending on laser power, laser beam scanning velocity, and H$_2$-pressure [6-8]. This change in physical properties within the laser-processed region is based on the local depletion of oxygen [9]. Laser-induced metallization has been studied in detail for electro-optical PLZT ceramics Multiple line output (337-356 nm) of a Kr$^+$ laser was used. The power dependence of the resistance per unit length of stripe on PLZT is shown in Fig. 6. Above the reduction threshold of P 70 mW a drastic decrease in resistance with increasing laser power is observed. The dependencies on scanning velocity and H$_2$ pressure show pronounced minima [7] in resistance for certain ranges of these processing parameters. This stresses the importance of the availability of H$_2$ and the laser beam dwell time for the metallization process.

Potential applications of laser-induced surface metallization can be envisaged in the production of electrodes. Laser-fabricated electrodes on PLZT have been shown to lead to higher contact capacitances than evaporated Au contacts [8], which in combination with the increased adherence of laser generated microstructures may be favourable for application to miniaturization problems.

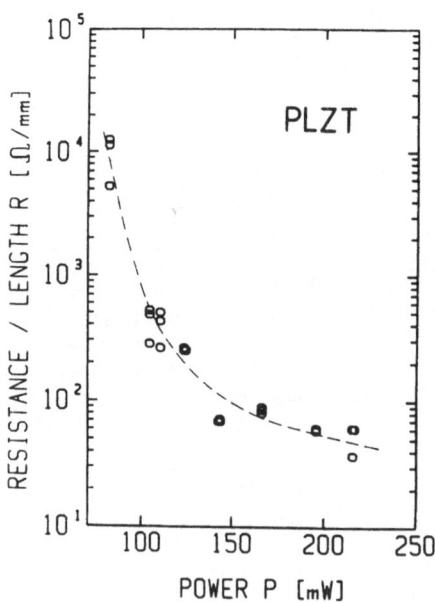

FIGURE 6. Resistance per length of laser-processed stripes on PLZT 9.5/65/35 as a function of laser power with v$_s$ = 25 μm/s and p(H$_2$) = 500 mbar.

4. ETCHING

In the following we shall briefly report on the laser-induced etching of ceramic PZT and single crystal Si.

4.1. Etching of PZT

When increasing the laser intensity above that used for metallization (Sect. 2), etching occurs. The most detailed investigations have been performed with UV and visible cw-Kr$^+$ or Ar$^+$ laser light [11], and more recently, with pulsed 308 nm XeCl excimer laser radiation [12].

With visible laser light, i.e. below band-gap radiation, regular structures can be produced only in a reducing atmosphere. On the other hand, continuous wave ultraviolet Kr$^+$ or pulsed XeCl excimer laser radiation with a photon-energy $h\nu > E_g$, permits regular etching also in air [11,12]. The role of H_2 for sub-band gap radiation is interpreted by the production of oxygen vacancies via reduction leading to locally enhanced and well defined absorption. Furthermore. a high concentration of oxygen vacancies favours a collapse of the perovskite lattice [9,11]. The drawback of using cw-laser radiation for PZT processing lies in a thermally damaged region around the etched pattern which can be seen in Fig. 7a. This region

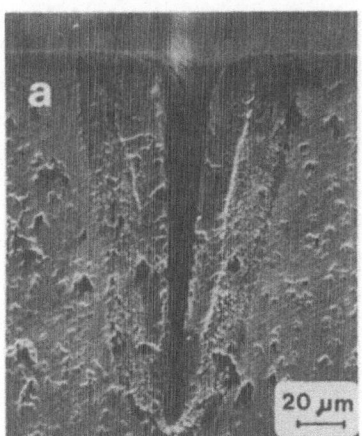

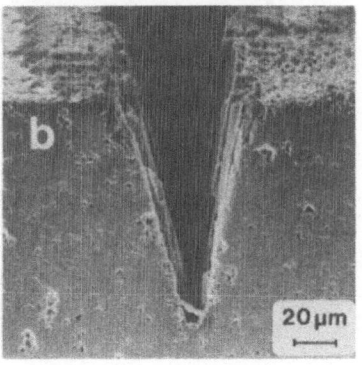

FIGURE 7. a) Scanning electron micrograph of a groove etched with cw Kr$^+$ laser light in an H_2 atmosphere: P = 0.72 W, λ = 647 nm, $2w_0$ = 18 μm, v_s = 8.4 μm/s, $p(H_2)$ = 90 mbar. b) Groove produced in air with a XeCl excimer laser using a stationary line focus of width 50 μm. Laser fluence Φ = 3.2 J/cm^2, N = 104 shots.

is porous and highly deficient in Pb and oxygen, but nevertheless still polarizable.

Thermal damage can be avoided by using pulsed lasers as, for example, excimer lasers. We have therefore investigated the etching of PZT in air using 308 nm XeCl excimer laser radiation [12]. Fig. 7b shows a groove produced in air by focusing the XeCl excimer laser light to a line of width w = 50 μm (FWHM). The dimensions of the groove are comparable to the one shown in Fig. 7a produced with cw Kr^+ laser light, but lacks the substantial damaged zone of the latter.

The etch rate of PZT as a function of fluence is shown in Fig. 8. A threshold fluence of about $\Phi_t \sim 2$ J/cm^2 is observed below which no etching occurs. The low fluence region $\Phi < \Phi_t$ can be divided into two ranges: first, a range up to a fluence Φ_1, where the laser pulse causes no changes in surface morphology, and second, a range $\Phi_1 < \Phi < \Phi_t$ where a pronounced change in surface morphology can be observed in scanning electron micrographs, similar to the surface regions next to the edges of the groove shown in Fig. 7b. For fluences $\Phi > \Phi_t$ the etch rate increases initially with ln (Φ/Φ_t).

In view of potential applications of this process we mention the possibility of defining the shape of the etched cross-section by a suitable choice of processing parameters. This makes the method versatile for fabricating piezoelectric devices based on PZT, such as ultrasonic arrays, jet printers etc.

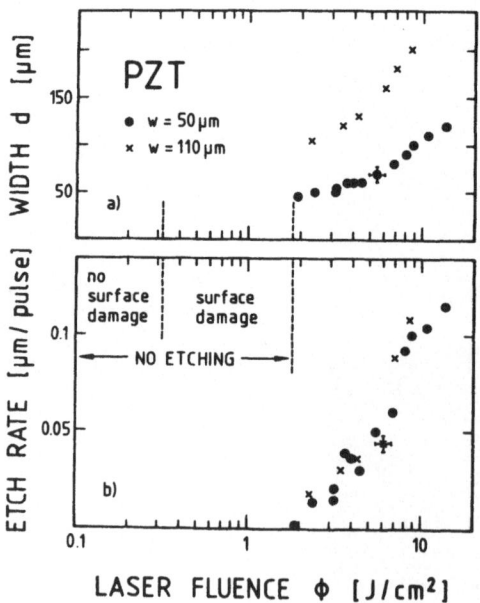

FIGURE 8. Width of grooves and etch rate as a function of the incident laser fluence. • w = 50 μm; x w = 110 μm.

4.2. Etching of Silicon

Laser-induced chemical etching of semiconductors is a powerful technique for processing electronic materials. Of particular importance is laser-induced dry etching of silicon which can be achieved in chlorine atmosphere. To obtain a better understanding of the microscopic mechanisms, the etch-rate dependence on energy density and wavelength of the laser light has been investigated for various chlorine pressures. A detailed description is given in [13].

Experiments were performed mainly with a XeCl-excimer laser (λ = 308 nm) and an excimer laser pumped dye laser (λ = 423 nm, 583 nm). At normal incidence to the Si substrate, the laser beam was collimated to a spot of, typically, 150-500 μm^2, resulting in fluences of up to 1.2 J/cm^2. The samples were lightly p-doped (100) Si wafers. Chlorine pressures up to 100 mbar were used. Fig. 9 shows the etch rate for excimer laser irradiation as as function of energy density and various Cl$_2$ pressures. The laser-induced maximum surface temperature which was calculated by numerically solving the one-dimensional heat equation is shown in the upper scale.

Fig. 10 shows the etch rate as a function of laser fluence for three

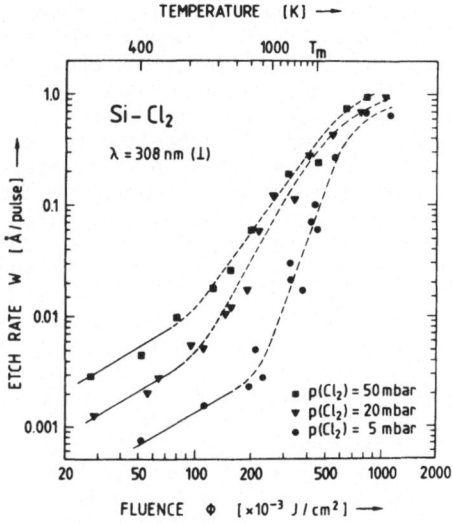

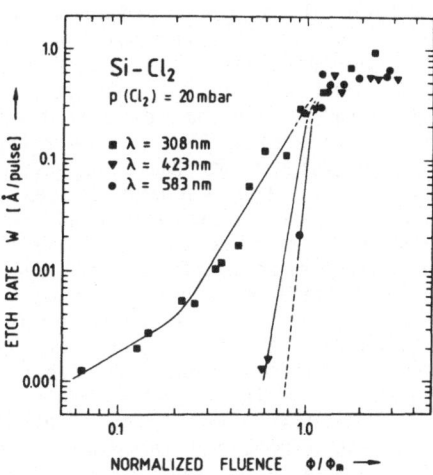

FIGURE 9. Etch rate of Si as a function of 308 nm XeCl-excimer-laser fluence for various Cl$_2$ pressures. The temperatures indicated on the upper scale were calculated.

FIGURE 10. Etch rate of Si for three different wavelengths. The fluence was normalized to the fluence necessary for surface melting. The Cl$_2$ pressure was kept constant at p(Cl$_2$) = 20 mbar.

different wavelengths. Fluences have been normalized to the fluences $\Phi_m(\lambda)$ necessary to melt the surface. For fluences $\Phi > \Phi_m$ the etch rate saturates and becomes nearly independent of the wavelength indicating a purely pyrolytic process. In this regime, the etch rate depends only slightly on chlorine pressure which shows that chlorine atoms or molecules adsorbed on the Si surface are involved in the etching process.

Below the melting point, the etch rate increases linearly with chlorine pressure. Therefore, the flux of chlorine atoms or molecules to the surface is a rate limiting factor in this regime. For $\Phi < \Phi_m$ probably no etching occurs with 583 nm radiation. With $\Phi < 0.9 \Phi_m$ the etch rate is definitely below the detection limit which was about 0.0008 Å/pulse. At 423 nm etching takes place, however the etch rate is much smaller than for 308 nm. Chlorine molecules can be dissociated only by laser light having a wavelength below 500 nm, with a strong increase in absorption cross section towards the near UV. Therefore. the results shown in Fig. 10 demonstrate that below the melting point, heating of the Si surface results in etching only in the presence of gas-phase chlorine atoms. Heating of the surface will facilitate penetration of these Cl atoms into the silicon lattice and enhance Si-Si bond breaking and desorption of $SiCl_x$ species.

For low laser fluences only a modest temperature rise takes place and etching is a photolytic process involving chlorine atoms generated in the gas phase and photocarriers in the semiconductor. Etching involves several reaction steps [14]: Irradiation of the semiconductor results in the generation of electron-hole pairs Photoelectrons will be transferred to chlorine atoms impinging onto the silicon surface thereby forming Cl^- ions. These ions can easily diffuse into the silicon lattice and break Si-Si bonds which leads to the formation of $SiCl_x$ species A comparison between the chlorine atom flux to the surface and the observed etch rates indicate that chlorine atoms impinging up to about 100 ns after the laser pulse are involved in the etching process. As for our experimental conditions, chlorine atom recombination takes place on a millisecond time scale, the chlorine atom flux to the surface during and immediately after the laser pulse is proportional to the laser fluence Φ. Due to the high laser fluence and the small absorption length of only 70 Å high photocarrier densities will be generated. Auger recombination will dominate which results in a photocarrier density being only slightly dependent on the laser fluence in the time interval mentioned above. Therefore an almost linear increase of the etch rate with laser fluence Φ can be observed in this regime

This mechanism was further investigated in another experimental setup by employing a combined irradiation scheme. An excimer laser beam parallel to the surface generated chlorine atoms in the gas phase, and a HeNe laser focused perpendicularly onto the sample provided the photolytic excitation of the semiconductor. Etching was observed only in a region around the HeNe laser spot. As can be expected from the model described above, the etch rate increased both with chlorine-atom concentration and HeNe laser intensity. As for the low HeNe laser intensity the photocarrier concentration will increase linearly with laser intensity.

REFERENCES

1. Bäuerle D: Chemical Processing with Lasers. Springer Series in
    Materials Science 1. Berlin, Heidelberg: Springer, 1986.
2. For recent reviews see: a) Laude LD. Bäuerle D, Wautelet M(eds.):
    Interfaces under Laser Irradiation. Nato ASI Series. Dordrecht:
    M. Nijhoff, 1987. b) Bäuerle D. Kompa KL, Laude LD(eds.): Laser
    Processing and Diagnostics II. Les Ulis: Les Editions de Physique,
    1986. c) Bäuerle D(ed.): Laser Processing and Diagnostics. Springer
    Series in Chemical Physics 39. Berlin, Heidelberg: Springer, 1984.
    d) Johnson AW, Ehrlich DJ, Schlossberg HR(eds.): Laser Controlled
    Chemical Processing of Surfaces. New York: North-Holland, 1984.
    e) Osgood RM, Brueck SRJ, Schlossberg HR(eds.): Laser Diagnostics and
    Photochemical Processing for Semiconductor Devices. New York: North-
    Holland, 1983.
3. Szörényi T, Piglmayer K, Zhang GQ. Bäuerle D: to be published
4. Zhang GQ, Szörényi T, Bäuerle D: J. Appl. Phys., 1987
5. Zhang GQ, Piglmayer K, Szörényi T, Bäuerle D: to be published
6. Otto J, Stumpe R, Bäuerle D: in Laser Processing and Diagnostics,
    Bäuerle D(ed.). Springer Series in Chemical Physics 39. Berlin,
    Heidelberg: Springer, 1984, p. 320
7. Kapenieks A, Eyett M, Bäuerle D: Appl. Phys. A 41, 331 (1986)
8. Kapenieks A, Stumpe R, Eyett M, Bäuerle D: in Proc. of the 6th IEEE
    Int. Symp. on Applications of Ferroelectrics, 1986, p. 696;
    Kapenieks A, Eyett M, Stumpe R, Bäuerle D: in Laser Processing and
    Diagnostics II, Bäuerle D, Kompa KL, Laude L(eds). Les Ulis: Physique,
    1986, p. 165
9. Perluzzo G. Destry J: Can J. Phys. 56, 453 (1978); Sprogis A. Dimza V:
    phys. stat. sol. (a) 72, K57 (1982); Bäuerle D, Wagner D, Wöhlecke M,
    Dorner B, Kraxenberger H: Z. Physik B 38, 335 (1980)
10. Eyett M, Bäuerle D: to be published
11. Eyett M, Bäuerle D, Wersing W, Lubitz K, Thomann H: Appl. Phys. A 40,
    235 (1986)
12. Eyett M, Bäuerle D. Wersing W, Thomann H: J. Appl. Phys. (1987)
13. Kullmer R, Bäuerle D: Appl. Phys. A 43 (1987)
14. Winters HF, Coburn JW, Chuang TJ: J. Vac. Sci. Technol. B1 (2),
    469 (1983)

HIGH TECHNOLOGY MANUFACTURING: CRITICAL ISSUES FOR THE FUTURE

E.L. HU

ECE Department and Center for Robotic Systems
      in Microelectronics
University of California
Santa Barbara, CA 93106

1. INTRODUCTION

The economy of fabrication integration has moved the semiconductor integrated circuit (IC) industry forward at an unparalleled pace. The steady move to greater integration of computational functions achieved by smaller minimum feature sizes and larger wafer and die sizes has characterized IC technology since its inception. Figure 1 details the increase in circuit complexity over the past ~ 25 years. For a large part of that time, circuit complexity doubled each year.[1] The continual decrease in device dimension and increase in wafer size, have allowed a greater number of more sophisticated devices to be produced per wafer. The level of circuit integration has proceeded from "small scale integration" (SSI, ~100 gates) to the threshold of "ultra-large scale

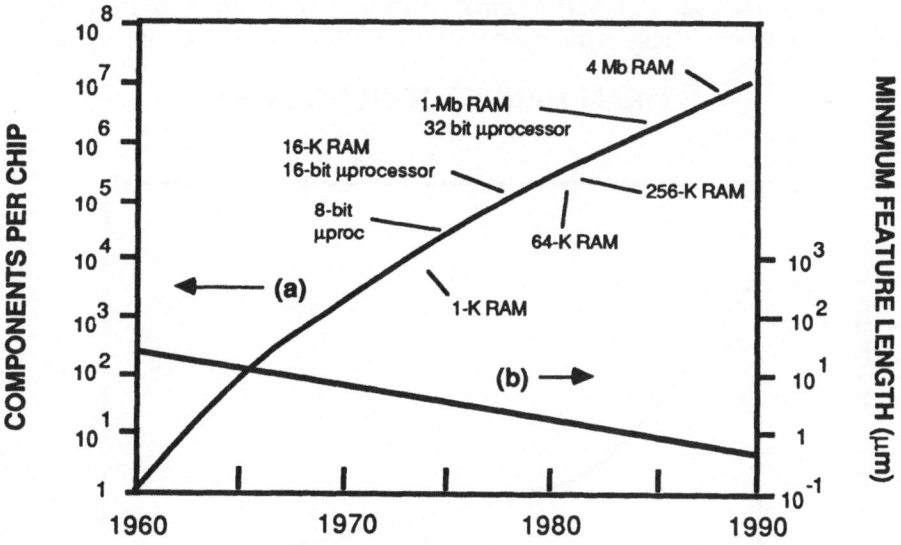

Figure 1. a) Increase in circuit complexity, or components per chip, as a function of time. b) Decrease in minimum device dimension as a function of time.

D. J. Ehrlich and V. T. Nguyen (eds.), Emerging Technologies for In Situ Processing, 45–54.
© *1988 by Martinus Nijhoff Publishers.*

integration" (ULSI, ~ 10^9 gates). This has been influential in achieving the reduction of cost per "bit", as shown in Figure 2. As in Figure 1, we note that the most recent data shows a leveling off of the rate of change of either reduction in device dimension or reduction in cost per bit. This levelling off reflects changes in and current constraints to device and processing technology, and suggests re-evaluation of "traditional" manufacturing strategies for continued economic viability in this area . Those traditional strategies have included extensive use of "batch processing" (simultaneous processing of large numbers of wafers), performed on increasingly sophisticated, disjoint equipment within extensive clean room facilities.

2. THE CLEAN ROOM ENVIRONMENT

As minimum device dimensions are reduced, reduced as well is the "critical" size of particles that can cause "fatal defects" or can render sections of the circuit inoperable by their presence (e.g. by electrically shorting together two metal lines). Current IC fabrication is carried out in specialized "clean rooms" of Class 100 or better. A Class 100 clean room ensures an environment having fewer than 100 particles/ft^3 of diameter 0.5 µm or larger. Standards have not yet been officially set for Class 10 or Class 1 clean rooms, although one may extrapolate a definition from the Class 100 specification. Clean room air is rapidly

DRAM MANUFACTURING COST

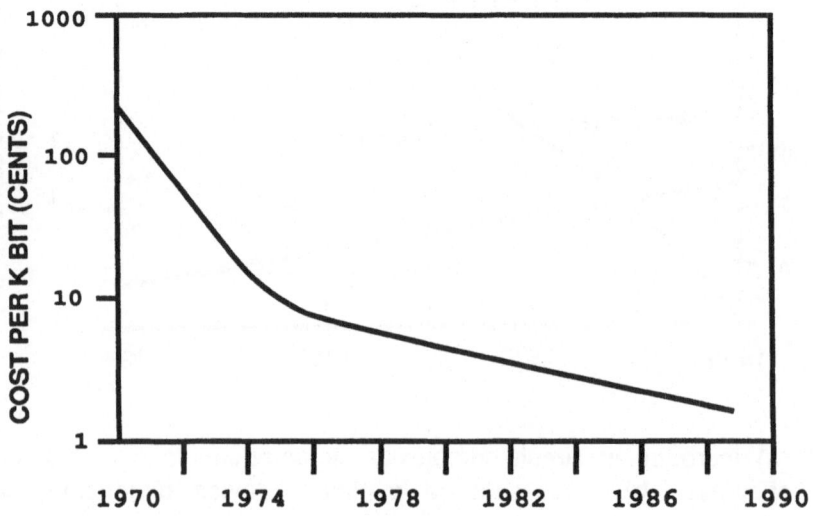

Figure 2. Reduction of IC cost as a function time.

circulated through High Efficiency Particulate (HEPA) filters at rates of a few hundred cubic feet per minute; temperature and humidity of the clean room environment are tightly controlled.   It is not surprising, therefore that the initial capital investment in clean room construction is high, as are the costs for continued operation of the clean room alone. So critical is cleanliness to the ability to manufacture IC's at reasonable yield, that costs notwithstanding, there is a   projected 40% rate of increase in square footage of Class 1 clean rooms between 1985 and 1990 [2]. Clearly, the rising costs of a fabrication environment having ever greater constraints on cleanliness will eventually pose questions of economic feasibility.

3. BATCH PROCESSING

Fundamental to the success of IC manufacture has been the extensive use of "batch" processing in which "lots" of ~20-40 wafers simultaneously go through the various processing steps involved.   The total processing costs can therefore be distributed over a large number of fabricated devices.   A typical sequence of processes to be carried out might be represented by the diagram in Figure 3.   In order to understand the efficiencies of the process, it is important to note those steps which are most commonly repeated in this sequence: i.e. lithography, inspection and cleaning.   It is also important to note that the various processes are carried out in a variety of machines which may be fairly widely distributed in space over the fabrication area.   Early in the history of IC manufacture, such reduction of critical dimensions was achieved fairly easily, and the benefits realized from such reductions were great.   The *continued* ability to realize such improvements and reduction in cost may be reaching some limits, as shown in Figure 2, by the flattening out of the curve of cost reduction as a function of time.   This is borne out by the data of Table I, which suggests that as we progress to higher levels of integration, incorporating micron and submicron device geometries, the fabrication technology alone is providing greater constraints to achievement of high yield. A previously acceptable variation of processing (such as etch rate, film thickness) over a single wafer or from wafer to wafer , that was inherent in batch processing is no longer tolerable. For example, if processing tolerances were a simple percentage of device size, say ±10%, then linewidths of 20µm could have a variation of ±2 µm, but current 1µm technology will only tolerate ±0.1 µm.   The former may be reasonably achievable, but the latter constraint may require more individualized monitor than can be obtained with batch processing.

As a specific illustration of the trends discussed above, consider the progression of pattern transfer (or etching)   techniques.   Original etch-transfer of patterns was accomplished by wet chemical etching, simply

TABLE 1. Cost reduction vs. Level of Integration[12]

Type of production line	Average annual cost reduction	Description
SSI/MSI	53%	equipment inexpensive, large benefits from device shrinkage
LSI	24%	innovations in equipment: projection aligners, plasma etchers
VLSI	20%	μm and sub-μm geometries provide technoligical problems in processing equipment expenses high defects severely limit yield

performed and amenable to batch processing (dipping cassettes of wafers into etch solutions). The implementation of dry or plasma etching techniques greatly improved the cleanliness and the resolution achievable. Wet etching inevitably takes place isotropically, or with an anisotropy determined by crystallographic planes of the substrate . The etchant baths will contain particles of the etch material which must be filtered out if they are not be deposited onto the surface of the wafers themselves. Use of reactive radicals or ions, formed from plasma discharges, allows a greater control, reproducibility and resolution to the process. Batch processing is still possible, using either planar reactors that will accomodate several wafers at a time or "hexode" etchers that will etch ~20 wafers at a time. In all reactors thus far designed, particularities in the reactor geometry, gas flow pattern or power distributions have led to small variations in etch rate and selectivities among all wafers etched. Given the ever tighter constraints on device fabrication, those variations are becoming increasingly insupportable - so much so that there is currently serious exploration of "single wafer at a time" etchers which will allow sensitive monitoring and adjustment of process conditions for each wafer. The cost of such an approach is reduced throughput - the compelling reason for its implementation is high yield.

4. AUTOMATION

As discussed previously, current micron and submicron device geometries place greater demands on the fabrication technology and

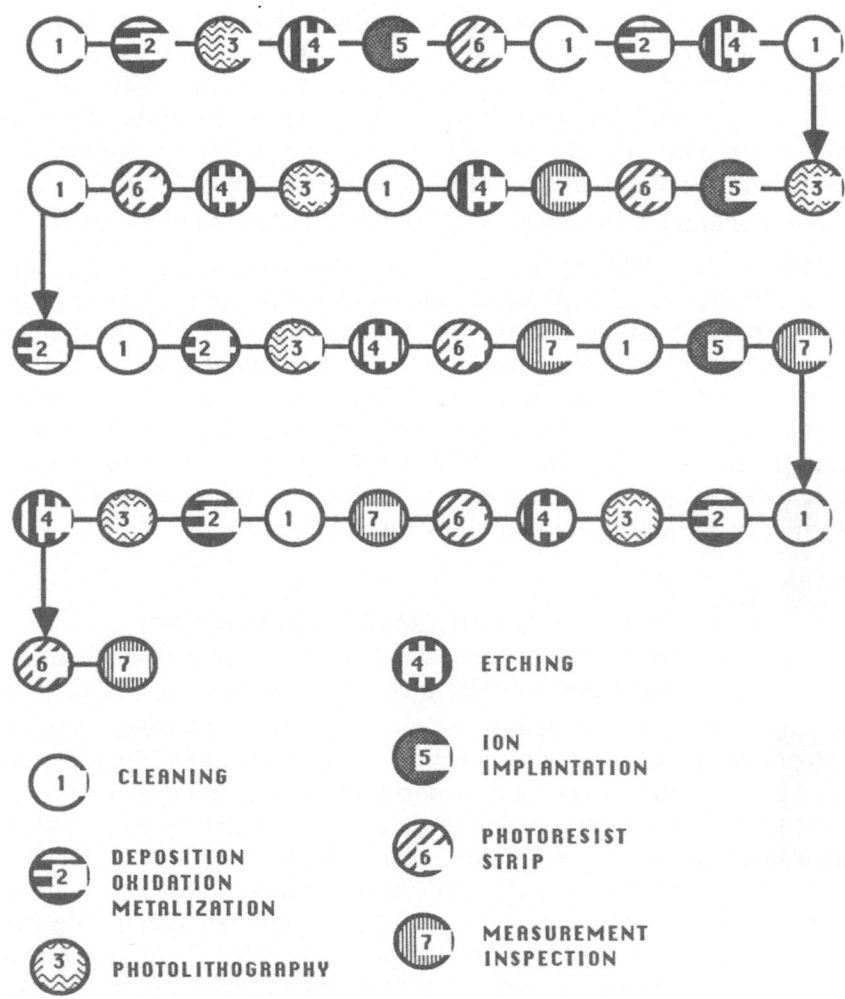

Figure 3. "Typical" process sequence for IC fabrication.

therefore on the equipment used to carry out that fabrication. Individual pieces of processing equipment have grown increasingly sophisticated, with a high degree of microprocessor control; however, the transport of cassettes of wafers, the loading and unloading of wafers from machine to machine, machine and process qualification, the all-important inspection steps and the critical exercise of judgement are still almost completely operator-performed. In order to ensure a viable, manufacturing technology, this is a situation that must change. Robotic, or automated IC processing would far better achieve (1) the necessary cleanliness and (2) process monitoring and control for high yield; and (3) effective scheduling and facility management for the most efficient utilization of process equipment to ensure maximum throughput and minimum cycle time (total required to process a lot of wafers from start to finish). Perhaps the paramount issue is *cleanliness*. Human beings are generators of particles ($\sim$6000/ft^3/minute, even in clean room garb[3]). In addition, current fabrication technology calls for a degree of control, accuracy, reproducibility and detailed analysis that is beyond the capabilities or patience of human beings. The predominant opinion has been that automation is not only highly desirable, it is mandatory for effective manufacturing. There are very few examples of its full implementation into a fabrication line, and many reasons might be cited for this. However, rather than explore the bottlenecks to automation in this paper, let us turn to two examples of fully automated lines as realize by Mitsubishi and NMB Semiconductors.

5. THE MITSUBISHI AND NMB FULLY AUTOMATED CLEAN ROOMS

The Mitsubishi Saijo factory consists of two building dedicated to production of 64K and 256K DRAMs, built at a total cost of $\sim$\$320 million [4]. The capacities of these plants are on the order of 10 million parts per month (estimated $\sim$20,000 wafer starts). The fabrication area consists of Class 10/Class 100 areas, and is designed with a "main street" , "side street" concept: trackless, optically guided automatic vehicles (AGV's) carry cassettes of wafers down the main streets to I/O stations at the junctures of the main streets and side streets. The side streets represent , specific, individual process areas, such as photolithography. Robots moving in the side streets transfer cassettes from I/O stations to the various processing machines. Inspection is done by closed circuit TV (CCTV), made by operators outside of the clean room area. The NMB automated facility has many points in common, but chooses to use an overhead rail system to transport the cassettes of wafers [5]. The clean room area is divided into a Class 1 fabrication area (total area $\sim$ 10,000 ft^2), Class 30 maintenance area and Class 300 operator/monitor area. Monitoring here is also done by CCTV. Mitsubishi claims a total fabrication to proble cycle time of 3 weeks. Although not disclosing the absolute device yields, the *relative increase in yield due to automation* was claimed as being 20%. Cycle time for the NMB 256K memories is $\sim$ 6 weeks.

6. APPLICATION SPECIFIC (ASIC) OR CUSTOM IC'S

The accomplishments of the fully automated fabrication lines, described above, have been formidable. Those lines were designed for the production of large quantities of memory chips, characterized by regular, repetitive arrays of memory cells on a single chip. Once the processes comprising the fabrication sequence have been finely tuned, such plants can efficiently and economically produce a high volume of

such chips; in fact such plants are designed to be operated at top rates for peak efficiency, and may not be easily scaled back to produce lower volumes of material. A problem situation then arises during periods of sluggish economic growth and lessened demand; the operation of such "monster" IC plants[6] is then no longer economical. In addition, an increasing part of the IC market is being captured by the custom, semi-custom or "application-specific" (ASIC) IC's. In a sense, this is a logical outcome of the increased sophistication of individual chip performance; as more functions are incorporated onto a single chip, it begins more and more to resemble a system itself, rather than an individual component. The pace of ASIC development has been facilitated by sophisticated computer aided design tools (CAD) that allow the rapid design and design verification of innovative custom chips. The fabrication requirements of ASICs are not the same as those applicable to the high volume "commodity" chips that have dominated the industry in the past. ASIC manufacture will emphasize a large variety of chips and perhaps of technologies; these custom chips are expected to be produced in much smaller volume than the commodity memory chips have been, they will place a premium on rapid cycle time producing a high yield on the individual wafer level (rather than averaged over a large number of lots). In other words, the ASIC manufacturing environment will have to be *flexible;* it will need to be low-cost at low-volume: the operating costs cannot now be averaged over many 1000's of wafers, and the "learning curve" for high yield manufacture must be reached in a much shorter time than is true presently.

These requirements re-emphasize a need to re-evaluate the more "traditional" means of IC manufacture. A large degree of intelligent automation in ASIC manufacture is mandatory, if we are to rapidly and continually implement new designs and technologies, achieving high yield and short cycle times. Large, expensive clean room facilities, even those that are fully automated, may prove too inflexible and too costly for ASIC manufacture. Lastly, as high yield and short cycle time are paramount in realizing economic advanatges to new chip designs, batch processing and high throughput may have lesser considerations. If single wafer at a time processing can accomplish the degree of process control and reliability that is required, and batch processing cannot, there is much less argument for these low volume products as to which method should be pursued.

7. IN SITU PROCESSING

There are other device technologies, in addition to the ASICs described above, where devices are to be produced in small volume, where device uniqueness, rather than quantity and cost are the important economic factors, and where a high degree of monitoring and control of sophisticated processes need to be performed within a well-controlled fabrication environment. Optoelectronic devices, such as lasers, modulators and detectors, fabricated from heterostructure compound semiconductors (e.g. GaAs, AlGaAs) fall within this category. In fact, so important is the control of the fabrication environment, that Optoelectronic Joint Research Laboratory (OJRL)[7] has made a strong investment in the realization of an ultra high vacuum (UHV) *in situ* processing station that would incorporate Molecular Beam Epitaxy (MBE) substrate growth chambers with chambers to perform ion beam modification (such as focussed ion beam, dry etching), metallization, analysis, all connected by transfer tubes, and all maintained at high vacuum conditions. A number of individual laboratories are also exploring various degrees of this *in situ* processing concept. Such an approach would allow the sequential execution of the requisite device processing steps to take place within a *compact*, controlled environment, with sufficient analytical capabilities that continual and adequate monitoring and inspection could also be performed in situ. Such an approach is generally recognized to be useful and necessary for many of the optoelectronic devices under consideration, where large volume production is not a critical consideration, and such in situ control determines whether or not the device can be fabricated at all. What implications does such an approach have for future silicon IC manufacture, which may include an ever larger share of ASICs?

8. FACTORY OF THE FUTURE?

Recognizing the need for a comparable degree of control in the process environment for certain steps that require a number of steps to be rapidly performed in succession, some semiconductor equipment manufacturers have already come out with modular, multi-chamber process tools. Examples of such equipment are the Drytek Quad System[8] and the Applied Materials Precision 5000 CVD[9]. An example of a multi-step process would be the dry etching of aluminum features, using photoresist masking. It has been found to be desirable to remove the photoresist immediately after having etched the aluminum features, without having exposed the wafer to the ambient atmosphere, i.e. to do an *in situ* removal of photoresist. By-products of the Cl_2 etching of aluminum, adsorbed onto the remaining photoresist, may react with water vapor in the air and form HCl which can then attack the aluminum itself[10]. Because of the sensitivity of dry etching (and other) processes to the history of the reactant chamber, it is very often unwise to perform a sequence of steps, utilizing different gases and process conditions within the same chamber. A multiple chamber configuration also improves cycle times by allowing several processes to be carried on simultaneously. We thus have the capability of utilizing the control possible with single wafer at a time processing, while achieving reasonable cycle times with the multiple processing technique. Such systems currently are cassette-loaded into load-locks. All processing and robotic transfer is done subsequently carried out under vacuum conditions.

The commercial availability of such machines both suggests a response and indicates a solution to the limitations of conventional IC fabrication. It also suggest that the notion of *in situ* processing, as applied to silicon IC's, may represent a realistic, viable means of producing the requisite chips of the future. If we look again at Figure 3, we note that each of the numbers corresponds to one or more separate machines or stations which may be widely dispersed across the factory floor. One may then add to the cycle time because of the delays and distances over which wafers must be transported in trying to follow the process flow. Some of that delay could be reduced if processes like step 7, measurement and inspection, could be integrated within the process itself, or be performed in situ. Some steps, such as the numerous cleaning steps, number 1, could be reduced if better control of the processing and of the wafer surface could be effected. If multiple-chamber, single wafer at a time machines are already gaining commercial credence, one can easily speculate that ultimately it would be possible to utilize these compact machines to perform the entire sequence of processes that comprise IC fabrication. Huge throughput is not a critical factor for custom circuits. In addition, "By putting in raw wafers at one end of of machine and getting out fully processed wafers at the other, the chance of contamination or particulate damage would be almost completely eliminated... This machine itself would be the clean room"[11].

9. CONCLUSION

The characteristics of IC manufacture that have ensured its success and phenomenal rate of growth may today no longer be appropriate. Escalating clean room costs, the need for flexibility and diversity in ever more sophisticated chips and the pacing limitations of fabrication technology call for new approaches in IC manufacture. These factors may make attractive, even necessary, approaches formerly regarded as being impractically slow or other uneconomical. With a future emphasis on a diversity of small volume, low-cost and high yield custom circuits, the factory of the future may need to be a self-contained, highly intelligently automated *in situ* station.

This work was performed at the Center for Robotic Systems in Microelectronics under the sponsorship of NSF grant CDR 8421415.

REFERENCES

1. Moore GE: Progress in Digital Integrated Electronics:" International Electron Devices Meeting, Dec. 1975
2. Class 1 Clean Room Construction to Grow: Semiconductor International: Sept. 1986
3. Parikh M and Bonora A: SMIF Technology reduces clean room requirements: Semiconductor International, May 1985
4. Grenier J: The Mitsubishi Saijo factory: a new and fully automated IC facility: Solid State Technology, Jan. 1986
5. Grenier J: NMB Semiconductor: a new automated Japanese semiconductor facility: Solid State Technology, Sept. 1986
6. Waller L:Automated 'Monster' IC plants may be an expensive mistake: Electronics, Feb. 5, 1987
7. Hayashi, I:these proceedings
8. Drytek: 16 Jonspin Road, Wilmington, MA
9. Applied Materials, Inc.: 3050 Bowers Avenue, Santa Clara, CA.
10. Kammerdiner L: Aluminum Plasma Etch Considerations for VLSI Production: Solid State Tech., Oct. 1981
11. Naryanamurti V: in Cole B: Here comes the billion-transistor IC: Electronics, April 2, 1987
12. Hutcheson, DG: Economics in wafer processing automation:Semiconductor International, Jan. 1985

ULTRA HIGH VACUUM PROCESSING: MBE

J. P. Harbison, P. F. Liao, D. M. Hwang, E. Kapon, M. C. Tamargo

Bell Communications Research, Red Bank, New Jersey, USA 07701-7020

G. E. Derkits, Jr.

AT&T Bell Laboratories, Murray Hill, New Jersey, USA 07974

J. Levkoff

AT&T Engineering Research Center, Princeton, New Jersey, USA 08540

1. Abstract

As an ever increasing variety of processing steps are being performed in vacuum or vacuum-based environments, and as pressures for finer control and higher reproducibility drive these vacuums toward the ultra high vacuum (UHV) range, the combination of UHV processing and UHV semiconductor crystal growth by molecular beam epitaxy (MBE) entirely in an interconnected UHV environment becomes increasingly attractive. In this paper we discuss three examples of such combinations of MBE and processing, each of which yields unique structures unobtainable by combinations of conventional processing and crystal growth. In the first, we combine MBE and refractory metal evaporation to produce buried metal layers in III-V materials for use as buried gates and metal interconnects. In the second, we add plasma etching to produce selective area growth on substrates by means of a technique called tungsten patterning which opens up the possiblity of growing different layer structures at different points on the wafer for possible integration of more than one type of device on the same optoelectronic integrated circuit. And in the third, we invert the order of growth and patterning to produce superlattice structures grown on patterned substrates which, due to the lateral variation in superlattice period, offer possibilities of varying index of refraction and effective bandgap laterally across the wafer.

2. Introduction

At the present time, the entire spectrum of semiconductor processing steps are either vacuum based or capable of being made vacuum compatible. Metal evaporation for forming contacts and interconnects, thermal vacuum annealing for implant dopant activation or ohmic contact formation, and ion implantation for selective area doping are examples of processes currently routinely performed in vacuum. Other steps such as plasma deposition of dielectrics and passivation layers, thermal oxidation, sputter deposition of metals and insulators, and plasma and reactive ion etching all require a vacuum environment free from air contamination as a starting point into which various process gases are introduced to form the necessary controlled atmosphere for the process. The semiconductor processing workhorse, wet chemical etching, a non-vacuum-compatible process, is being superseded by the vacuum-compatible gas phase etching processes just mentioned. Even the position of polymer based photolithography,

D. J. Ehrlich and V. T. Nguyen (eds.), Emerging Technologies for In Situ Processing, 55–60.
© *1988 by Martinus Nijhoff Publishers.*

inherently non-vacuum in its execution, is now being challenged by a whole new class of UHV compatible contact-mask-free schemes for patterning, including focused ion and electron beam writing, laser direct-write schemes, and even projection eximer laser and projection ion beam systems of lithography.[1] These exciting alternatives allow the possibility of replacing even wet photolithographic processing with entirely vacuum compatible steps.

The demand for higher reproducibility and finer linewidth control is driving these vacuum based processing technologies in the direction of better vacuum conditions, and as they currently move into the UHV regime, an obvious next step is to begin integrating these steps sequentially in vacuum without exposing the wafer to air between steps. With such a scheme in place, we can then integrate these processing steps with the initial "processing" step of semiconductor epitaxial crystal growth in the form of the UHV based technique of MBE, and by doing so we open an important new possibility. Conventional device processing proceeds by forming all the epitaxial semiconductor layers first, with all the processing steps being performed subsequently. Overgrowth after processing steps, particularly in the case of MBE, is virtually precluded due to the air contamination problem, but with vacuum integration of growth and processing, regrowth becomes feasible. Already Ishida et al.[2] have demonstrated successful buried heterostructure stripe laser fabrication using an integrated UHV MBE growth/focused ion beam lateral patterning system. The ability to deposit additional semiconductor layers at various points in the processing sequence opens new vistas for device structures.

With UHV technology progressing at a rapid pace, and UHV manufacturers already offering commercial systems which allow one in theory to integrate many of these functions entirely in UHV, the exciting challenge becomes that of finding processing steps and techniques made possible by this integration which yield new structures previously unattainable by conventional processing. There are a number of examples of such work in the MBE literature including in situ formed Schottky barriers, barrier height modification, in situ formed ohmic contacts, epitaxial insulators, and selective area doping and growth.[3] These combinations will in turn accelerate the transition to such integrated systems, offering more than just an evolutionary improvement in factors such as yield or linewidth control by going to all UHV schemes, offering instead quantum leaps in our ability to produce new structures and devices.

In this paper we would like to present three examples of work in our laboratory that has the potential for opening up such new opportunities by breaking with the conventional sequence of semiconductor growth subsequently followed by processing steps. In the first, we integrate refractory metal deposition and MBE growth to achieve buried metal layers, one application of which is a perforated metal base transistor, which we discuss briefly. In the second, we further integrate plasma etching with MBE growth and metal deposition to achieve a lateral tungsten patterning technique which has the potential of allowing selective area growth of different device layer structures on different parts of the wafer. And in the final example, we invert the initial MBE growth and the traditional processing step of patterned etching to perform MBE growth of superlattices on previously etch-patterned substrates to provide structures with laterally varying properties such as index of refraction and conduction band minimum, leading to in situ grown waveguides and potential lateral quantum confinement devices.

3. GaAs/W/GaAs Buried Metal Structures

One structure which would be useful in device design is the placement of metal layers within the semiconductor layer structure. Such a structure, for example, is at the heart of the operation of the permeable base transistor.[4] Although a number of metals such as aluminum, silver and iron have been successfully grown epitaxially on GaAs in MBE system,[3] these non-refractory metals are not stable with respect to the underlying GaAs at GaAs growth temperatures (typically 550-700 °C). Tungsten on the other hand is relatively unreactive with GaAs in this temperature range and experiences minimal interdiffusion into the GaAs at those temperatures, but its use in an epitaxial GaAs/W/GaAs structure is precluded by an inability to grow the W epitaxially on the GaAs.[5,6] To get around this constraint, we made use of a morphological feature of the polycrystalline W films which we deposited, namely the appearance of an array of perforations, apparent in Figure 1 which shows a planar transmission electron microscope (TEM) micrograph of a 10 nm thick W layer deposited at 700 °C on GaAs in situ in the same chamber in which the GaAs is grown. By overgrowing GaAs, again in the same chamber, free from air contamination which traditionally precludes high quality overgrowth, we get epitaxial seeding through the holes in the W from the underlying GaAs.[7] This is shown graphically in the TEM cross section in Figure 2 which shows the thin W layer across the middle of the micrograph, with the first and second layers of GaAs below and above it, respectively. As the TEM specimen becomes progressively thinner when moving to the right side of the micrograph, the holes and the lateral seeding through them become more apparent. Upon coalescence of these multiply-seeded areas, we end up with a complete epitaxial GaAs upper layer in registry with the underlying GaAs. The importance of this technique is that we are able to choose the metal layer material based solely on reactivity and interdiffusivity considerations, freed from the severely confining constraint of the need for epitaxial match to the semiconductor material.

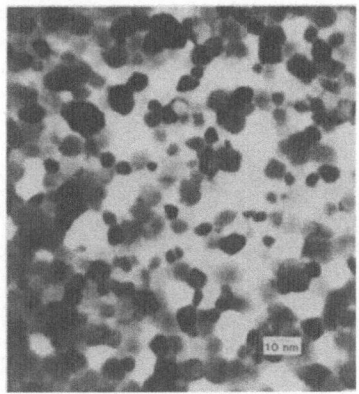

Figure 1. Planar TEM micrograph of 10 nm thick tungsten layer grown over GaAs. Note the perforations which serve as seeds for subsequent GaAs overgrowth.

Figure 2. Cross-sectional TEM view of a GaAs/W/GaAs structure as described in the text showing seeded growth through the perforations in the tungsten film.

One prime application for such a buried metal layer is in a vertical-transport transistor structure employing the metal layer as a base. We have fabricated such a device from this GaAs/W/GaAs structure.[8] The W point-to-point current-voltage (I-V) characteristics are quite linear and ohmic and show continuous conductivity within the W layer despite the perforated nature of the layer's structure. Resistivities lying in the range of 90 to 300 $\mu\Omega$-cm indicate that the grain structure is indeed sufficiently above the planar two-dimensional percolation threshold to allow contact to it across the entire base. I-V characteristics from both the W base to the lower GaAs layer and the W base to the upper GaAs layer show reasonable Schottky barrier characteristics. And indeed the device does work in three terminal operation as a metal gate transistor with $\alpha \sim 0.4$-0.6 and $\beta \sim 0.2$-1.4.

We believe that the physical basis of the transistor action in this device is by modulation of the electrical current flowing through the pores shown in Figures 1 and 2. The pores are substantially smaller than the depletion width in the material, and therefore conduct well before they go into accumulation. This is in marked contrast to the permeable base transistor which is essentially a vertical field effect transistor, relying on changes of depletion widths between W fingers to control the current. It is also fundamentally different from the true metal base transistor which consists of back-to-back Schottky diodes close enough due to the thin metal base to allow ballistic transport from one to the other. In the perforated base transistor structure we are looking at here, similar in principle to the metal gate transistor originally proposed by Lindmayer,[9] the current essentially never enters the metal. The perforated base device we have made shares in common with the other two vertical transport transistors just mentioned a short emitter to collector dimension on the order of 10 nm or so, achievable in thin deposition which makes it attractive in terms of ultimate high speed operation.

4. Tungsten Patterning

Though a number of processes have been demonstrated to date which integrate one or more processing steps with the semiconductor crystal growth steps, one key element needed for such an integration to be useful is the ability to perform these integrated operations <u>selectively</u> across the wafer. The work we will discuss in this section specifically addresses the problem of performing growth of the semiconductor layers selectively across the substrate which then opens the possibility of integrating a variety of devices that each require a different layer structure all on the same chip. The process used, which we refer to as tungsten patterning,[10] relies at its heart on the in situ processing combination of MBE growth, refractory metal tungsten metallization, and plasma etching, all within the same set of UHV vacuum enclosures.

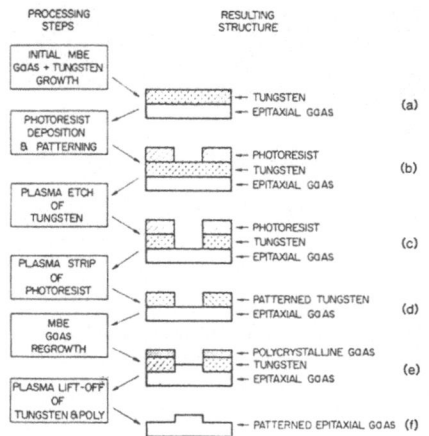

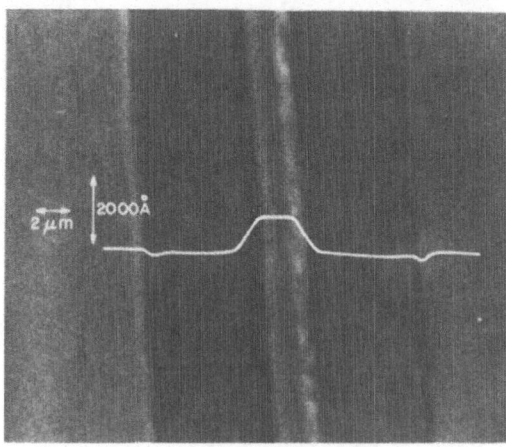

Figure 3. Details of the processing steps involved in performing selective area epitaxy using tungsten patterning.

Figure 4. Nomarski optical micrograph of a free-standing selectively grown using tungsten patterning. The white overlay represents a height vs. distance trace across the center of feature.

The tungsten patterning process is shown schematically in Figure 3. The sequence of steps to achieve selective area III-V growth are as follows: First, the underlying epitaxial GaAs is grown by conventional MBE. Next, in the same UHV chamber, a 1000 Å thick layer of W is evaporated to completely seal the UHV-clean as-deposited GaAs layer (Fig 3a). This then allows us to remove the wafer from the UHV environment for conventional spin-on organic photoresist deposition and patterning (Fig 3b), though it is eventually envisioned that this patterning step will be achievable entirely in the UHV environment using some of the techniques described elsewhere in this volume such as focused or projected ion or electron beam lithography schemes. Plasma etching in a fluorine based plasma completely removes the parts of the W that are exposed through the openings in the photoresist, stopping abruptly at the underlying semiconductor layer which is not attacked at all by such a plasma (Fig 3c). Once the photoresist has been stripped by an O_2 plasma (Fig 3d), the sample is returned to the in situ processing/MBE system. As a final step before regrowth, the wafer is lightly etched in a chlorine containing plasma, which attacks III-V's but not W, to remove any surface contamination from the GaAs. It is then transferred directly into the MBE growth chamber without breaking UHV vacuum, and a layer of GaAs is grown. Since the W is relatively unreactive with respect to the GaAs even at elevated growth temperatures, it serves as an ideal mask layer. The GaAs that grows over the exposed semiconductor part of the wafer grows as a single crystal, whereas the GaAs growing over the W mask layer grows polycrystalline (Fig 3e). The final step involves another fluorine based plasma which attacks and undercuts the W leading to the complete lift-off of the W mask and the polycrystalline GaAs grown over it. The resulting structure then consists of selectively grown single crystal GaAs features (Fig 3f). Figure 4 shows a micrograph of a ridge of GaAs material grown by this technique, with a Dektak surface profilometer trace superimposed. The regrown stripe is 2-3 μm wide and 1200 Å high with a submicron step profile.

These regrown layers of course need not be just GaAs. Any heterostructure capable of being deposited by MBE is possible due to the fact that the selectivity of the W and the GaAs to fluorine and chlorine based plasmas respectively extends to the entire range of III-V materials. The GaAs lying under the W layer throughout the process stays UHV clean throughout. Our results show that it is as smooth as the initial epitaxial layer once the W is removed in the last step. And presuming that the last plasma lift-off takes place in UHV, it is clean and ready for yet another regrowth step to occur above it. This opens the possibility of integrating a number of devices on the same wafer even though they have radically different layer structures. Such a capability is impossible without the regrowth capability opened up for us by in situ processing.

5. Growth of Superlattices on Patterned Substrates

As a final example we present work we have done[11] which combines processing and growth by inverting the traditional order of initial MBE growth followed by conventional etching to achieve pattern definition, and perform MBE growth of superlattices on substrates which have been first patterned by selective etching techniques. Though previous studies of growth on patterned substrates exists, this work focuses for the first time on very thin layer structures (~ 10 nm). The patterned grooves develop a series of crystal facets, as delineated by the labels in Figure 5 showing a transmission electron micrograph cross-section of the structure. The

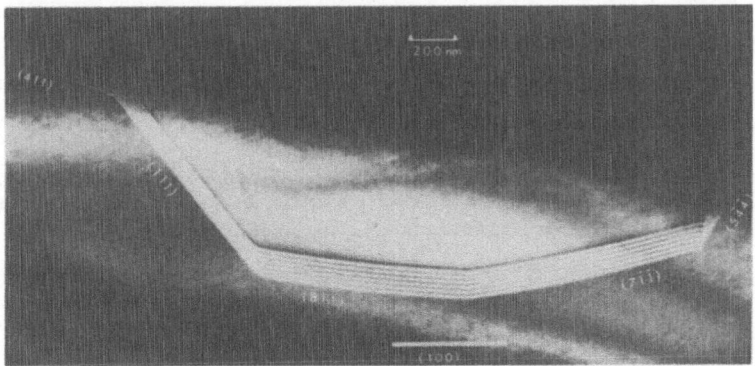

Figure 5. TEM micrograph of a (01$\bar{1}$) cross section of a GaAs/Al$_{0.3}$Ga$_{0.7}$ superlattice grown in a groove etched in a GaAs substrate. The white numbers indicate the crystal planes of the various facets formed. The dark stripes represent GaAs layers.

details of the individual facets which are formed are discussed in more detail in reference 11. The key point to note from Figure 5 is that the period spacing of the superlattice grown over this structure varies by up to a factor of two depending on which facet it is growing on. To first approximation this variation can be explained by the non-normal incidence of the Ga, Al and As atoms with respect to the various facets, leading to a $1/\cos\theta$ thickness dependence.

A lateral variation in superlattice periodicity leads to a lateral variation in any material properties that depend on that periodicity. One example is the index of refraction. The index is higher in the longer period superlattice at the bottom of the groove than it is in the shorter period sidewalls, leading to a waveguiding effect.[12] A second example is the effective bandgap of the material which increases as the superlattice period becomes smaller due to quantum confinement effects. Since the period becomes shorter on the sidewalls of the groove, there is a lateral confinement barrier of $\sim$ 50 meV associated with such a structure. If the lateral dimensions can be further decreased, lateral quantum confinement effects might be observable in such structures, leading to a whole spectrum of interesting physics and device applications.

6. Conclusions

So in summary we have demonstrated three combinations of processing and MBE growth which not only allow better control and reproducibility by combining growth and processing steps, but also open up new capabilities for fabrication of structures unobtainable by conventional separate growth and processing schemes. In the first case, we can bury metal layers which can serve as gates or buried interconnects within the semiconductor epitaxial layers. In the second case, we can selectively deposit different layer structures to be used for entirely different circuit elements on different parts of the same wafer. And in the third case, we can provide lateral variations in properties such as index of refraction or effective bandgap which can give us lateral as well as depth control of confinement of carriers and light. Thus the combination of UHV processing and MBE growth can open a number of new vistas in our ability to tailor properties for semiconductor device design.

REFERENCES

1. See for example other papers in this volume.

2. K. Ishida, T. Takamori, K. Matsui, T. Fukunaga, T. Morita, E. Miyauchi, H. Hashimoto, and H. Nakashima, Jap. J. Appl. Phys. 25, L738 (1986).

3. We refer the reader to our previous brief review surveying this literature:
   J. P. Harbison, J. Vac. Sci. Technol. A4, 1033 (1986).

4. C. O. Bozler and G. D. Alley, IEEE Trans. Electron Devices ED−27, 1128 (1980).

5. J. Bloch, M. Heiblum, and Y. Komen, Appl. Phys. Lett. 46, 1092 (1985).

6. G. E. Derkits, Jr., J. P. Harbison, and D. M. Hwang, Appl. Phys. Lett. 47, 19 (1985).

7. J. P. Harbison, D. M. Hwang, J. Levkoff, and G. E. Derkits, Jr., Appl. Phys. Lett. 47,1187 (1985), and J. Vac. Sci. Technol. B4, 662 (1986).

8. G. E. Derkits, Jr., J. P. Harbison, J. Levkoff, and D. M. Hwang, Appl. Phys. Lett. 48, 1220 (1986).

9. J. Lindmayer, Proc. IEEE 52, 1751 (1964).

10. J. P. Harbison and G. E. Derkits, Jr., J. Vac. Sci. Technol. B3, 743 (1985).

11. E. Kapon, M. C. Tamargo, and D. M. Hwang, Appl. Phys. Lett. 50, 347 (1987).

12. E. Kapon and M. C. Tamargo, to be published.

EPITAXIAL GROWTH OF III-V MATERIALS ON IMPLANTED III-V SUBSTRATES

P.N. FAVENNEC, H. L'HARIDON, M. SALVI, L. HENRY, A. LE CORRE, D. LECROSNIER
M.A. DI FORTE POISSON*, J.P. DUCHEMIN*
CNET LAB/ICM/TOH LANNION FRANCE - *LCR TH-CSF ORSAY FRANCE

I - INTRODUCTION

In conjunction with ion implantation, the epitaxial processes can be very attractive technologies for achieving monolithic integration of microwave and electro-optical devices. We present results on the epitaxial growth of III-V compounds, molecular beam epitaxy (MBE) and metalorganic chemical vapor deposition (MO - CVD), on III-V substrates previously implanted either on the whole surface or on selected areas. More especially, the following combinations are studied : (i) GaAs growth by MBE on implanted GaAs substrates (ii) InP growth by MO - CVD on implanted InP substrates (iii) $In_{0.53}Ga_{0.47}As$ growth by MBE on implanted InP substrates. We will determine the conditions for an overgrowth of a good quality crystal and also the conditions for overgrowth of a polycrystalline growth. Then, we will demonstrate that the use of ion implantation can lead to selective monocrystalline and polycrystalline growth of III-V compounds and a semi-insulating electrical isolation between conducting areas.

II - GaAs GROWTH BY MBE ON IMPLANTED GaAs SUBSTRATES

<001> and <111> oriented, semi-insulating GaAs substrates were chemically etched in a solution of H_2SO_4 : H_2O_2 : H_2O (1 : 1 : 100) to remove any mechanically damaged layer. Using a conventional photolithographic technique for masking, only a part of the substrate was exposed to bombardment, in order to compare the layer quality on the implanted wafer to the one simultaneously grown on the unimplanted section for reference. The implanted ion was oxygen (but the ion species does not seem significant). Its energy was 300 KeV and the beam intensity was no more than 500 nA/cm^2. The implantations were tilted 7° from the <100> or <111> axis. After implantation and after mask removal, the substrates had no postimplant anneal, but were directly located in a MBE system. The epitaxial growth was performed at a substrate temperature of 600°C.

Microscopic inspection revealed no difference between implanted and unimplanted areas in the surface morphology of the epitaxial layer as long as the implanted doses did not exceed the critical amount. In fact, we had a single crystalline epigrowth on GaAs substrates implanted with doses less than 5 x 10^{14} O^+ cm^{-2} at room temperature. When the doses approach the upper limit of about 5 x 10^{14} O^+ cm^{-2}, a significant change in the growth behaviour takes place for RT implants. Examples are given in Fig. 1, which demonstrates the rough deposition on the oxygen-implanted region bordering on an unimplanted area with a smooth epilayer. This figure shows micrographs of an epitaxial sample. Two regions are distinct :

61

D. J. Ehrlich and V. T. Nguyen (eds.), Emerging Technologies for In Situ Processing, 61-70.
© 1988 by Martinus Nijhoff Publishers.

62

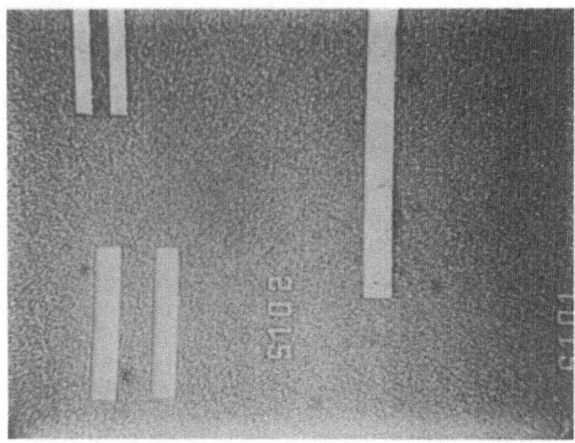

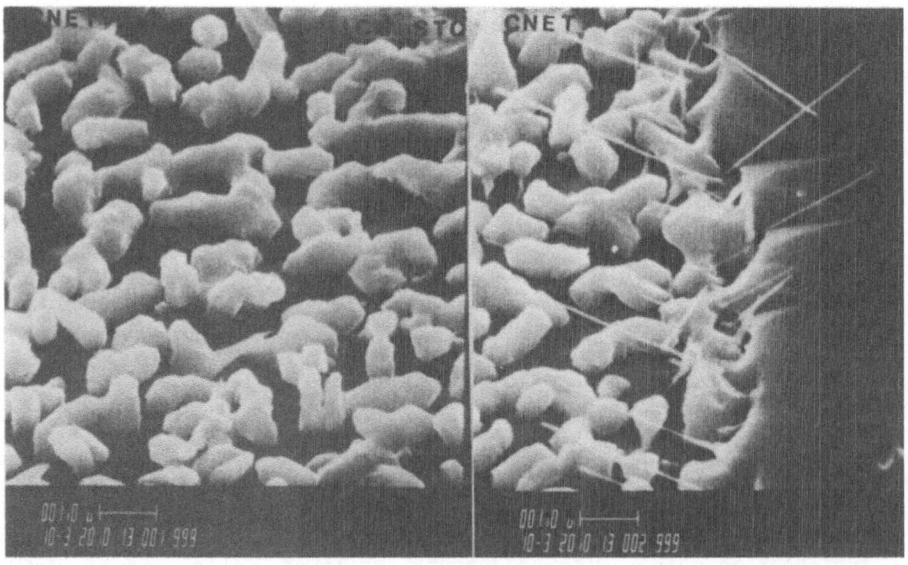

Figure 1 – Micrographs showing GaAs deposition (a) monocrystalline stripes (x 260) in a nonmonocrystalline sea (b) small blocks on implanted areas (c) implanted region bordering on an unimplanted area.

(i) a monocrystalline region. This area was masked during implantation and then was unimplanted. The epitaxial thickness was 2000 Å.

(ii) a second region with islands of polycrystallites. This area was implanted before the epitaxy process. The implantation conditions were : oxygen ions, 300 KeV, $10^{15}O^+cm^{-2}$ at RT. We emphasise that the molecular epitaxy was performed directly on a substrate.

Fig. 1b shows a scanning electron micrograph (x 2000 magnification) of the nonmonocrystalline region. The polycrystalline small blocks are about 5000 Å in height when the thickness of the monocrystalline epilayer was 2000 Å. Between these deposited blocks there is no deposition of GaAs. Electrically, this area has a high resistivity (1 - 3 x 10^5 Ω-cm). As a consequence, we can have conductive monocrystalline GaAs regions in a nonconductive sea. Thus we obtained monocrystalline stripes (Fig. 1a) intentionally doped at 2 x 10^{17} electron cm^{-3}. The final stripe width was that of the mask window used during the implantation.

In conclusion, we have shown that : (a) for low-dose-implanted substrates, the epitaxial layers are comparable to those obtained on unimplanted substrates ; (b) for high-dose-implanted substrates, the parts of implanted substrates are dotted with small islands.

The island formation of GaAs can find an application in the selective growth of GaAs for achieving monolithic integration of microwave and electro-optical devices. The described method seems advantageous to other MBE selective-growth classical methods. [1 - 3]. In addition, after longer periods of GaAs deposition on the implanted parts, from these islands, there is a growth of whiskers [4].

III - InP GROWTH BY LP-MOCVD ON IMPLANTED InP SUBSTRATES

Chemically etched (100) oriented undoped and iron dor semi-insulating InP substrates have been used. Using a conventional photolithographic technique for masking, regions of the substrates were exposed to ion bombardment in order to compare the layer quality on an implanted area with that on an unimplanted area, for reference. The implanted ion was oxygen. Several energies have been used varying from 300 to 1000 KeV. The implantations were performed at room temperature and the samples were tilted 7° from the $\langle 100 \rangle$ axis. After implantations and after mask removal, the substrates did not have a post-implant anneal, but were directly placed in an MOCVD reactor in order to grow InP layers. The deposition temperature was 550°C [5].

In Fig. 2 we show scanning electron micrographs of the deposited InP on the implanted substrate at various doses. All the results presented in this figure are obtained for 300 KeV implantation. Microscopic inspection revealed, on the surface morphology of the epitaxial layer, no difference between an area implanted at a dose of $10^{14}O^+cm^{-2}$ and an unimplanted area. For the dose of $10^{15}O^+cm^{-2}$, the coverage of the substrate is incomplete. When the implantation dose is increased, the surface area covered by epitaxial growth decreases, and for a dose of 2 x $10^{15}O^+cm^{-2}$, only nuclei appear on the surface. They are approximately hemispherical in shape. And the density of these nuclei becomes zero for a dose of 5 x $10^{15}O^+cm^{-2}$.

64

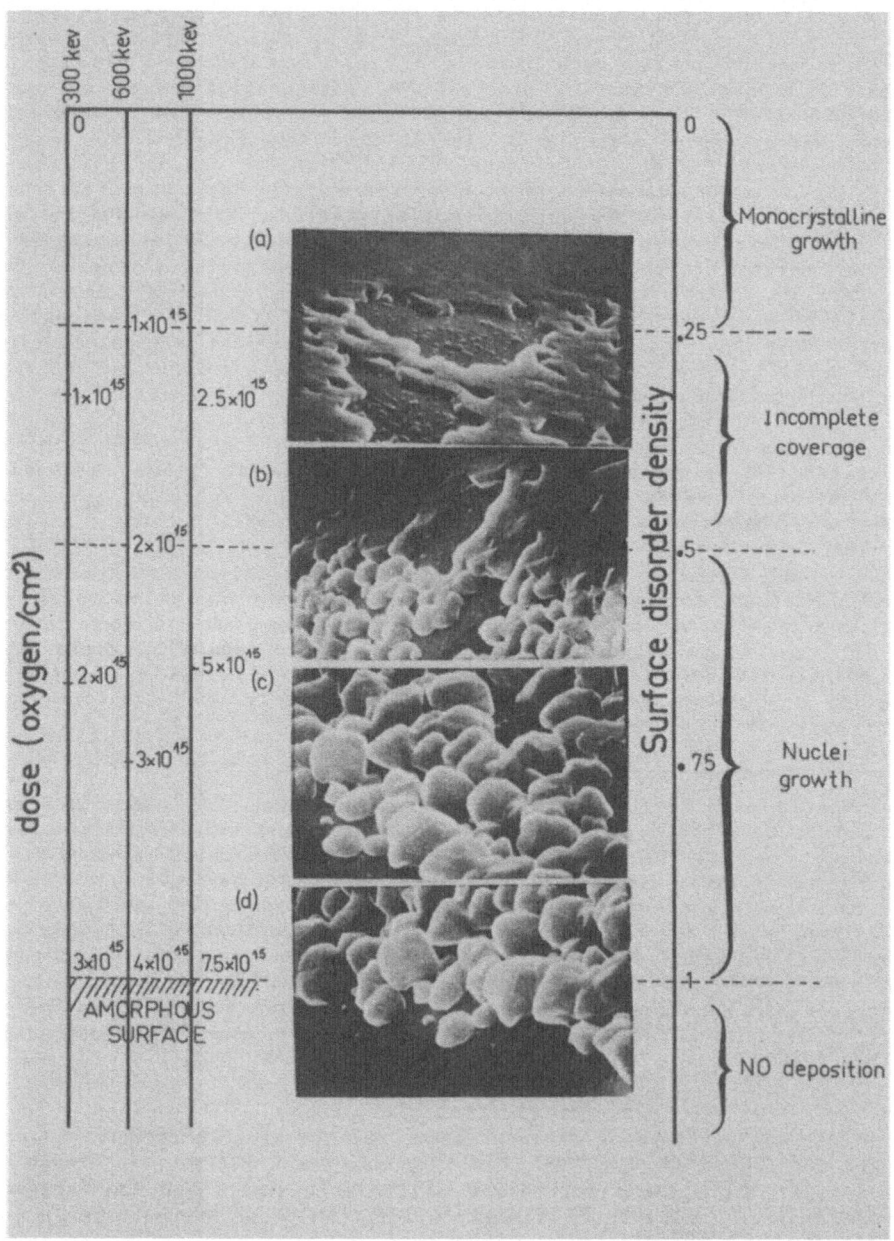

Figure 2 – Micrographs of the InP deposited on the implanted substrate at various doses for 300 KeV, 600 KeV and 1000 KeV. The surface morphology depends on the surface disorder density.
(a) x 5000   (b) x 8000   (c) x 10000   (d) x 10000

Similar results have been observed for other implantation energies.

In Fig. 3 we show a scanning electron micrograph (SEM) of the sample surface obtained after a selective area implantation at 600 KeV and with a dose of $5 \times 10^{15} O^+ cm^{-2}$. The implant pattern used is similar to that shown in Figure 1(a). In the implanted area (lower part of the micrograph), there is no InP deposition despite the fact that the whole surface was exposed to the chemical vapor flow. However, in the unimplanted area (upper part of the micrograph), we have effectively demonstrated an InP monocrystalline deposition. For the two areas (implanted and unimplanted), the boundary is not well defined. In this region, the deposition is thicker, and the edge is rounded. A rough estimation shows that the volume of material deposited in the unimplanted part of the sample has the same volume (surface x thickness) as if we had deposited material uniformly on the total surface. This shows that the total volume of material deposited was limited by mass transport in the gas phase, and that locally there was material transport from the implanted area towards the unimplanted area. This mass transport phenomenon observed during the deposition process would be similar to the transport of InP observed by Liau and Walpole [6], after a heat treatment in H_2 and PH_3 atmospheres. Ion implantation having disoriented the surface, the extra thickness observed near the edge of the monocrystalline layer is due to the material initially deposited on the amorphized area which moved to the preferential crystallographic orientation at the surface of the sample [5].

The ion implantation modifies the substrate surface and therefore the interaction between the material deposited and the surface substrate. When the surface is not too perturbed by the ion implantation, MOCVD deposition gives a monocrystalline layer. When the perturbation increases, there is no longer complete coverage of the surface. The coverage decreases with the dose, then the flat growth areas separate into irregular shapes producing islands which are approximately hemispherical in shape. It could appear that In and P atoms skate on the implanted surface up to an obstacle (in the example shown in Fig. 3, the obstacle was the edge of the epitaxial layer) producing excess InP growth on the edge.

The quality of the InP deposition or its absence on the InP substrate is a function of the damage density of the surface exposed during the MOCVD process. It is not a function of the total damage produced by the bombardment. As an example, an implantation at 300 KeV of $2 \times 10^{15} O^+ cm^{-2}$ produces less defects than an implantation at 1000 KeV of $2 \times 10^{15} O^+ cm^{-2}$, but at the surface, the concentration of displaced atoms is higher for the 300-KeV implantation than the 1000-KeV implantation. In Fig. 4 we show an estimation of the variation of the surface damage density, calculated from the nuclear energy loss at the surface, versus the oxygen dose for several energies used. Arbitrarily, the surface is defined as amorphized for a surface disorder density of 1, and the assumption is made that no self-annealing takes place at room temperature during the ion bombardment. From these curves, clearly we see (i) when the surface is amorphized, there is no deposition ; (ii) for the surface damage densiy lying from 0.5 to 1, there is nuclei growth ; (iii) for the damage density between 0.25 to 0.5, there is an incomplete coverage of the crystalline layer ; (iv) for the surface disorder density smaller than 0.25, the growth layer is monocrystalline and the coverage is complete.

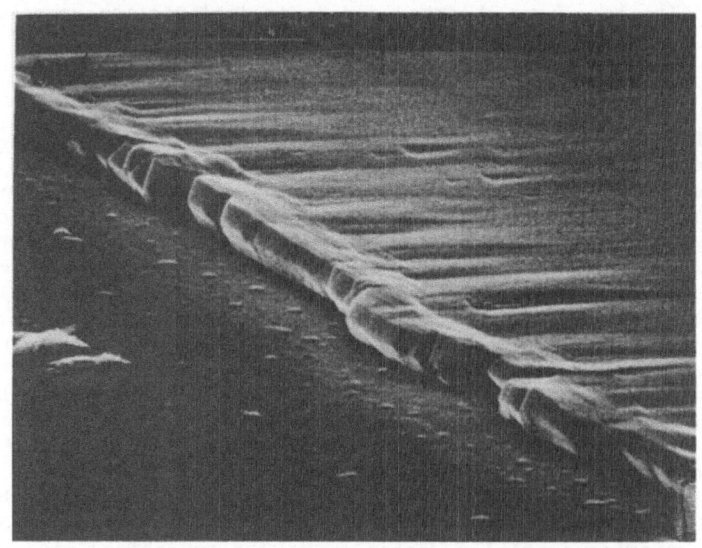

Figure 3 – Micrograph (x 10200) showing a selected area epitaxy.
Unimplanted area : epitaxy layer of 6000 A InP
Implanted area at 600 KeV and 6 x $10^{15}O^+cm^{-2}$: no deposition of material.

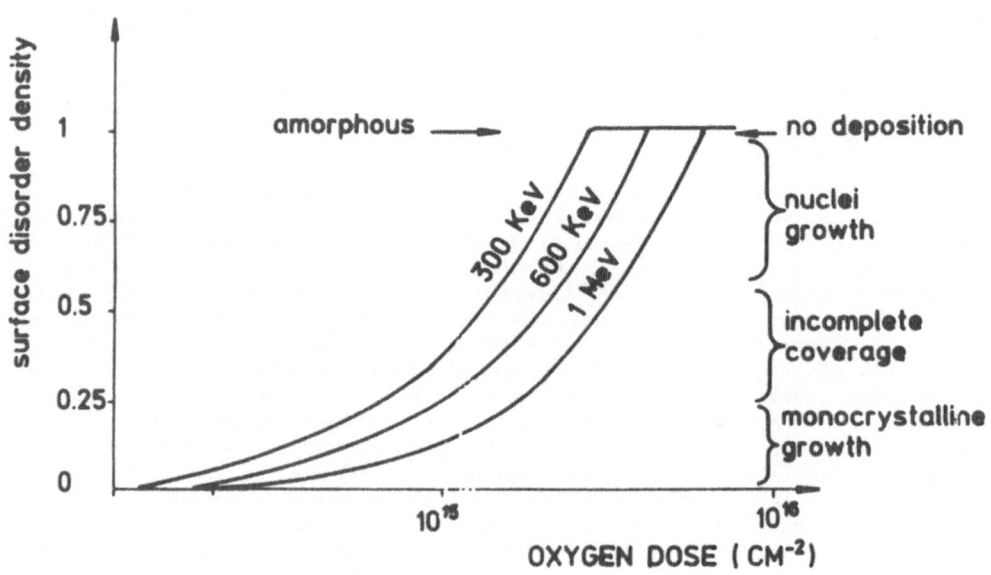

Figure 4 – Surface disorder density versus the implanted oxygen dose.

Experiments have been made using argon and neon instead of oxygen ions, and the results are similar to those described here, except for a shift in the dose. Whatever the ions are we have the same phenomenon, if we have the same surface disorder density. This means that if we required a selected area epitaxy for a given substrate we must choose the nature of the ions, the energy and the dose for having an amorphous surface.

IV - $In_{053}Ga_{047}As$ GROWTH BY MBE ON IMPLANTED InP SUBSTRATES

The previous results suggest that similar results would be obtained if we grew a ternary or quaternary material on ion implanted substrate. We show some results related to the $In_{053}Ga_{047}As$ growth by MBE. The InP substrates were annealed or unannealed after implantation and the MBE growth was performed under a substrate temperature of 600°C.

At first, let us consider the case of the substrate surface not too disturbed by ion implantation. For that, the semi-insulating InP substrates were implanted with 150 KeV silicon ions at various doses then annealed at 700°C to get silicon activation. The electron carrier profiles, as measured by POLAPON, before and after MBE epitaxy, are shown in Figure 5. The silicon implantation leads to a n-type layer at the substrate surface with the same electrical profile before and after epigrowth and the ion implantation has no influence on the electrical characteristics of the epigrowth layer. In addition, as observed by SIMS, the silicon impurities do not diffuse during the epitaxy process, either towards the In GaAs layer or towards the substrate. This process permits structures such as n-In GaAs / n^+. InP / S.I. InP to be obtained.

If the substrate surface is very disturbed by ion implantation, the $In_{053}Ga_{047}As$ is nonmonocrystalline. We show an example in Figure 6. This sample was ion implanted at a high dose on selected areas and unannealed. The surface disorder density was 1. As seen on the micrograph, the GaInAs deposition was a smooth epilayer on the unimplanted areas and a rough epilayer on the implanted areas. In fact, a polycrystalline layer grows on the damaged regions. Using enhanced etching of polycrystalline regions, selective polycrystalline growth also leads to the idea of self-aligned etching [7].

V - CONCLUSION

All the results obtained concerning the epitaxial growth of III-V material on implanted III-V substrates show a strong dependence of the layer quality on the surface disorder density (sdd) of the substrate. These results have been verifyed for the GaAs/GaAs, InP/InP and GaInAs/InP couples.

By using a conventional ion implantor and an epitaxial system (MBE or MOCVD) : (i) if the sdd is low, there is an epitaxial growth (ii) if the sdd is high, there is no monocrystalline deposition.

When pattern doping in the substrate is performed by ion implantation, the growth of the material on this substrate leads to (i) a homogeneous monocrystalline layer if the implanted areas have a low sdd (ii) an inhomogeneous layer if the implanted areas have a high sdd : monocrystal-

electron carrier concentration (cm^{-3})

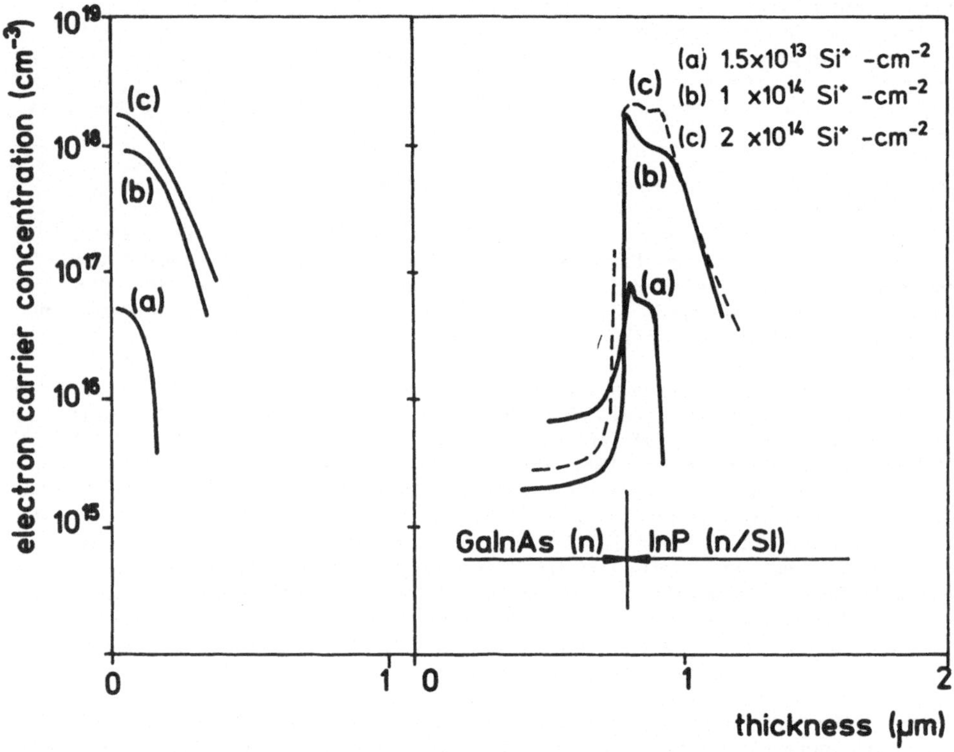

10^{19}

10^{18}

10^{17}

10^{16}

10^{15}

(c)

(b)

(a)

(c)

(b)

(a)

(a) 1.5×10^{13} Si$^+$ –cm^{-2}
(b) 1×10^{14} Si$^+$ –cm^{-2}
(c) 2×10^{14} Si$^+$ –cm^{-2}

GaInAs (n) | InP (n/SI)

0   1 0   1   2

thickness (µm)

Figure 5 – Electron carrier profiles (a) after implantation and annealing (b) after MBE epitaxy of In$_{0.53}$Ga$_{0.47}$As

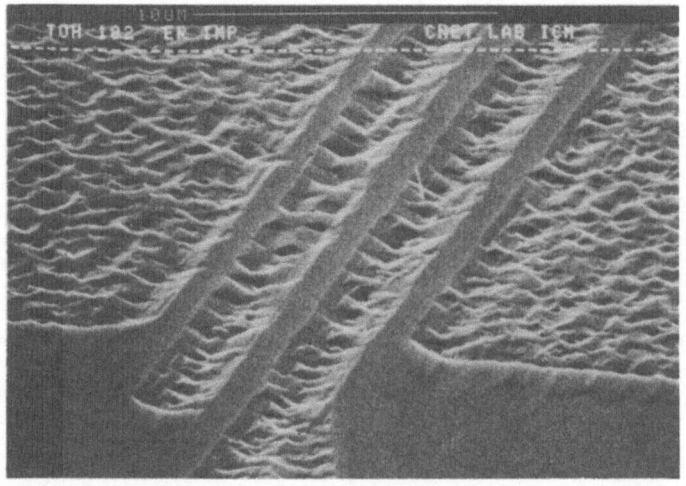

Figure 6 – Micrograph showing a selected area epitaxy of InGaAs. Unimplanted area : monocrystalline area ; implanted area : polycrystalline area.

line areas on the unimplanted areas and polycrystalline (or no deposition) areas on the implanted areas, then, we have a selected area maskless monocrystalline growth.

This work has been performed with a conventional implantor using broad ion beams. But, the results can be directly applied if we use a focused ion beam implantor (FIBI).

The use of a FIBI system would enable us to have a complete process without mask and this technology, coupled to MBE or MO-CVD growth, would offer a powerful device fabrication tool. The main disadvantage of this FIBI technique is the restricted choice of the implanted ions (liquid source).

REFERENCES

1 - TSANG W.T. and ILEGEMS M., Appl. Phys. Lett., 1977, 31, 301.

2 - TSANG W.T. and CHO A.Y., Appl. Phys. Lett., 1978, 32, 491.

3 - METZE G.M., LEVY H.M., WOODARD D.W., WOOD C.E.C. and EASTMAN L.F., Appl. Phys. Lett., 1980, 37, 628.

4 - FAVENNEC P.N., HENRY L., REGRENY A. and SALVI M., Electron. Lett., 1982, 18, 933.

5 - FAVENNEC P.N., SALVI M., DI FORTE POISSON M.A. and DUCHEMIN J.P., Appl. Phys. Lett., 1983, 43, 771.

6 - LIAU Z.L. and WALPOLE J.N., Appl. Phys. Lett., 1982, 40, 568.

7 - HASHIMOTO H. and MIYAUCHI E., Nucl. Instr. Meth., 1987, B19, 381.

MECHANISMS OF LASER-INDUCED DEPOSITION FROM THE GAS PHASE

R. L. JACKSON, T. T. KODAS, G. W. TYNDALL, T. H. BAUM, and P. B. COMITA
IBM Almaden Research Center, San Jose California 95120

1. INTRODUCTION

Laser-induced deposition from the gas phase has been extensively studied in the last several years. Because a high degree of spatial localization can be achieved via deposition employing a tightly focused laser beam, a number of applications for laser deposition have been found in the microelectronics field. For example, highly localized deposition induced by a scanning, focused cw laser beam has been successfully used for direct writing of interconnection lines in integrated circuits (1-4). Related applications have been found in integrated circuit modification and customization (5), direct-writing of waveguides (6), and localized deposition for repair of clear defects in lithographic masks (7-9). A mask repair system based on laser deposition of a metal repair patch is now commercially available (10).

Despite the importance of laser deposition, understanding of the elementary processes involved is only beginning to emerge. In this chapter, we will discuss work performed in our laboratory as well as a number of other laboratories that has addressed the details of both thermal and photochemical laser deposition processes. We will try to summarize current knowledge of laser deposition mechanisms.

2. GENERAL CONSIDERATIONS

Laser deposition from a gas phase precursor can be performed via two fundamentally different approaches. In the photothermal approach, laser light heats the substrate, inducing decomposition of the precursor by a process closely related to conventional chemical vapor deposition. In the photochemical approach, laser light is absorbed by the precursor, which decomposes photochemically, thereby inducing deposition. Laser-induced photochemical deposition is typically performed using uv lasers at power densities well below those used in laser-induced photothermal deposition, which minimizes heating of the substrate during the deposition process. By tightly focusing the laser beam, deposits less than 1 μm in width have been produced by both photothermal (11) and photochemical (12) deposition.

In the remainder of this chapter, we will discuss mechanisms of photothermal and photochemical deposition driven by cw lasers focused directly onto the substrate. Photochemical deposition has also been performed, however, using pulsed uv excimer lasers and dye lasers, either with the beam incident on the substrate or directed parallel to it, to induce deposition over large areas. More extensive discussion of pulsed laser-induced photochemical deposition can be found in reference (13).

3. LASER-INDUCED PHOTOTHERMAL DEPOSITION

Photothermal deposition processes may be broken down into several elementary steps, which include 1) migration of reactants to the heated region of the surface, i.e. the reaction zone, 2) adsorption of reactants, 3) reaction on the surface, 4) desorption of the reaction products, and 5) migration of the products away from the reaction zone. Two critical elements that can influence these steps are the surface temperature and gas phase mass transport. Reactant adsorption and product desorption are critically dependent on the surface temperature. The surface reaction rate is also a function of the surface temperature. Surface reaction pathways may vary with surface temperature, thus determining the purity of the deposited material. Migration of reactants to the reaction zone and of products away from the reaction zone are governed by gas phase mass transport. In the next subsections, we will examine surface temperature and gas phase mass transport contributions to laser deposition kinetics.

3.1 Surface temperature profile

Several authors have calculated the temperature profile obtained on a solid surface illuminated

D. J. Ehrlich and V. T. Nguyen (eds.), Emerging Technologies for In Situ Processing, 71–81.
© *1988 by Martinus Nijhoff Publishers.*

by a Gaussian laser beam. The heat conduction equations can be solved analytically for semi-infinite homogeneous solids (14, 15) and multilayer solids (4, 16, 17). These cases are not appropriate to describe surface temperature profiles during laser-induced photothermal deposition, however, since the deposition process alters the substrate reflectivity and thermal properties. Thus, the heat conduction equations must be solved numerically. To model photothermal deposition processes, accurate theoretical models of the surface temperature profile as a function of time during deposition must be developed. Experimental measurements of surface temperatures must be also obtained as verification of the accuracy of the model calculations.

We have employed thin-film thermocouples deposited onto various substrates to measure surface temperature distributions obtained upon illumination with the focused beam of an argon ion laser (514.5 nm) (18). For the case we will discuss here, thin-film thermocouples consisting of a 200 nm thick gold line overlapping a 100 nm thick nickel line were deposited onto a silicon substrate having a 200 nm overlayer of thermally grown SiO_2. The thermocouples were fabricated by standard lithographic lift-off techniques. These structures are particularly useful for measuring surface temperatures under conditions similar to those encountered in laser-induced photothermal deposition of gold lines onto integrated circuits consisting of a silicon wafer covered with a SiO_2 dielectric layer. The metal lines of the thermocouple serve not only to measure the surface temperature accurately, but also to simulate the thermal conduction and optical properties of a line deposited using a scanning laser.

Temperature profiles were measured by scanning the laser beam across the thermocouple junction. The measured temperature profiles were compared to profiles calculated using a numerical finite difference technique, accounting for the changing reflectivity and thermal properties of the substrate with increasing temperature. The results for thermocouple line widths of 8 μm and a focused beam diameter (1/e) of 9 μm show that the surface temperature is strongly influenced by the presence of the metal lines. Measured and calculated peak surface temperatures for the case of the laser focused directly onto the gold top line of the thermocouple junction are significantly higher than temperatures for the case of the laser focused onto the SiO_2 layer without metal lines. This is because the lines are insulated from the thermally conductive Si substrate by the presence of the relatively non-conductive SiO_2 layer. Heat conduction away from the thermocouple junction occurs predominantly through the metal lines. In contrast, when the laser is focused away from the metal lines, absorption of the laser energy occurs in the Si layer and heat conduction occurs in three dimensions. In addition, the reflectivity of the gold lines in our thermocouples is much lower than the reflectivity of the silicon substrate (18), which further increases the temperature when the laser is focused directly onto the thermocouple junction. The calculated temperature distributions agree well with our measured distributions. When the laser is scanned over the top of the thermocouple junction, the measured temperature profile should match the profile calculated for the laser focused onto the gold line. The agreement between the calculated and the measured temperature distributions for this case is illustrated in Fig. 1, considering a radial distance from the center of the thermocouple junction of less than 5 μm where the laser is incident predominantly on the gold line. When the laser is scanned past the junction onto the surrounding substrate, i.e. beyond ~13 μm, the measured temperature distribution should more closely match the profile calculated for the case of the laser focused onto the substrate without the metal lines. The agreement between calculated and measured temperature profiles for this case is again demonstrated in Fig. 1.

The thickness of the metal lines can also strongly affect the surface temperature. Fig. 2 shows the calculated maximum temperature as a function of line thickness for the case of the laser focused directly onto the gold line at the junction. The temperature drops dramatically with increasing line thickness. This further demonstrates that precise models for the surface temperature are required to calculate laser-induced photothermal deposition rates. Indeed, the effect of line thickness on surface temperature may explain our observation that cross sections of lines deposited via argon ion laser-induced photothermal deposition of copper from copper bis-(hexafluoroacetyl acetonate) onto SiO_2/Si substrates (19) are strongly affected by scan rate. At low scan rates, the lines have rounded cross sections, while at high scan rates, the lines have "volcano" type profiles, i.e. the deposited lines are significantly thinner in the middle than at the edges. The deposited lines are thinner at the faster scan rates; this results in a higher surface temperature in the center of lines deposited at high scan rates. High surface temperatures have been shown in some cases

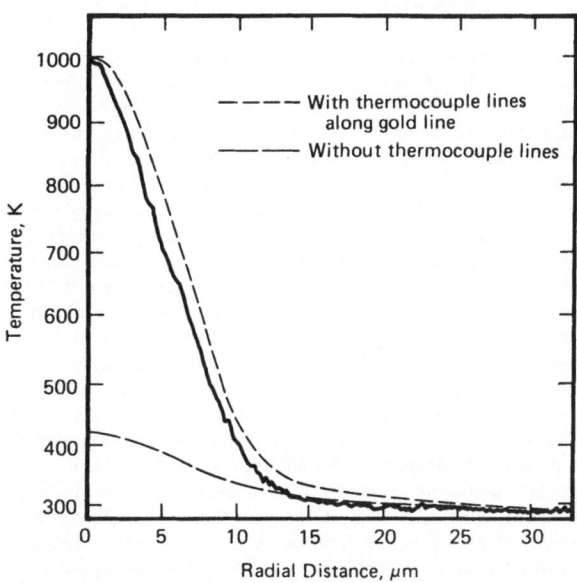

Fig. 1: Calculated (dashed lines) and measured (solid line) temperature profile for 0.5 W incident power and a 1/e laser beam diameter of 9 μm.

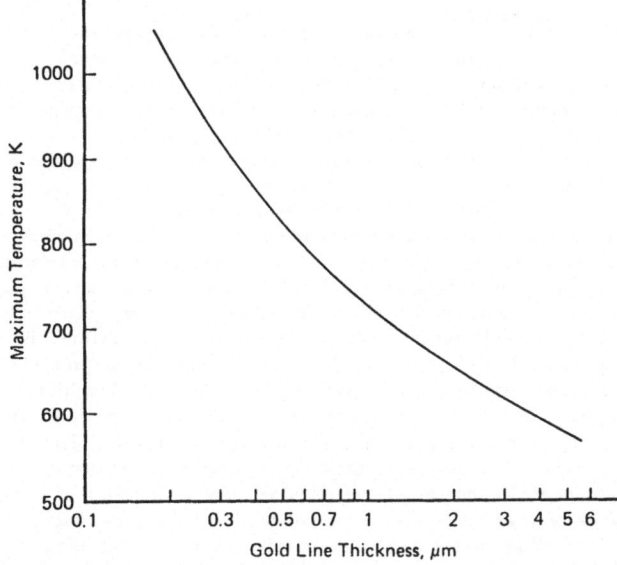

Fig. 2: Calculated influence of thermocouple line thickness on temperature with the laser incident on the center of the thermocouple junction. Laser power is 0.5 W and laser beam diameter is 9 μm.

to result in a reduced deposition rate (19, 20) (see below), thus accounting for the observed line cross sections.

For laser-induced photothermal deposition processes where the growth rate is governed by the surface reaction kinetics, rather than mass transport, knowledge of the surface temperature profile provides a good start toward building a model of the deposition process. Bauerle (21) has modeled laser-induced photothermal deposition from $Ni(CO)_4$ by considering the surface temperature profile. Although his model neglects the temperature dependence of the deposit and substrate thermal properties, he was able to gain insight into deposit shapes as a function of the ratio of the thermal conductivity of the substrate to that of the deposit. A quantitative fit of deposition rate as a function of laser power at high $Ni(CO)_4$ pressure was obtained, where mass transport effects on the surface reactant and product concentration can be largely neglected, using a simple Arrhenius model for the dependence of the surface reaction rate on temperature. More recently, Allen et al. (22) have modeled cw CO_2 laser (10.6 μm) induced photothermal deposition of spots from $Ni(CO)_4$. Temperature profiles were calculated numerically by a finite difference technique, accounting for the dependence of substrate thermal and optical properties upon temperature and deposit thickness. Allen et al. were able to obtain a reasonably quantitative fit to measured deposition rates and deposit dimensions as a function of laser power, using Carlton and Oxley's model (23) for the dependence of the surface reaction rate on temperature and treating surface reactant and product concentration as an adjustable parameter. Because the substrate in this case has a lower reflectivity than the deposit, the surface temperature drops with increasing deposit thickness and the deposition rate reaches a maximum when the deposit is very thin. By incorporating into their model the changing reflectivity of the surface as deposition proceeds, the expected maximum in surface temperature and deposition rate is accounted for. This effect is opposite to the effect we observed for gold on SiO_2/Si substrates. In that case, the temperature increases when gold is present, since the gold is less reflective than the underlying substrate. As a result, the surface temperature is expected to increase dramatically during the early stages of gold deposition (24, 25) onto SiO_2/Si.

3.2 Gas phase mass transport effects

Modeling of laser photothermal deposition processes over a wide range of deposition conditions requires accurate solution of the reactant and product mass transport equations as well as the surface temperature equations. For sufficiently high surface reaction rates, the deposition rate is limited by gas phase mass transport, independent of the surface temperature (11, 26, 27). Recently, Skouby and Jensen (28) have modeled deposition from $Ni(CO)_4$ onto SiO_2 induced by a CO_2 laser, considering gas phase mass transport effects and surface temperature effects in detail. Surface temperature profiles were calculated accounting for changes in substrate properties with temperature and with increasing deposit thickness. Coupled mass transport and surface temperature equations were solved by a numerical finite element technique to obtain growth rates using various models for the surface reaction kinetics. The value of this approach is in predicting the cross-sectional shape of deposited spots as a function of irradiation time and various deposition parameters. Because they accurately model both surface temperature and gas phase mass transport, Skouby and Jensen are able to model deposits with "volcano" type cross sections that have been observed at high laser powers by Allen (20) for CO_2 laser-induced deposition from $Ni(CO)_4$ onto SiO_2, and by us (19) for argon ion laser-induced deposition from copper bis-(hexafluoroacetylacetonate) onto SiO_2/Si. Several possible explanations for these profiles have been considered, revolving around the high surface temperature achieved in the center of the deposit. The reduced deposition rate at high temperatures could be due to depletion of reactants at the center of the deposit, reduced transport of reactants to the center of the reaction zone as a result of convection, and reduced trapping coefficients of reactants above a critical temperature. The model of Skouby and Jensen shows that under conditions where deposition is mass transport controlled, a reduced reaction rate can be obtained in the center of the deposit due to depletion of reactants. The reduction in deposition rate is small, however, and is not consistent with the depth of the volcanoes observed by us and by Allen. Skouby and Jensen showed that reduced reactant adsorption with increasing temperature can accentuate the depth of the volcano structure, however.

We have recently shown that deposition of gold from dimethyl gold hexafluoroacetylacetonate $(Me_2Au(hfac))$ onto Al_2O_3 substrates operates in the mass-transport limited growth regime over a wide range of surface temperatures (27). The gold deposits are highly conductive and largely

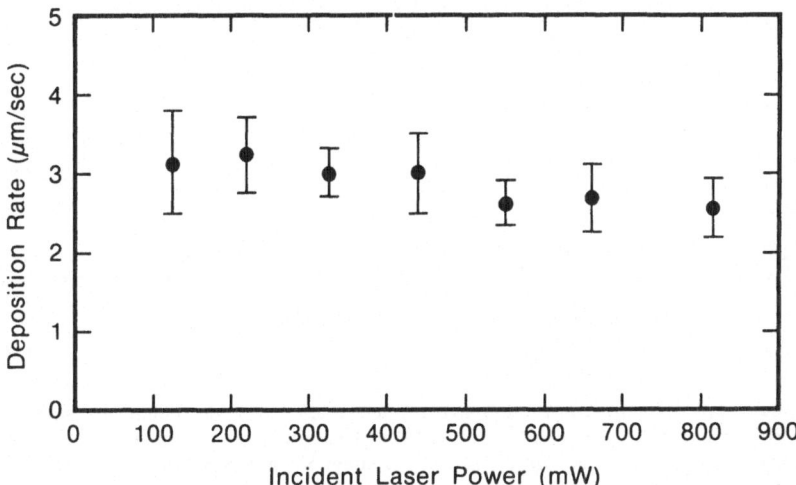

Fig. 3: Deposit vertical growth rate vs. laser power for Me$_2$Au(hfac) pressure of 0.35 torr and laser beam diameter of 7 μm.

free of contaminants (>96 at.% gold) (29). Evidence that the process is mass-transport limited is provided by the lack of dependence of the deposition rate on laser power, as shown in Fig. 3. Deposition rates in this study were measured by a new general optical technique, which we have termed optical profilometry (30). This technique measures the vertical growth rate of a deposit by monitoring the transmitted intensity vs. time for a probe laser beam directed parallel to the substrate as the deposit grows up through its focal point.

Further evidence that the rate of deposition from Me$_2$Au(hfac) is mass transport limited comes from theoretical consideration of the growth rate vs. Me$_2$Au(hfac) pressure and buffer gas pressure. In the regime where the mean free path is much greater than the deposit radius, the vertical growth rate for a hemispherical deposit, which to a good approximation describes the geometry of deposits obtained from Me$_2$Au(hfac) under a wide range of conditions, is given by (27)

$$\frac{dr}{dt} = \frac{\alpha v_1 p_1}{(2\pi mkT)^{1/2}},$$
(1)

where r is the radius of the deposit, α is the fraction of collisions of the precursor molecule with the heated region of the substrate that result in deposition, v_1 is the volume that a gold atom occupies in the deposit, p_1 is the partial pressure of gold precursor vapor, m is the mass of a precursor molecule, k is Boltzmann's constant, and T is the gas phase temperature. The reaction probability, α, is a combination of a number of parameters, including the precursor trapping coefficient and the probability of reaction vs. desorption. If the reaction rate and product adsorption rate are sufficiently high, α approaches the trapping coefficient. Since the trapping coefficient is a weak function of temperature and often approaches unity (31), the value of α for mass transport limited deposition processes will be near unity and will depend weakly on the surface temperature, as observed for deposition from Me$_2$Au(hfac). A fit of the deposition rate vs. Me$_2$Au(hfac) pressure yields a value of 0.6 for α, consistent with the value expected for a mass transport limited deposition process.

As buffer gas is added, the mean free path is reduced and eventually becomes smaller than the deposit radius. For a wide range of buffer gas pressures, the vertical growth rate for a mass transport limited process can be written as (27)

$$\frac{dr}{dt} = \frac{\alpha v_1 P_1}{(2\pi mkT)^{1/2}} \left[\frac{4Kn/3\alpha}{(1 + 4Kn/3\alpha)} \right]. \tag{2}$$

Kn is the Knudsen number defined by

$$Kn = \frac{3D_{AB}}{r(8kT/\pi m)^{1/2}}, \tag{3}$$

where D_{AB} is the binary diffusion coefficent of the precursor/buffer gas system. Kn is the ratio of the gas phase mean free path to the deposit radius. For Kn << 1, equation (2) reduces to

$$\frac{dr}{dt} = \frac{D_{AB}v_1 P_1}{rkT}, \tag{4}$$

which is the equation for the diffusion-controlled deposition rate under conditions where the mean free path is much less than the deposit radius (27). Deposition rates calculated from equation (3) were compared to measured deposition rates as a function of argon buffer gas pressure, using α = 0.6, as previously determined. The value of D_{AB} was calculated using the empirical method of Fuller et al. (32), kinetic theory (33), and Chapman-Enskog theory (33). These methods give values for D_{AB} between 0.01 and 0.04 cm^2/sec at 760 torr and 25°C. We used the mean value of 0.02 to obtain D_{AB} as a function of pressure. The agreement between the calculated and measured deposition rates is excellent, as shown in Fig. 4, supporting the conclusion that buffer gas reduces the deposition rate by reducing the rate of precursor transport to the surface.

4. LASER-INDUCED PHOTOCHEMICAL DEPOSITION

Photochemical deposition processes induced by a focused cw laser beam incident on a substrate

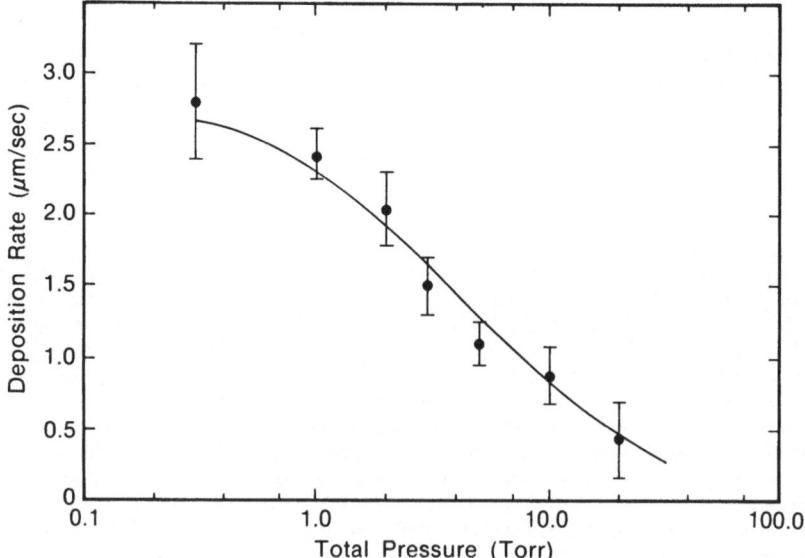

Fig. 4. Deposit vertical growth rate vs. argon buffer gas pressure for Me$_2$Au(hfac) pressure of 0.3 torr, laser beam diameter of 7 μm, and laser power of 127 mW. Solid line is calculated from equation (3) using α = 0.6, r= 20 μm, and D_{AB} = 14.5/P, where P is the argon pressure in torr.

surface can be divided into two basic types. In the first, the deposition kinetics are controlled by photochemical reactions of gas phase species; in the second, the deposition kinetics are controlled by photochemical reactions of surface-adsorbed species. Unfortunately, it is not often clear which process is dominant or if both may be important. Much of the discussion of photochemical deposition mechanisms has thus centered around ways of determining whether precursor photodissociation occurs in the gas phase or on the surface.

A strong indication that reactions of surface-adsorbed molecules are important in a number of deposition processes is the presence of ripples in the deposits. The ripples have a regular spacing on the order of 0.5-1λ. Rippled deposits have been observed upon frequency-doubled argon ion laser (257 nm) induced deposition from metal alkyls (34), metal carbonyls (35) (see Fig. 5), and copper bis-(hexafluoroacetylacetonate) (36). The ripples, oriented perpendicular to the electric field vector of the polarized laser beam, appear to be due to interference of the incoming beam with surface plasma waves having the same period as the observed ripples (37). For the deposit to grow with a rippled structure, the deposition process must faithfully reproduce the intensity variations of the light at the surface. This can only be accomplished if deposition occurs by photodissociation of surface-adsorbed species, since the mean free path in the gas phase is typically very much larger than the ripple spacing.

Other evidence points to the importance of gas phase photoreactions in deposition from $Cd(CH_3)_2$. Wood et al. (38) have developed a simple model for the dependence of the deposit growth rate on laser beam focal diameter at the surface. The growth rate is defined as the increase in deposit thickness per unit time. The model predicts that the deposition rate for a photochemical deposition process dominated by single-photon dissociation of surface-adsorbed species should depend on r^{-2}, where r is the radius of the laser beam at the surface. This is because the photodissociation rate is proportional to the laser intensity, which in turn is proportional to r^{-2} for a given laser power. The deposition rate for a photochemical deposition process dominated by single-photon dissociation of gas phase species should be approximately proportional to r^{-1}, however. Although the photodissociation rate is proportional to r^{-2}, photoproducts formed a distance r away from the surface contribute to the growing deposit, resulting in an overall dependence of roughly r^{-1}. Wood et al. found the deposition rate for frequency-doubled argon ion laser-induced deposition from $Cd(CH_3)_2$ to be proportional to $r^{-0.7}$, which they took to indicate that the deposition processes is controlled by photodissociation of $Cd(CH_3)_2$ in the gas phase.

Fig. 5. Scanning electron micrograph of a small area of a deposit produced from $Cr(CO)_6$ using a polarized, frequency-doubled argon ion laser (257 nm). The laser electric field vector is oriented perpendicular to the ripples. The white bar at the bottom left is 1 μm.

Another way to determine whether a laser-induced photochemical deposition process involves photoreactions of gas phase or surface-adsorbed species is the dependence of the deposition rate on precursor pressure. In either case, the deposition rate is expected to be proportional to precursor concentration over a wide range of conditions, but the dependence on precursor pressure is different for the two cases. For deposition under conditions where the rate-determining step involves photoreactions of gas phase species, the rate should increase linearly with precursor pressure. This has been observed for deposition from a number of metal alkyls (39, 40) and metal carbonyls (35, 41-43). For deposition under conditions where the rate-determining step involves photoreactions of surface-adsorbed species, the rate will increase with the precursor adsorption isotherm. For many systems, a Type II isotherm (44) characteristic of physical adsorption is expected. Type II adsorption behavior has been observed for $Cd(CH_3)_2$ (45), $[Al(CH_3)_3)]_2$ (45), and $TiCl_4$ (6). Deposition rates proportional to the adsorption isotherm have in fact been observed for frequency-doubled argon ion laser-induced deposition from $TiCl_4$ (6). This is strong indication that deposition proceeds via photodissociation of adsorbed $TiCl_4$.

To gain some insight into laser-induced photochemical deposition mechanisms, we recently studied frequency-doubled argon ion laser-induced deposition from the Group 6 metal carbonyls, $Cr(CO)_6$, $Mo(CO)_6$, and $W(CO)_6$ (35, 43). In these experiments, we employed a quartz crystal microbalance (QCM) for the first time in a laser-induced deposition process to measure absolute deposition rates in real time (46). By considering the equations for deposition of a small spot onto the microbalance crystal surface, we were able to obtain absolute deposition rates in terms of the mass deposited per unit time. A QCM employing AT-cut 10 MHz crystals operated on the third overtone was used to measure deposition rates from the hexacarbonyls of chromium, molybdenum, and tungsten as a function of metal carbonyl pressure and laser power. For all three metal carbonyls, the deposition rate increases linearly with metal carbonyl pressure (Fig. 6). The dependence of log deposition rate on log laser power is linear with a slope near unity, as summarized in Table I. The deposits are not purely metallic, being significantly contaminated by carbon and oxygen, as shown by the results of Auger analysis of the films, summarized in Table 2. Similar deposit compositions were observed by Gluck et al. (47) for deposition from the Group 6 carbonyls under similar conditions.

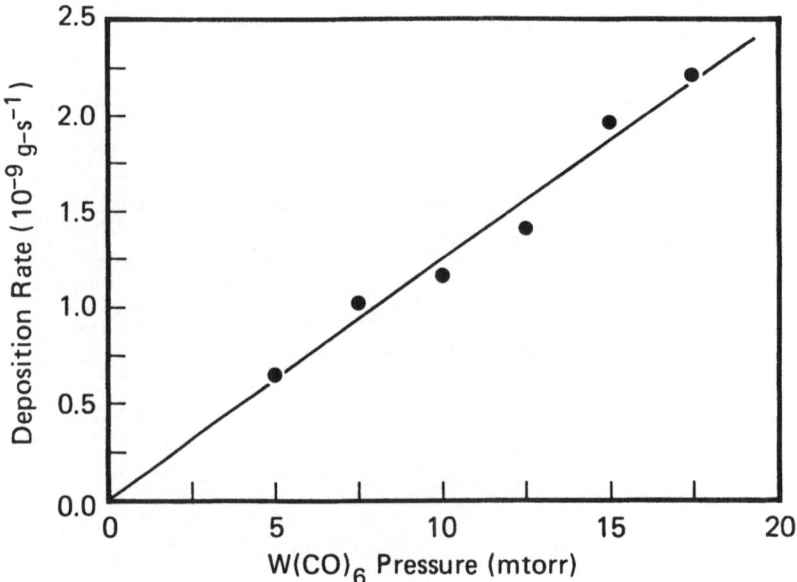

Fig. 6. Deposition rate vs. $W(CO)_6$ pressure from 257 nm laser-induced photochemical deposition. The line is a least squares fit to the data. Similar linear dependences of deposition rate on metal carbonyl pressure were observed for $Cr(CO)_6$ and $Mo(CO)_6$.

Table I. Slope of deposition rate vs. metal carbonyl pressure and the exponent in the dependence of deposition rate on laser power for 257 nm deposition from the Group 6 metal carbonyls.

	Deposition rate $(g \ s^{-1} \ torr^{-1})$	Exponent
$Cr(CO)_6$	1.9×10^{-7}	0.86
$Mo(CO)_6$	1.4×10^{-7}	0.89
$W(CO)_6$	1.2×10^{-7}	0.96

Table II. Auger spectral analysis of deposits (at. %) formed via 257 nm laser-induced deposition from the Group 6 metal carbonyls. Deposits were cleaned by argon ion sputtering to remove surface contaminants prior to analysis.

	M	C	O
$Cr(CO)_6$	39	<2	61
$Mo(CO)_6$	54	15	31
$W(CO)_6$	36	37	27

As we discussed above, the linear dependence of deposition rate on precursor pressure indicates that the deposition process involves photodissociation of the gas phase metal carbonyls. This is consistent with our observation that the metal carbonyls are not adsorbed (<0.05 monolayers) onto the metal surface of the crystal electrode, even at pressures approaching the metal carbonyl vapor pressure (48). The presence of ripples in the deposits, as shown in Fig. 5, is not consistent with a deposition mechanism involving purely gas phase photoreactions, however. Thus it is clear that photoreactions of surface-adsorbed species must also be occurring during deposition.

In addition, the simplest explanation of the linear dependence of deposition rate on laser intensity is that deposition proceeds via one-photon dissociation of the metal carbonyl to form a photoproduct that condenses on the surface, but it is not possible to account for the deposit composition by this mechanism. One 257 nm photon (4.82 eV) could remove up to four CO ligands from $Cr(CO)_6$ (49, 50), but only two or three from $Mo(CO)_6$ (51) and $W(CO)_6$ (52). A deposit composed of photofragments with a CO to metal atom ratio of 2-4 is not consistent with the Auger analysis. The metal content of the deposits can only be explained if deposition is a multiple-photon process. One possibility is that the gas phase photodissociation process is partially saturated, causing the deposition rate to show a linear dependence on laser intensity, even though the overall photoprocess involves several photons. We have studied gas phase multiple-photon dissociation of each of the Group 6 metal carbonyls using a KrF* laser (248 nm), and we see no evidence of saturation at laser powers well above those used in our deposition experiments (53, 54) . Thus, it is clear that photodeposition is a multiple-photon process, but it is not possible that this process occurs solely in the gas phase.

To account for our observations, we propose the following deposition mechanism. The first step is single-photon gas phase dissociation of the metal carbonyl to yield a photofragment, $M(CO)_x$, where x<6. This is consistent with the known gas phase photochemistry of metal carbonyls (55-59). In the next step, $M(CO)_x$ diffuses to the QCM crystal surface. Since $M(CO)_x$ is unsaturated, it should adsorb rather strongly to the surface of the QCM crystal or to the surface of the deposit. That fraction of the $M(CO)_x$ that adsorbs within the laser beam or that diffuses along the surface into the laser beam can absorb additional photons in the last step to form the observed deposit. Because the last step involves photodissociation of adsorbed species, ripples are formed (Fig. 6) and the deposit is confined to the laser beam focal diameter, even under conditions where the gas phase mean free path is much larger than the laser beam focal diameter. Surface diffusion of $M(CO)_x$ is likely to be much slower than further photodissociation at the laser intensities used in our experiments, so the probability of $M(CO)_x$ dissociation will approach unity and will be independent of laser intensity over a relatively broad range. A combination of gas phase dissociation of $M(CO)_6$ and gas phase transport of $M(CO)_x$ will thus be the rate-determining steps and the deposition rate will depend approximately linearly on laser intensity and $M(CO)_6$ pressure.

An important point of our study of 257 nm deposition from the Group 6 carbonyls is that laser-induced photochemical deposition processes do not have to proceed by photodissociation of purely gas phase or surface-adsorbed species. For the Group 6 metal carbonyls, both processes play a critical role, although the gas phase photoreactions dominate the deposition kinetics. A mechanism of this kind may be fairly general. For most of the species that have served as deposition precursors, the energy of one photon at 257 nm is not sufficient to allow complete dissociation to metal atoms via a single-photon dissociation process. Gas phase multiple-photon dissociation of a precursor to yield exclusively metal atoms is unlikely at the laser intensities used in deposition experiments. Both $Cd(CH_3)_2$ and $Zn(CH_3)_2$ can be completely dissociated to metal atoms via single-photon dissociation at 257 nm (60), but $M(CH_3)$ fragments as well as metal atoms are formed upon uv photodissociation of both species in the gas phase (60). Thus 257 nm laser-induced deposition from most precursors, including $Cd(CH_3)_2$ and $Zn(CH_3)_2$ to some extent, may proceed by a mechanism similar to the one proposed above for deposition from the Group 6 metal carbonyls, i.e. gas phase photodissociation to produce partially dissociated fragments, adsorption of the fragments on the surface, followed by photodissociation of the adsorbed fragments. For the metal alkyls, this is consistent with the linear dependence of the deposition rate on metal alkyl pressure (38, 39), laser power (38, 39), and the inverse of the focal spot size (37), as well as the presence of ripples in the deposits (33), as we discussed above for the Group 6 metal carbonyls.

5. CONCLUSION

We have summarized currently available knowledge on the mechanisms of laser-induced photothermal and photochemical deposition. Substantial advances have been made toward complete quantitative modeling of photothermal deposition by performing accurate calculations of the surface temperature distribution (18,22) during deposition, measuring surface temperature profiles with thin-film thermocouples (18), and calculating mass transport contributions to the deposition rate (11, 26-28). Less progress has been made in modeling photochemical deposition. Mechanistic work has concentrated on determining whether gas phase or surface phase photodissociation processes dominate the deposition process (34, 37, 38, 42). Recent work on the kinetics of photochemical deposition from the Group 6 metal carbonyls has shown that photochemical deposition processes may not fit this simple picture; gas phase and surface phase photoreactions can operate simultaneously (34, 42).

6. REFERENCES

1. J. Y. Tsao, D. J. Ehrlich, D. J. Silversmith, and R. W. Mountain, IEEE Elect. Dev. Lett., EDL-3, 164 (1982).
2. B. M. McWilliams, I. P. Herman, F. Mitlitsky, R. A. Hyde, and L. L. Wood, Appl. Phys. Lett., 43, 946 (1983).
3. I. P. Herman, B. M. McWilliams, F. Mitlitsky, H. W. Chin, R. A. Hyde, and L. L. Wood, "Laser-Controlled Chemical Processing of Surfaces", A. W. Johnson, D. J. Ehrlich, and H. R. Schlossberg, ed., Elsiever, New York, 1984, pp. 29-34.
4. I. P. Herman, Springer Ser. Chem. Phys., 39, 396 (1984).
5. D. J. Silversmith, D. J. Ehrlich, J. Y. Tsao, R. W. Mountain, and J. H. C. Sedlacek, "Laser-Controlled Chemical Processing of Surfaces", A. W. Johnson, D. J. Ehrlich, and H. R. Schlossberg, ed., Elsiever, New York, 1984, pp. 55-60.
6. J. Y. Tsao, R. A. Becker, D. J. Ehrlich, and F. J. Leonberger, Appl. Phys. Lett., 42, 559 (1983).
7. D. J. Ehrlich, R. M. Osgood, Jr., D. J. Silversmith, and T. F. Deutsch, IEEE Elect. Dev. Lett., EDL-1, 101 (1980).
8. J. N. Randall, D. J. Ehrlich, and J. Y. Tsao, J. Vac. Sci. Technol. B, 3, 262 (1985).
9. M. M. Oprysko, M. W. Beranek, and P. L. Young, IEEE Electr. Dev. Lett. EDL-6, 344 (1985).
10. Quantronix Corp., Smithtown, New York.
11. D. J. Ehrlich and J. Y. Tsao, J. Vac. Sci. Technol. B, 1, 969 (1983).
12. D. J. Ehrlich, R. M. Osgood, Jr., and T. F. Deutsch, J. Vac. Sci. Technol., 21, 23 (1982).
13. F. A. Houle, Appl. Phys. A, 41, 315 (1986).

14. M. Lax, Appl. Phys. Lett., 33, 786 (1978).
15. J. E. Moody and R. H. Hendel, J. Appl. Phys., 53, 4364 (1982).
16. J. P. Colinge and F. Van de Wiele, J. Appl. Phys., 52, 4796 (1981).
17. I. D. Calder and R. Sue, J. Appl. Phys., 53, 4357 (1982).
18. T. T. Kodas, T. H. Baum, and P. B. Comita, J. Appl. Phys., 61, (1987), in press.
19. C. R. Moylan, T. H. Baum, and C. R. Jones, Appl. Phys. A, 40, 1 (1986).
20. S. D. Allen, J. Appl. Phys., 52, 6501 (1981).
21. D. Bauerle, Springer Ser. Chem. Phys., 39, 166 (1984).
22. S. D. Allen, J. A. Goldstone, J. P. Stone, and R. Y. Jan, J. Appl. Phys., 59, 1653 (1986).
23. H. E. Carlton and J. H. Oxley, Am. Inst. Chem. Eng. J., 12, 86 (1967).
24. T. H. Baum and C. R. Jones, Appl. Phys. Lett., 47, 583 (1985).
25. T. H. Baum and C. R. Jones, J. Vac. Sci. Technol. B, 4, 1187 (1986).
26. I. P. Herman, R. A. Hyde, B. M. McWilliams, A. H. Weisberg, and L. L. Wood, "Laser Diagnostics and Photochemical Processing for Semiconductor Devices", R. M. Osgood, S. R. J. Brueck, and H. R. Schlossberg, ed., Elsiever, New York, 1983, pp. 9-18.
27. T. T. Kodas, T. H. Baum, and P. B. Comita, J. Appl. Phys., in press.
28. D. C. Skouby and K. F. Jensen, Proc. Soc. Photo-Opt. Instrum. Eng., 797 (1987), in press.
29. T. H. Baum, J. Electrochem. Soc., 134 (1987), in press.
30. P. B. Comita and T. T. Kodas, J. Appl. Phys., in press.
31. W. H. Weinberg and R. P. Merrill, J. Vac. Sci. Technol., 8, 718 (1971).
32. E. N. Fuller, P. D. Schettler, and J. C. Giddings, Ind. Eng. Chem., 58, (5), 18 (1966).
33. R. B. Bird, W. E. Stewart and E. N. Lightfoot, "Transport Phenomena", Wiley, New York, 1960, p. 510.
34. R. M. Osgood and D. J. Ehrlich, Opt. Lett., 7, 385 (1982).
35. G. W. Tyndall and R. L. Jackson, to be published.
36. F. A. Houle, R. J. Wilson, and T. H. Baum, J. Vac. Sci. Technol. A, 4, 2452 (1986).
37. S. R. J. Brueck and D. J. Ehrlich, Phys. Rev. Lett., 48, 1678 (1982).
38. T. H. Wood, J. C. White, and B. A. Thacker, Appl. Phys. Lett., 42, 408 (1983).
39. D. J. Ehrlich, R. M. Osgood, Jr., and T. F. Deutsch, IEEE J. Quant. Electr., QE-16, 1233 (1980).
40. M. S. Chiu, K. P. Shen, and Y. K. Ku, Appl. Phys. B, 37, 63 (1985).
41. D. J. Ehrlich, R. M. Osgood, Jr., and T. F. Deutsch, J. Electrochem. Soc., 128, 2039 (1981).
42. M. S. Chiu, Y. G. Tseng, and Y. K. Ku, Opt. Lett., 10, 113 (1985).
43. R. L. Jackson and G. W. Tyndall, J. Appl. Phys., 62, (1987), in press.
44. A. W. Adamson, "Physical Chemistry of Surfaces", Fourth Ed., Wiley, New York, 1982, p. 534.
45. D. J. Ehrlich and R. M. Osgood, Jr., Chem. Phys. Lett., 79, 381 (1981).
46. A QCM was used to monitor relative deposition rates from $Cr(CO)_6$ induced by a pulsed laser beam oriented parallel to the crystal surface by T. M. Mayer, G. J. Fisanick, T. S. Eichelberger IV, J. Appl. Phys., 53, 8462 (1982). A QCM was also used to monitor arc lamp deposition from $Pb(C_2H_5)_4$ over a large area by L. J. Rigby, Trans. Faraday. Soc., 65, 2421 (1969).
47. N. S. Gluck, G. J. Wolga, C. E. Bartosch, W. Ho, and Z. Ying, J. Appl. Phys., 61, 998 (1987).
48. Lack of $Cr(CO)_6$ adsorption on a QCM crystal was also noted by Mayer et al. (46).
49. The average Cr-CO bond dissociation energy is 1.11 eV. See ref. 50.
50. G. Pilcher, M. J. Ware, and D. A. Pittam, J. Less-Common Met., 42, 223 (1975). See also K. E. Lewis, D. M. Golden, and G. P. Smith, J. Am. Chem. Soc. 106, 3905 (1984).
51. The average Mo-CO bond dissociation energy is 1.57 eV. See ref. 50.
52. The average W-CO bond dissociation energy is 1.84 eV. See ref. 50.
53. G. W. Tyndall and R. L. Jackson, J. Am. Chem. Soc. 109, 582 (1987).
54. G. W. Tyndall and R. L. Jackson, to be published.
55. T. R. Fletcher and R. N. Rosenfeld, J. Am. Chem. Soc., 107, 2203 (1985).
56. T. A. Seder, S. P. Church, and E. Weitz, J. Am. Chem. Soc., 108, 4721 (1986).
57. W. Tumas, B. Gitlin, A. M. Rosan, and J. T. Yardley, J. Am. Chem. Soc., 104, 55 (1982).
58. T. A. Seder, A. J. Ouderkirk, and E. Weitz, J. Chem. Phys., 85, 1977 (1986).
59. G. Nathanson, B. Gitlin, A. M. Rosan, and J. T. Yardley, J. Chem. Phys., 74, 361 (1981).
60. P. J. Young, R. K. Gosavi, J. Connor, O. P. Strausz, and H. E. Gunning, J. Chem. Phys., 58, 5280 (1973).

TIME RESOLVED MEASUREMENTS IN THE THERMALLY ASSISTED PHOTOLYTIC
LASER CHEMICAL VAPOR DEPOSITION OF PLATINUM

D. BRAICHOTTE AND H. VAN DEN BERGH
Laboratoire de Chimie Technique
Ecole Polytechnique Fédérale (ETH)
CH-1015 Lausanne, Switzerland

ABSTRACT:

   The laser chemical vapor deposition (LCVD) of platinum is studied with a cw
argon ion laser near 350 nm. Deposition rates are studied in real time by measuring the laser light transmitted through the depositing metal film and the
substrate. At low power densities the transmitted light decreases monotonically with time in a purely photolytic process leading to low quality films. At
higher power densities two quite different time phases are observed: A slow
initial photodeposition followed by a rapid pyrolytic deposition. The high power density deposits contain less impurities and show good electrical conductivity and surface adherence. The low power photolysis takes place mainly in the
layers of organometallic molecules adsorbed on the surface, whereas at higher
powers photolytic LCVD is a predominantly gas phase process.

INTRODUCTION:
   Laser chemical vapor deposition of metals is one way to make small ohmic
contacts for possible applications in microelectronics like circuit repair and
circuit prototyping (1-3). Other applications of metal deposition may include
the production of Schottky diodes (4,5) optical gratings (6) and mask repair
(7). Lasers have a wide variety of other applications in micro- and opto-elec-
tronics like highly localized chemical surface modifications (8), deposition
of composit substances (9), etching (10), and doping (11). All these can be
done, at least in principle, with a resolution of better than 1 micron due to
the small focal points of laser beams. Also some particular aspects of the
laser-induced processes themselves allow for surface modifications which are
even smaller than the laser beam waist at the focal point. Laser CVD or direct
writing has the advantage of being a simple one step process which contrasts
it with more classical lithographic processes.
   In the present paper we are concerned with the LCVD of platinum which is a
convenient model system for metal deposition, and which furthermore may have
some special applications (4,5). Generally, two mechanisms have been invoked
in LCVD, pyrolytic and photolytic deposition. In the former the surface is
locally heated by the focussed laser beam to a temperature where the metalorga
nic molecules which hit the hot surface pyrolyse leaving behind the metal atom
In the latter, higher energy photons are used to photodissociate the metalorga
nic molecules, either in the gas phase or on the surface, but in this case the
surface may remain cool. We have found in the LCVD of several metals (6,8,9)

D. J. Ehrlich and V. T. Nguyen (eds.), Emerging Technologies for In Situ Processing, 83–91.
© 1988 by Martinus Nijhoff Publishers.

that pure <u>photolysis</u> leads to metal films which include a relatively high percentage of ligand material. This implies bad electrical conductivity and surface adherence. <u>Pyrolysis</u> however leads to significantly less inclusion of ligand material, and these films can show excellent electrical conductivity (less than twice the bulk value) and good surface adherence. Unfortunately in pyrolysis the surfaces have to be heated significantly (depending on the metalorganic used) for acceptable decomposition rates, and this may damage the surface in some way. Furthermore, laser induced surface heating depends on surface optical properties like absorption and reflection as well as on the surface thermal conductivity, all of which may vary from one point to another along the surface where the metal contact is to be deposited. Thus the morphology and the electrical properties of such a contact made by pyrolytic LCVD will vary with local changes in the substrate which is undesirable. There may be several solutions to this problem. First one may try to use successive laser beams: A UV laser first "seeds" the surface with a fine stripe of metal clusters. This seeding would be photolytic and independent of substrate characteristics. The seeding step can be fast as the deposition of only very little material is required. Then a second visible (or IR) laser beam would strongly heat the area with the metal clusters inducing pyrolytic deposition. As seeding with metal clusters may enhance the deposition, a highly localized deposition exclusive to the area of cluster seeding, may be found. A second solution is the use of a so-called hybrid deposition mechanism which we have called thermally assisted photolytic LCVD (4,5,8). In this deposition we use a single laser at a wavelength at which photolytic LCVD is efficient (two lasers, one in the UV and one in the visible, with coincident beams may also be used). The laser beam power is chosen to be sufficient to cause a fair amount of surface heating. At intermediate temperatures this heating may help to "cure" the metal film from the undesired ligand impurities. At higher light intensities, eventually pyrolysis may set in. Again, one may speculate that the light absorption in the initially photodeposited thin metal layer is so high as compared to the surrounding substrate that the heating, and hence the pyrolysis, are localized by the limits of the well defined photolytic deposit The transmission experiments described below actually confirm this hypothesis At first, a thin metal film is deposited by a slow photolytic process. When the metallic layer gets thick enough to absorb a significant fraction of the incoming light it heats up considerably, and pyrolytic LCVD sets in. One of the main goals of the present work is to study this hybrid mechanism in some detail. We clearly show at which conditions the photolytic deposition takes mainly place in the "ad-layer" or on the contrary in the gas phase. In a future communication we will show how this information may be transferred to practical direct writing problems on different substrates such as semiconductors, plastics, ceramics etc.

EXPERIMENTAL:

A schematic of the setup is shown in Fig. 1. The photolysis cell is a stainless steel cylinder with 4 cm internal diameter and 6 cm optical depth. A jacket through which a thermostatted liquid is circulated controls the cell temperature Tc which is equal to the temperature of the photolysis substrate. A small reservoir in the back of the cell contains the metalorganic solid powder which is separately thermostatted at a temperature Tu which is generally well below the cell temperature to avoid condensation on the windows. Tu

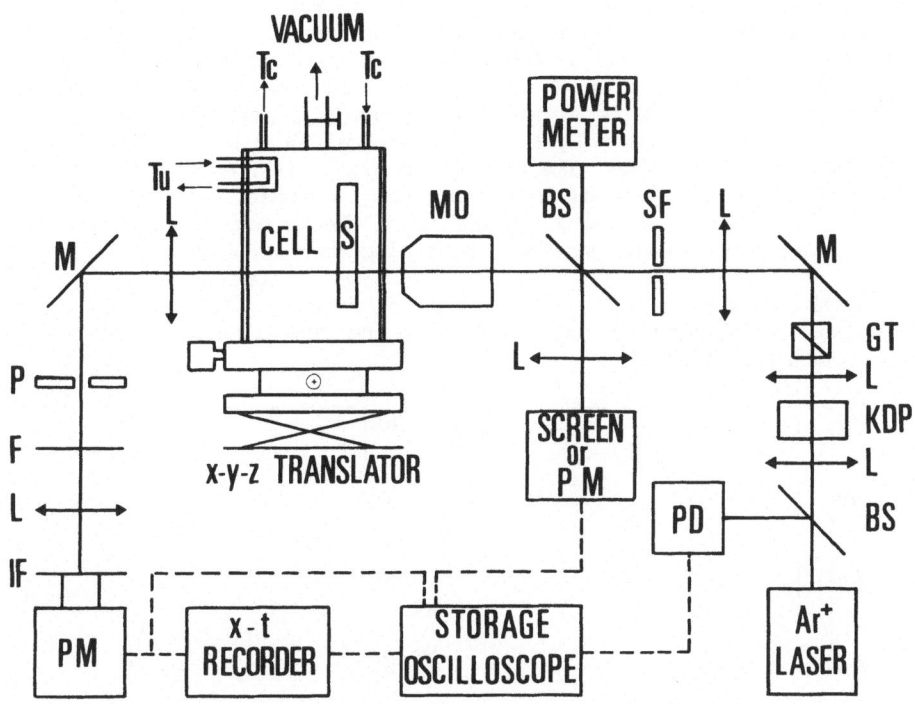

FIG. 1: The experimental setup used to measure LCVD rates by measuring the light transmitted through the depositing metal film in real time. BS=beam splitter, F=filter, GT=Glan-Thompson prism, IF=interference filter, KDP=frequency doubler, L=lens, M=mirror, MO=microscope objective, P=pinhole, PD=photodiode, PM=photomultiplier, S=substrate, SF= spatial filter, Tc=cell temperature, Tu=organometallic solid temperature of the reservoir.

controls the metalorganic vapor pressure in the cell. In the absence of laser radiation the ratio Tc/Tu controls the number of metalorganic adlayers on the substrate. The cell is evacuated with a two stage mechanical pump through a liquid nitrogen trap. The cell position is controlled by an X,Y,Z translator. A line tunable argon ion laser is used which can emit up to 1 W of UV light in the range 351-363 nm. For some applications the 514.5 nm line may be frequency doubled with a KDP temperature tuned crystal. In these cases the green visible beam is separated from the UV beam at 257 nm with a Glan-Thompson prism and a reflection filter (60%T at 257 nm). The beam is then widened and passed through a pinhole and the power is measured continuously by a fixed % of the beam being reflected from a beam splitter onto a power meter. Different microscope objectives are used to focus onto the substrate. The light which is back-reflected from the surface of the substrate is passed in reverse through the microscope objective, and reflected off a beam splitter onto a screen. This projection on a screen of the back-reflected image of the focal point permits accurate focussing. This screen may be replaced by a photomultiplier which measures the re-

flected light as a function of the time. This is particularly useful for real
time measurements of LCVD on surfaces which are not transparent. The light
transmitted through the cell is collected by a lens and passed through a pinhole and some filters onto a PM. The PM signal is fed into an oscilloscope for
the case of fast signals and into an X-t recorder in the case of slow signals.
The deposition is started by opening a mechanical shutter (not shown) which is
placed just after the laser. The photodiode signal is then used to trigger the
oscilloscope. The platinum bis hexafluoroacetylacetonate was prepared as before
(4,5) and its vapor pressure as a function of the temperature and its UV-Vis
absorption spectrum in the gas phase have also been reported (4,5).

RESULTS AND DISCUSSION:
Figures 2 and 3 show <u>typical</u> raw data on the time dependent transmission decreases observed in LCVD.

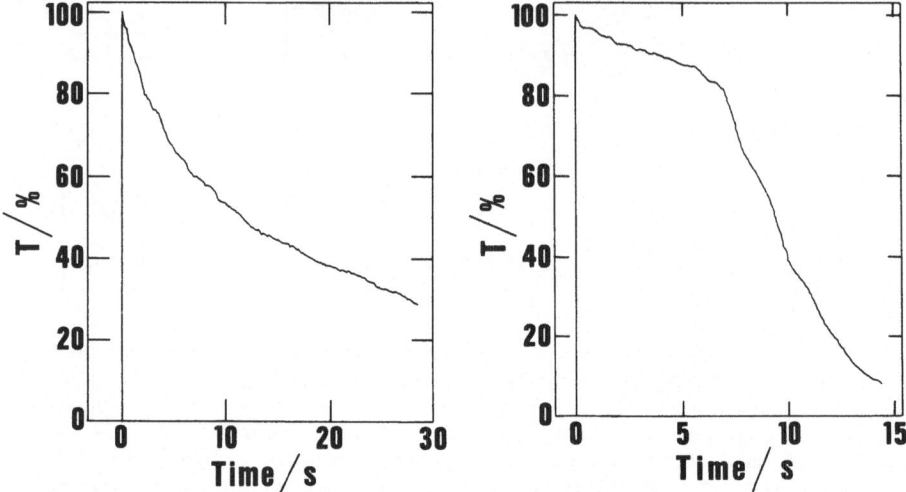

FIG.2: Real time transmittance decrease in purely photolytic LCVD of
platinum. Laser power is 0.08 mW.
Further conditions as in Fig. 4 below.

FIG.3: Real time transmittance decrease in hybrid LCVD. Laser power
is 20 mW. Conditions as in Fig. 4
below. Note the two distinct phases.

Fig. 2 shows the case of purely photolytic deposition at low light intensity
with a monotonically decreasing transmission. Fig. 3 shows a hybrid deposition, at somewhat higher laser beam intensity which starts with a slow photolytic phase, followed by a fast pyrolytic phase. Data similar to that of Figs.
2 and 3, obtained over five orders of magnitude change in laser light intensity, are compiled in Fig. 4. The relative photolytic deposition rates are defined as the inverse of the time necessary for the transmitted light to decrease
from 90%T to 70%T. In case of the pyrolysis data the inverse of the time necessary for the decrease from 80%T to 60%T was used. These relative rates in
units of inverse seconds may be approximately converted to absolute deposition
rates in units of Angstroms per second by multiplying them by a factor of 20.
Between I=10^{-2} and I=5 kW cm^{-2} the photolytic LCVD rate increases, after which

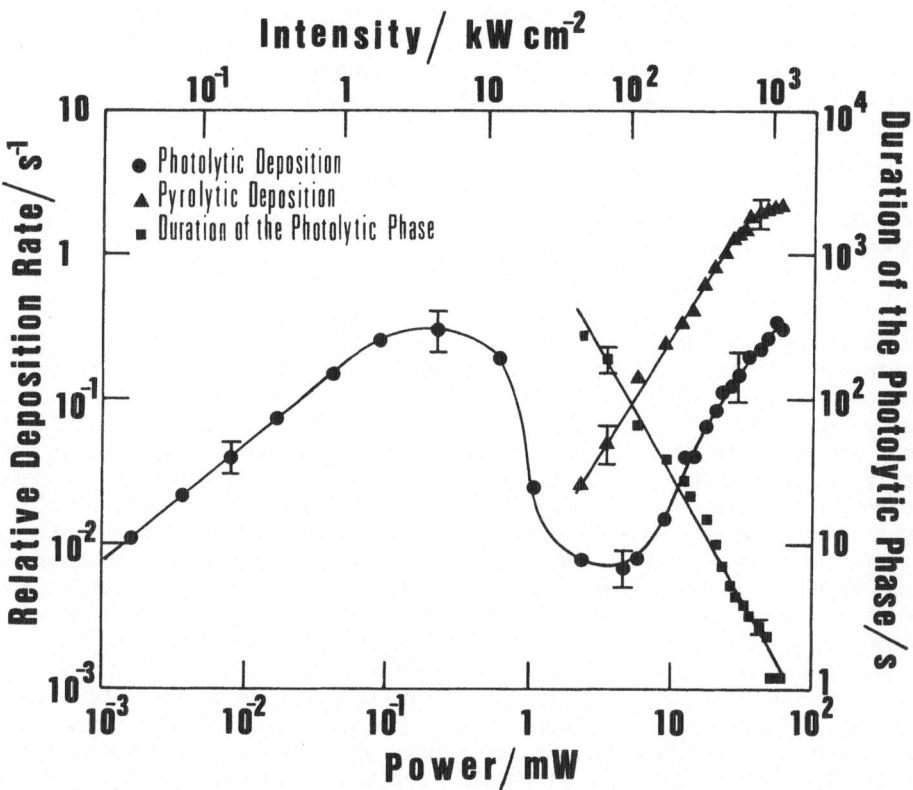

FIG. 4: Relative deposition rates in the LCVD of platinum from Pt-bis hex-afluoroacetylacetonate near 350 nm. (●) are data for photolytic deposition (▲) are data for pyrolytic deposition. (■) represent the duration of the photolytic phase in the case of hybrid deposition. The beam waist at the focal point on the surface is about 2.5 microns (1/e). Tc=50°C, Tu=40°C. The substrate is Pyrex glass, the microscope objective used had N.A.=0.2.

It passes through a maximum and decreases up to I=80 kW cm^{-2}. At intensities a-bove 80 kW cm^{-2} two deposition phases in time are observed rather than one, the first of which is again photolytic deposition. Above 80 kW cm^{-2} we see again an increase in this photolytic LCVD rate with increasing power density up to the maximum intensity of about 1 Megawatt cm^{-2}. At these very high power densities the photolytic deposition rate is equal to that observed at much lower intensities of about 5 kW cm^{-2}. Our interpretation of these observations is that at the lowest intensities photolysis takes place predominantly from the molecules adsorbed on the surface. At higher intensities the surface starts to heat up and the adsorbed molecules evaporate and "ad-phase" photolysis is no longer possible. At significantly higher power densities photolysis from the gas phase starts to dominate. It should be noted that apparently photoly-sis from the "ad-phase" is significantly more efficient than the gas phase photolysis at the applied conditions. To arrive at these conclusions we follow

an argument established by T.H. Wood and coworkers (12) who proposed that one
may distinguish between gas phase photolytic LCVD and "ad-phase" photolytic
LCVD by measuring the deposition rate as a function of the laser beam spot
size on the surface. These authors derived expressions for the deposition rate
from the gas phase (R_g)

$$R_g = C_g \cdot P_L \cdot \frac{1}{w_o} \quad\quad (1)$$

where C_g is a constant and P_L is the laser power, w_o is the beam radius. For
the deposition rate from the "ad-phase" (R_a)

$$R_a = C_a \cdot P_L \cdot \frac{1}{w_o^2} \qu\quad (2)$$

The additional factor of w_o in R_g is due to the metal atoms created within a
distance of the order of w_o from the surface which contribute to the deposition. Hence, if we plot $\log(R)$ against $\log(w_o)$ at constant P_L we should find
a slope of -1 for gas phase photolysis and -2 for "ad-phase" photolysis. Such
plots are shown in Fig. 5. (●) are points taken at $P_L=0.09$ mW, i.e. in the
purely photolytic LCVD range. The observed slope is -1.7 which is strongly
indicative of adlayer photolysis. That the slope does not exactly equal 2.0
may in part be due to the fact that there are some approximations in the de-
rivation of Eqns. 1 and 2. (■) shows the data taken at $P_L=60$ mW which corres
ponds to the photolysis part of the hybrid LCVD. The observed dependence of
$\log(R)$ on $\log(w_o)$ is near the expected value of -1 at low w_o. At higher w_o
however this curve flattens out and even a small increase in R is observed at
larger spot sizes. A possible explanation for this effect is that one increases w_o without changing P_L which causes the temperature of the substrate to
decrease. This may cause a change in the photolysis mechanism, as the temperature decrease of the substrate may enable an "ad-phase" to re-establish
itself. The data of Fig. 5 confirm this hypothesis as the flattening out is
observed at ($P_L=60$ mW, $w_o=10$ micron) $I= 20$ kW cm^{-2}, which in Fig. 4 correspond
exactly to the power where we see an increase in the deposition rate due to a
change in deposition mechanism. The flattening out of the curve obtained at 6C
mW in Fig. 5 may thus well be due to two compensating effects: first the normal decrease in R with increasing w_o, and second an increase in R due to a
temperature decrease which enables the more efficient "ad-phase" photolysis to
start playing a role. Hence the method proposed by Wood et. al. has some disadvantages that can be eliminated by rearranging equations 1 and 2 in the following way:

$$R_g = C_g \cdot \frac{P_L}{w_o} \quad\quad (3)$$

$$R_a = C_a \cdot \frac{P_L}{w_o} \cdot \frac{1}{w_o} \quad\quad (4)$$

If we take the logarithm on both sides and experimentally now keep the ratio
P_L/w_o constant we obtain

$$\log(R_g) = \log(C_g \cdot \frac{P_L}{w_o}) = \text{constant} \quad\quad (5)$$

$$\log(R_a) = \log(C_a \cdot \frac{P_L}{w_o}) - 1 \cdot \log(w_o) \quad\quad (6)$$

Now a plot of $\log(R)$ against w_o will give a slope of O for the case of gas

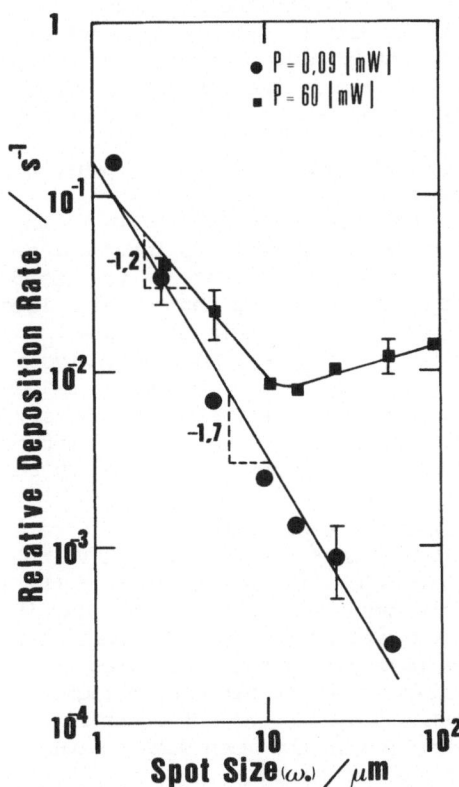

FIG. 5: Relative photolytic LCVD
rates as a function of the laser
spot size on the surface, in order
to distinguish between "ad-layer"
and gas phase photolysis. Conditions
as in Fig. 4.

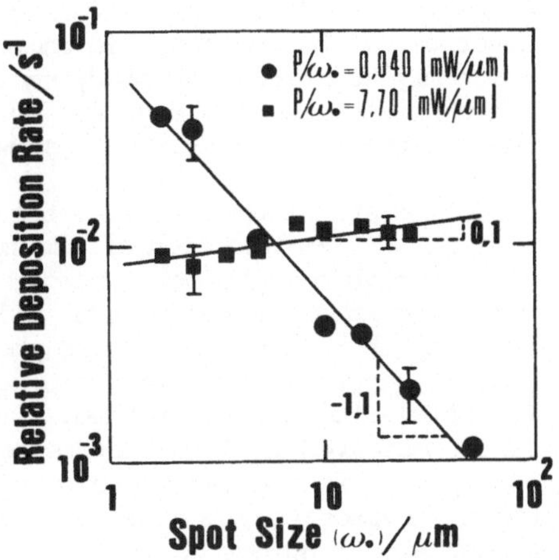

FIG. 6: Relative photolytic LCVD
rates as a function of the laser
spot size, but now keeping P_L/w_O
constant (rather than keeping P_L
constant as in Fig. 5). Conditions as in Fig. 4. A clearer
distinction between "ad-phase"
and gas phase photolysis is obtained.

phase photolysis, whereas for the case of "ad-phase" photolysis a slope of -1
should be observed. This method has the advantage that if we keep P_L/w_O constant, the temperature (at least in the center of the deposit) will stay
approximately constant (13). The results of such plots are shown in Fig. 6.
(●) are points taken at low P_L/w_O which corresponds more or less to the low
power part of Fig. 4. A slope of -1.1 is observed which is very close to the
predicted value for "ad-phase" photolysis. (■) are data obtained at high
P_L/w_O which corresponds to the high power data of Fig. 4. Here a slope of 0.1
is observed, again very close to the predicted zero slope for gas phase photolysis.

The data of Fig. 4 lead to several other observations which should be considered. First of all why should photolysis in the adsorbed phase be so much
more efficient than in the gas phase? Among several possible explanations one
might mention:

1) A spectral shift in the "ad-phase" as compared to the gas phase absorption spectrum of the metalorganic compound leading to enhanced absorption. This point may be checked by studying the same phenomena at other wavelengths where spectral shifts may lead to a decrease in the absorption coefficient.
2) Energetically more favorable pathways being open to photolysis on the surface.
3) More favorable geometry, i.e. the atoms generated in the gas phase need to diffuse in the direction of the surface.
4) If more than one photon is needed to dissociate the molecules , the lower diffusion rates in the adlayer will prevent a photolysis intermediate produced by absorption of a first photon from diffusing out of the laser beam (as may happen in the gas phase) prior to the second photon which liberates the metal atom being absorbed. It is not clear at this point whether the slopes observed in Fig. 4 of respectively 1 in the adphase photolysis data and about 2 in the gas phase photolysis data can be interpreted as being due to respectively one photon and 2 photon rate limiting processes.
5) A much higher number of organometallic molecules in the "laser-surface interaction region" in the case of adlayer photolysis as compared to the gas phase photolysis.

Another set of data plotted in Fig. 4 are the observed pyrolysis rates in the hybrid deposition. Between 50 and 10^3 kW cm^{-2} the pyrolytic LCVD rates are about 10 times faster than the photolytic deposition rates found in the initial phase. The pyrolytic data indicate the onset of a possible plateau in the deposition rates near 1 MW cm^{-2} which may well be due to diffusion limiting. It will be interesting to compare these pyrolysis deposition rates in the hybrid deposition with pyrolytic deposition rates obtained with light at for instance 514.5 nm where one is sure that photolysis can not contribute significantly to the deposition. i.e. the question is how much does photolysis contribute in the pyrolytic part of the hybrid mechanism.

The duration of the photolytic phase in the hybrid deposition is also shown in Fig. 4. A linear decrease of this photolytic initiation period of the pyrolysis is found with increasing power density. A decrease is expected as pyrolysis sets in when a certain temperature T is reached. T will depend on both the laser power and the absorption of this power in the metal film. As power is increased the thickness of the metallic layer which gives rise to T is reached earlier.

CONCLUSIONS:

Thermally assisted photolytic laser chemical vapor deposition may well be an optimal approach for making microscopic ohmic contacts on surfaces that can not support strong heating. Real time deposition studies show that deposition occurs by an initial (slow) photolytic process which is followed under certain conditions by a fast pyrolytic deposition. Photolytic LCVD at low intensities is very efficient and occurs (at least at the applied conditions) from the surface adsorbed organometallic molecules. At higher intensities a less efficient gas phase photolysis predominates.

AKNOWLEDGEMENTS:

The authors are grateful to the Swiss Fonds National for financial support, and to C. Garrido for help with the data evaluation.

REFERENCES:

1. D.J. Ehrlich, Solid State Technology, December 1985, pp. 81–85.
2. R.M. Osgood, Jr., Ann. Rev. Phys. Chem. 1983, 34, pp. 77–101.
3. D. Bäuerle, Laser und Optoelektronik 1985, No. 1, pp. 29–36.
4. D. Braichotte and H. van den Bergh, Laser Processing and Diagnostics, D. Bäuerle et. al. Eds., Les Editions de Physique, Les Ulis, France 1986, pp. 95–99.
5. D. Braichotte and H. van den Bergh, Society for Optical and Quantum Electronics, Proc. International Conf. on Lasers '85, pp. 688–696.
6. M. Qiu, R. Monot and H. van den Bergh, Scientia Sinica 1984, A27, p. 531.
   K.H. Richter, W. Güttler and M. Schwoerer, Appl. Phys. 1983, A32, pp. 1–11.
7. M.M. Oprysko, M.W. Beranek, D.E. Ewbank, and A.C. Titus, Semiconductor International January 1986, pp. 90–100.
8. D. Braichotte, K. Ernst, R. Monot, J.-M. Philippoz, M. Qiu and H. van den Bergh, Helv. Phys. Acta 1985, 58, p. 879.
   M.C. Joliet, C. Antoniadis, R. Andrew and L.D. Laude, Appl. Phys. Lett. 1985, 46(3), pp. 266–267.
   T. Sugii, T. Ito and H. Ishikawa, Appl. Phys. Lett. 1984, 45(9), pp. 966–968.
9. D. Braichotte and H. van den Bergh, Helv. Phys. Acta 1986, 59, pp. 1014–1017.
   G.A. West, A. Gupta and K.W. Beeson, Appl. Phys. Lett. 1985, 47(5), pp. 476–478.
10. P.D. Brewer, G.M. Reksten and R.M. Osgood, Jr., Solid State Technology, April 1985, pp. 273–278.
11. T. F. Deutsch, D.J. Ehrlich, R.M. Osgood, Jr., and Z.L. Liau, Appl. Phys. Lett. 1980, 36(10), pp. 847–849.
    E.P. Fogarassy, D.H. Lowndes, J. Narayan and C.W. White, J. Appl. Phys. 1985, 58(6), pp. 2167–2173.
12. T.H. Wood, J.C. White and B.A. Thacker, Appl. Phys. Lett. 1983, 42(5), pp. 408–410.
13. M. Lax, J. Appl. Phys. 1977, 48, p. 3919.

EXCIMER LASER PROJECTION PATTERNING

M. ROTHSCHILD AND D. J. EHRLICH
Lincoln Laboratory, Massachusetts Institute of Technology
Lexington, MA 02173

1. INTRODUCTION

Photolithographic techniques, in particular those based on optical projection with visible/near-UV radiation, have attained a major role in integrated circuit manufacturing because they provide a unique combination of high precision, high volume and low cost. In these methods an opaque pattern on a trans-illuminated mask is imaged onto a photoresist-coated wafer. The intensity distribution in the image plane is imprinted on the wafer in two steps: photoinduced modification of the resist, followed by development. Conventionally, the photon sources in projection systems have been mercury lamps, operating at 436 nm, and more recently at 405 nm and 365 nm (the G, H, and I lines, respectively, in the emission spectrum of Hg atoms) [1].

The overall resolution obtained to date with the more refined conventional projection systems of this kind is in the 0.7- to 1.0-μm range [1], and further incremental improvements may lower it asymptotically to the vicinity of ~0.5 μm. Further reduction in photolithographic dimensions requires a more revolutionary modification of projection printers. This change is now widely accepted to be the replacement of the mercury lamps with excimer lasers as the photon source [2]. Such a change would have several benefits. The immediate, and more obvious one, is improved resolution due to shorter wavelength, since from very general considerations the smallest resolvable feature in optical imaging is proportional to the wavelength. Excimer lasers operate at several wavelengths, all of them shorter than the G, H, or I mercury lines. The most common ones are 351 nm (the lasing species is XeF), 308 nm (XeCl), 248 nm (KrF) and 193 nm (ArF).

Another property of excimer lasers which has important potential applications to projection lithography is their high peak intensity. Mercury lamps are cw devices with intensities in the 1 W cm^{-2} range. Excimer lasers, on the other hand, operate in the pulsed mode (10-25 ns pulse width), with peak intensities which are ~10^7 times higher. This dramatic difference in intensities, coupled with the shorter wavelength of the excimer lasers, has important implications to integrated circuit processing with resists. Nonlinear processes can now be induced in numerous materials with great efficiency. Therefore, the concern with resist photosensitivity, which is dominant in lamp systems, becomes of secondary importance. Instead, new resists can be employed with properties such as dry processibility, self-development, bleaching and contrast, better tailored for specific applications.

Moreover, the short wavelength of excimers may open up opportunities for a new class of processes, namely, the direct

93

D. J. Ehrlich and V. T. Nguyen (eds.), Emerging Technologies for In Situ Processing, 93–104.

(resistless) dry patterning technology. This comes about because at these new, higher photon energies numerous compounds exhibit intense dissociative absorption. Because of the resultant rich and diverse UV photochemistry, the excimer laser may go beyond the traditional role of the lamp, namely that of sensitizing the photoresist and thereby preparing the wafer for processing with other methods.

The great potential of excimer lasers in photolithography is at present tempered by a number of not yet fully solved challenges. Foremost of these is the availability of aberration-free projection optics in the UV. The design of such optics must take into account the specific output properties of the lasers, such as spatial coherence and uniformity, and spectral linewidth. The high peak intensity may cause catastrophic damage to optical elements, and the high average power may generate absorptive color centers in nominally transparent materials. Also, direct processes must take place at low values of peak power in order to avoid surface damage. This may reduce the throughput to unacceptably low levels, unless changes are introduced in present excimer laser technology. Although the processing and diffraction considerations mentioned above favor switching to the shorter excimer wavelengths, the difficulties encountered in the laser engineering and optical design end of such a lithographic system also increase rapidly as the wavelength is reduced.

2. OPTICAL CONSIDERATIONS

The characterization of an optical projection system involves several fundamental parameters, chiefly the wavelength, λ, and the numerical aperture, NA. Assuming the (by no means trivial) accomplishment of the design and production of an aberration-free system, its resolution is determined by λ and NA in two ways. First, diffraction dictates the size of the smallest resolvable feature, d_{co}, through the relationship $d_{co} = \lambda/4NA$. The smallest feature that can be imprinted in the photoresist, d_{pr}, is larger than and proportional to d_{co}, with the constant of proportionality depending on the material photoresponse. With currently available photoresists, $d_{pr} \approx 3d_{co}$. Second, diffraction also determines the tolerance in defocusing, $z_d = \lambda/2(NA)^2$. In practice, defocusing is unavoidable, because of focusing errors, wafer nonflatness and imperfect planarization. In current practice the smallest useful values of z_d are in the 0.8- to 1.0-μm range. The quantities d_{co} and z_d are therefore two parameters which determine the range of acceptable λ and NA of a particular system. Figure 1 represents the interrelationship between these quantities. In it, a point whose coordinates are z_d and d_{co}, determines λ and NA, and vice versa. The solid segments represent the operating values of present G-line and I-line projection systems with numerical apertures in the 0.30-0.45 range. It is evident from Fig. 1 that reducing d_{co} is compatible with practical values of z_d only with a simultaneous reduction in λ and NA. For instance, a lower limit of 1 μm on z_d imposes $d_{co} = 0.23$ μm at 436 nm and 0.18 μm at 248 nm, corresponding to $d_{pr} = 0.69$ μm and 0.54 μm, respectively, and NA = 0.47 and 0.35, respectively.

Thus, for a fixed z_d, switching from the mercury G line to the KrF laser implies increased resolution, as well as a reduction in NA, which in turn has the benefits of simplifying the optical design and enabling a larger field of view.

The transition from Hg lamps to excimer lasers is, however, not trivial. First, chromatic aberrations due to the laser spectral linewidth must be eliminated. The output of free-running excimer lasers has significant bandwidth. Its effect is further compounded by the fact that the dispersion of most optical materials increases rapidly with decreasing wavelength. For fused silica, for instance, the refractive index at 193 nm varies across the 100-cm^{-1} width of the ArF laser by $\delta n_{ArF} \approx 6.2 \times 10^{-4}$, while at 248 nm it varies across the 50 cm^{-1} of the KrF laser by $\delta n_{KrF} \approx 1.7 \times 10^{-4}$. If a fused silica lens is used for projection, these values of δn cause a spread in focal length, δf, which is proportional to the focal length, f. Even with a conservatively short $f = 1$ cm, δf assumes the values $\delta f_{ArF} \approx 11$ μm, $\delta f_{KrF} \approx 3.4$ μm, which are much larger than the defocusing tolerance $z_d > 1$ μm, discussed above. With refractive optics, δf can in principle be reduced by either spectral narrowing of the excimer laser output, or by achromatization of the projection lens. Both approaches have been recently applied to the development of steppers at 248 nm [3-6]. The first method introduces a degree of complexity in the laser-end of the projection system. For instance, the bandwidth of KrF lasers can be reduced to <1 cm^{-1} with the use of intracavity etalons or with injection locking in an oscillator-amplifier configuration. In either case there are certain drawbacks, such as reduced energy output or lower pulse repetition rate. The second approach employs free-running lasers, at the price of a more complex and more sensitive projection lens. The design and production of achromats in the visible is by now a routine task because of the availability of glasses with varying refractive indexes and dispersive properties. At the excimer laser wavelengths, however, there is a limited selection of transparent materials. These include fused silica and several metal fluorides, mainly calcium, magnesium and lithium. A projection lens which is achromatic over the laser bandwidth can indeed be designed using two or more of these materials. They all, however, have certain undesirable properties. For instance, LiF is hygroscopic and MgF$_2$ is birefringent. All develop color centers when exposed to high doses of laser radiation. It should also be noted that both the difficulties encountered in line narrowing the laser and the propensity toward developing color centers quickly increase at the shorter wavelengths. Thus, overcoming chromatic aberrations at 193 nm is a complex task, in comparison with 308 nm or 248 nm.

A radically different way to eliminate chromatic aberrations is the use of a projection lens made of reflective elements, since mirrors are intrinsically achromatic at all wavelengths. Reflective optics have indeed been employed in ring field scanning projection systems with 1:1 imaging using a XeCl laser [7]. For finer patterning, steppers with a magnification of 5X or higher seem to be more appropriate. An example of such a lens is the Schwarzschild microscope objective which we have used extensively in our own studies of projection lithography [8-10], at a reduction of up to 36X. The field of view, however, was more than an order of magnitude smaller than the > 10 mm field of most step-and-repeat systems. At present, reflective optics are not-yet-fully-developed candidates for projection lenses in commercial excimer-based steppers.

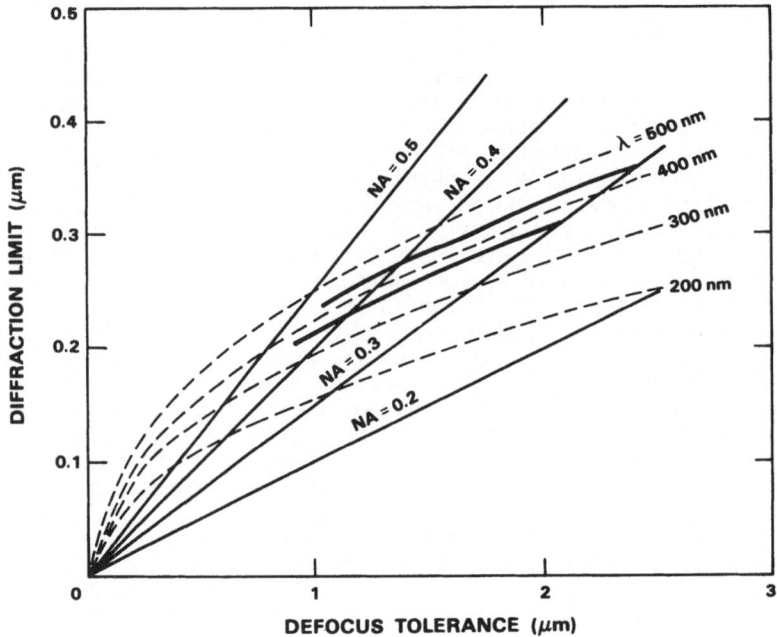

Figure 1. The diffraction limited linewidth, d_{co}, and the defocusing tolerance, z_d, as functions of the wavelength, λ, and numerical aperture, NA. For any imaging system, d_{co} and z_d are the coordinates of the point determined by the intersection of the appropriate λ-curve and NA-curve. The solid line segments represent the operating parameters of existing Hg-lamp steppers at $\lambda = 436$ nm and $\lambda = 365$ nm, with NA's in the range 0.30 to 0.45.

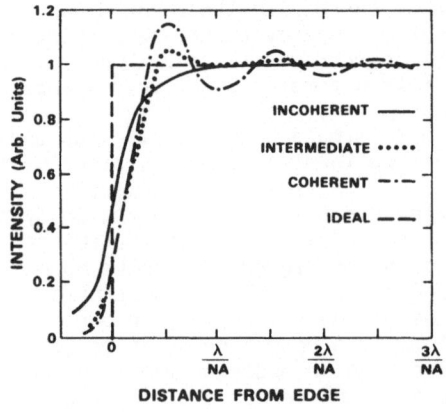

FIGURE 2. The effect of coherence on the image of a sharp edge.

A second topic in which lamp-based imaging systems differ from exicmer-based ones is control of spatial coherence. The degree of spatial coherence is an important parameter in determining the image quality. A highly coherent beam has a diffraction-limited cutoff feature size, $d_{co} = \lambda/2NA$, whereas under highly incoherent illumination the cutoff is half this value, $d_{co} = \lambda/4NA$. In fact, Fig. 1 and the related discussion above implicitly assumed incoherent illumination. While the smaller value of the diffraction-limited cutoff linewidth seems to favor incoherent imaging, it is noted that incoherent imaging is more sensitive toward defocusing than coherent imaging. There are further differences between the two cases. If the object is an isolated edge the image is sharper with coherent imaging, but it also exhibits intensity "ringing" in the illuminated region, as well as a lateral shift of the image from its geometrical position (Fig. 2).

Since lamps are for all practical purposes incoherent, a degree of coherence is determined by the ratio of the numerical apertures of the condenser and the objective as measured in the object (mask) space. Typically, an intermediate degree of coherence is chosen, so as to take full advantage of the positive features of both types of illumination.

The incorporation of excimer lasers in projection printers adds a new dimension to the control of the degree of spatial coherence. Unlike lamps, excimer lasers possess a certain degree of intrinsic spatial coherence, which is determined by the number of oscillating cavity modes. Since the number of modes in excimer lasers is very high, the degree of coherence is low. Nevertheless, it may have a profound impact on image quality. It can be shown that if no condenser is placed between the laser and the mask, the degree of coherence is determined by the divergence of the laser beam, the NA of the objective and its magnification ratio. For a 10X lens with low numerical aperture (NA = 0.2) employing a free-running excimer laser, the degree of coherence is within the range of that of conventional printers. Similarly, for a high-magnification lens such as the 36X, 0.5 NA Scharzschild objective employed in many of our studies, the illumination approaches the incoherent limit. However, the more common projection lenses have a high NA and lower magnification, and the resultant illumination is too coherent. Furthermore, if line narrowing is used, as discussed above, the number of spatial modes is typically reduced as well. This manifests itself in a reduced divergence and increased degree of coherence, beyond its optimal value. In most instances it is therefore necessary to reduce the spatial coherence of the excimer laser before propagating the beam through the rest of the optical system.

Several methods to accomplish this have been reported. At the longer wavelengths, a quartz diffuser can be placed in the beam to transform the laser output into that similar to an incoherent lamp [7]. A comparable effect may be achieved with a randomly oriented fiber bundle. These are, however, lossy processes, and their application to shorter wavelengths (248 nm, 193 nm) may be impractical because of the low energy throughput of diffusers and fibers. At 248 nm, a coarse diffuser has recently been used, namely, a set of closely spaced lenslets ("fly's eye") [11]. A different approach, which is applicable only to multiple-pulse exposures, was implemented at 248 nm [3]: the laser beam was focused onto a spot in the object

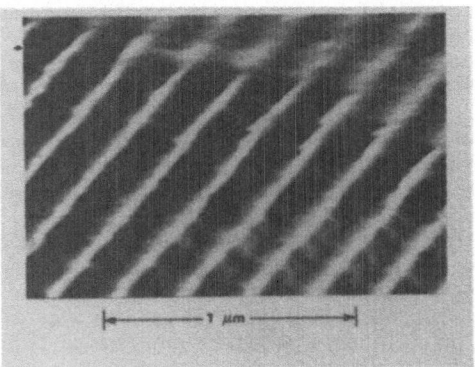

Fig.3   Scanning electron micrograph of nominal C.13-µm lines and spaces
etched in projection in O.15-µm-thick PMMA on Si. Exposure was
performed with one 15-ns laser pulse at 193 nm, and development
was for 5 min in MIK/isopropanol.

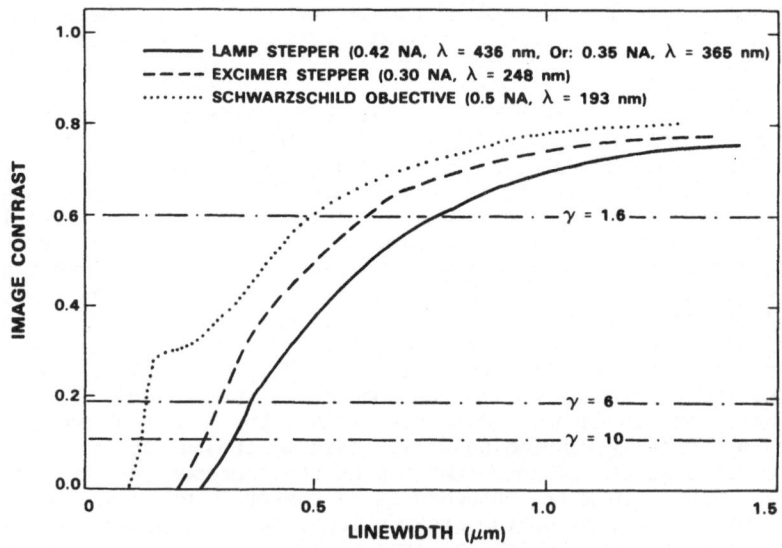

FIGURE 4.   Image contrast vs. image linewidth for three projection
systems, under aberration-free, incoherent illumination. The smallest
printable linewidth is determined by the intersection of the contrast
curve and the appropriate γ of the material response.

plane of the condenser, i.e., the source plane, and this spot was moved in its plane from pulse to pulse. Individual pulses are still coherent, but they do not contain enough energy to affect the photoresist; and the cumulative effect of multiple pulse exposure is the random superposition of individual speckle patterns, which average out to a constant value.

3. LASER-MATERIALS INTERACTION

The application of laser projection imaging to microfabrication is ultimately determined by the interaction between laser and the irradiated surface in ways which facilitate formation of desired patterns or structures. Conventional lamp-based photolithographic techniques are generally based on localized modification of organic or inorganic resists by photoinduced defects, such as bond scission (for positive resists) or crosslinking (negative resists). Following exposure, these defects are amplified in a development step, which usually involves preferential dissolution of the exposed (positive resists) or unexposed (negative resist) areas. These well-characterized processes have direct analogs when the lamp is substituted with an excimer laser. Excimer lasers with even moderate energy output can be used to generate desired photodefects in appropriate resist materials. Indeed, a 1:1 scanning projection system incorporating a XeCl laser (308 nm) [7], and several 5X and 10X steppers employing KrF lasers (248 nm) [3-6] have been reported recently, all using positive, wet developed organic resists. Although these systems are not yet optimized, linewidths as low as 0.35 μm were obtained with them. In our own laboratory we have demonstrated the highest resolution reported in wet-developed photoresists. Figure 3 shows an electron scanning micrograph (SEM) of 0.13-μm lines and spaces in polymethylmethacrylate (PMMA). This result was obtained with a single-pulse exposure at 193 nm in projection, followed by 5 min of wet development.

It should be noted, however, that excimer lasers are effective in promoting numerous types of laser/material interaction, in addition to generation of photoinduced defects. Their potential use in microfabrication will be more fully realized with the incorporation of the more diverse excimer-induced processes into the projection printing technology. For purposes of further discussion these processes can be grouped into two broad categories. They are laser-induced nonlinear transformations, such as photoablation, changes in stoichiometry and changes in bond structures; and laser-induced interfacial chemical reactions, including photodeposition and photochemical etching.

3.1. Laser-Induced Nonlinear Transformations

These processes are characterized by highly nonlinear material response to laser fluence. The order of nonlinearity, γ, of the material response is an important factor in determining the resolution achievable with a given projection system. Figure 4 demonstrates this fact for three systems: Hg-line steppers, a KrF laser stepper with refractive optics and an ArF-based laboratory stepper employing a high-NA, all-reflective Schwarzschild lens. In Fig. 4 the contrast in the image plane is plotted as a function of linewidth, assuming a high-contrast object and diffraction limited incoherent imaging. The absolute cutoff linewidth, $d_{co} = \lambda/4NA$, corresponds to zero contrast. In practice, the minimum printable feature, d_{pr}, is that at which the aerial image contrast is high enough to induce resolvable changes in the irradiated material. The lower limit on aerial image contrast thus depends on the order of nonlinearity of the material response to fluence variations, and it in turns determines d_{pr} via a curve such as those in Fig. 4. In Fig. 4 the effect of γ on d_{pr} is shown for $\gamma = 1.6$, $\gamma = 6$ and $\gamma = 10$. Clearly, all other parameters being equal, higher γ enables resolution closer to d_{co}. For wet-developable photoresists commonly used with Hg-lamps, $\gamma \approx 1.5\text{-}4.5$. However, excimer induced self-developing solid transformation processes have significantly higher γ in the range 10-25. This effect enables higher resolution with excimer steppers even beyond the changes in d_{co} discussed in Section 2.

Another aspect of nonlinear transformations is that they frequently produce results in a single pulse. Single-pulse exposure has distinct advantages for applications to step-and-repeat systems. First, the throughput can be extremely high. Assuming a 10-mm field of view, such as that of the recently developed steppers, a 4-inch wafer can be processed in less than one second even at the modest pulse repetition rate of 100 Hz. Second, the presently stringent requirements for long-term mechanical stability of the projection system can be significantly relaxed, since the wafer is translated in between pulses and alignment is performed "on-the-fly". A large variety of such nonlinear transformations with excimer lasers both in resists and in direct patterning has been demonstrated in recent years and some of these are detailed below.

3.1.1. Organic Systems. The use of organic bilayers, in which the top layer is reactively decomposed by 157- or 193-nm irradiation, has permitted single-pulse printing of 0.4-μm lines and spaces [8]. The bilayer is formed by overcoating a 1.5-μm-thick soft-baked PMMA layer with several UV absorption lengths of AZ 1350J resist. The top layer can be removed by ablation with a single pulse (~80 mJ/cm^2) without disrupting the underlying PMMA. The image is then developed using UV flood exposure and wet etching. A primary advantage of the bilayer is that thin AZ-resist imaging layers can be used. Importantly, a thin contamination or remnant of this layer left after ablation does not ultimately interfere with clean development to the substrate surface. The highly nonlinear intensity response of the overlayer to ablation produces high resolution and sharply defined edges.

Another organic system which is efficiently etched in projection is polyimide. Thin films of polyimide on oxidized silicon are cleanly etched with 1-10 pulses of 193 nm, at fluences of ~1 J/cm^2. Equal lines and spaces 0.9-μm wide are obtained in these studies. In separate experiments, a series of 6-μm via holes is etched in 1-μm-thick polyimide in projection (Fig. 5). The polyimide serves as intermetal dielectric in a GaAs circuit, and the excimer-generated via holes enable electrical contact between first- and second-level metals [12].

3.1.2. Solid-State Chemistry: Al/O Cermet. In principle, a variety of transformations can be induced by laser irradiation of metastable solid films. An example is the a solid-state transformation of evaporated Al/O cermets. This system and other metal/dielectric composites are particularly interesting since both chemical (e.g., etch resistance) and electrical properties can be altered with a heat pulse, and since they are immediately compatible with subsequent

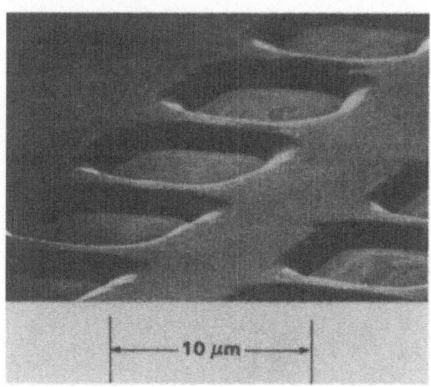

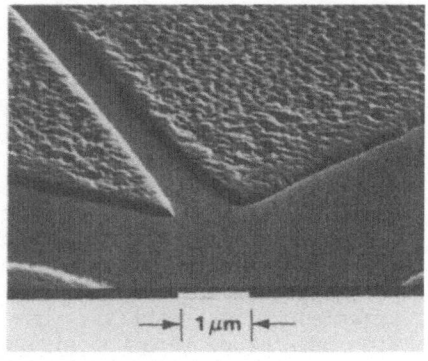

FIGURE 5. Scanning electron micrograph of a pattern of 6-μm via holes etched in projection in 1-μm-thick polyimide. Exposure was in air with 5 pulses of 3 J/cm^2 fluence.

FIGURE 6. Scanning electron micrograph demonstrating the excellent submicrometer resolution obtainable in projection patterning of an Al/O cermet film at 193 nm. One 120 mJ/cm^2 pulse was used to induce solid-state transformations in the 30-nm-thick film. Development was performed by wet etching in nitric acid/phosphoric acid followed by reactive ion etching in CHF$_3$.

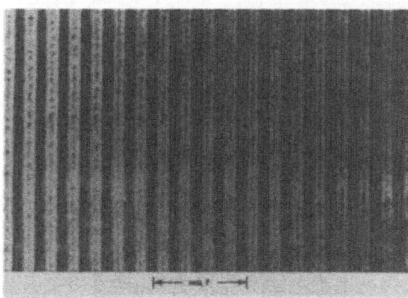

FIGURE 7. Scanning electron micrograph of 0.13-μm lines and spaces, patterned in hard carbon film on GaAs. The film thickness was 100 nm, and the projection patterning was performed with one 193-nm pulse (130 mJ/cm^2) in air.

high-temperature processing.

In our studies [8], a 30-nm-thick Al/O cermet was deposited on oxidized Si wafers by Al evaporation under a 3 x 10^{-6} Torr O_2 ambient. As evaporated, these films have a sheet resistance of 1000 $\Omega/\square$, and are smooth and shiny in appearance. After low-power single-pulse excimer laser irradiation, the films turn black; after higher-power irradiation the films are silvery. These color changes correspond to the creation of new phases with markedly different chemical-etching properties and electrical conductivities, e.g., the black phase has a sheet resistance of > 10^6 $\Omega/\square$.

The conductivity change induced by excimer pulses is of potential use for defining conductor patterns without any further treatment. Alternatively, the transformed material can be used as an etch mask to pattern underlying materials. To demonstrate the second application, images were transferred into SiO_2 by wet etching followed by reactive-ion etching. Typical results obtained with single-pulse excimer-laser exposure are shown in Fig 6. The projected image shows edges defined to a small fraction of a micrometer.

3.1.3. Patterning "Diamond-Like" Carbon Resist. For a study of the resolution capabilities of our system we have chosen hard "diamond-like" carbon resists, 100- to 200-nm thick, which are exposed by single 20-ns ArF laser pulses in a self-developing response [10]. Figure 7 is an SEM of 0.13-µm lines and spaces patterned by laser projection onto hard carbon resist on GaAs. The fluence, ~0.13 J/cm^2, was near threshold for ablation, and its effect was "blistering" of the carbon, which is nevertheless sufficient for further preferential dry etching of the GaAs substrate. The projected 0.13-µm-wide lines approach the absolute diffraction-limited cutoff width, $\lambda/4NA$ = 0.097 µm. In separate studies we determined that high contrast, γ = 22, characterizes the response of 100-nm-thick carbon films under self-development conditions. Our results show that, even with limited control of the aberrations in the projection optics, 0.13-µm lines and spaces can be generated in projection with the appropriate choice of system parameters. Key to optimal performance are, single-pulse exposure, a high contrast response, and built in "bleaching" via saturation of the resist response at high exposure.

3.2. Laser-Induced Interfacial Chemistry

This category includes the majority of laser-driven photochemical etching and deposition. The essential feature of these processes is the interaction between the laser and some external-phase (vapor-phase or adsorbed) species, and the subsequent reaction between the products of this interaction and the surface. Naturally, the laser may at the same time interact with the solid phase as well. These processes are limited fundamentally by the material coverage of an adsorbate, or by surface-collision frequencies from the external phase. Because of the low duty factor (~10^{-5}) of current excimer lasers these limitations imply long exposure times. Furthermore, for the case of substrate/gas-phase reactions, the resolution may be severely limited by lateral diffusion of the photogenerated species. This effect can be reduced by the addition of buffer gases, at the cost, however, of a reduction in the overall reaction rate. As a result of these considerations the rates and resolution achieved in photochemical projection etching and deposition have been significantly lower than those of the nonlinear processes detailed in the previous section.

3.2.1. Photochemical Etching. Excimer laser projection etching of GaAs was demonstrated at 193 nm [13]. The process is initiated by the vapor-phase photolysis of 0.5-Torr HBr. Linewidths of 4 μm were achieved with 10-Torr buffer gas (Ar or H$_2$), and the etch rate was ~0.25 A/pulse at a fluence of 20 mJ cm^{-2}/pulse.

Another III-V compound, InP, has been etched in projection with 1-Torr Cl$_2$ and no buffer gas [14]. The laser was operated at 193 nm, 20-Hz pulse repetition rate, and etch rates of ~3 A/pulse were achieved at fluences of 100 mJ cm^{-2}/pulse. The resolution was ~2.5 μm. In both instances the etching process is initiated by photolysis of the halogen-containing molecule (HBr or Cl$_2$) and generation of a high density of reactive halogen atoms (Br or Cl, respectively). These in turn react with the surface, leading to volatile halide molecules.

3.2.2. Photochemical Deposition. Projection patterning by photochemical deposition from gaseous precursors has been reported for several systems. An example is the patterned deposition of thin films of gold from dimethylgold acetylacetonate and from its fluorinated derivatives. These compounds have vapor pressures in the 10- to 300-mTorr range at room temperature, and they decompose photolytically at the excimer wavelengths. The detailed photochemistry of these compounds is still being investigated. In one set of experiments [15], 3000-A-thick, 2-μm-wide lines of Au were deposited with 193-nm, 248-nm and 308-nm pulses. Carbon contamination in the film, originating from the precursor organic molecule, varied from 11% to 73%, depending on laser wavelength and number of pulses. In a series of independent studies [14], gold films with linewidths of 1.6 μm were deposited in projection at 248 nm. The time-averaged deposition rate was ~0.13 A/pulse at ~60 mJ cm^{-2}/pulse.

4. CONCLUSIONS

Excimer lasers have emerged in recent years as effective and versatile tools for projection photolithography. Their power, coherence and wavelength characteristics make them well suited for microelectronics applications, particularly as design rules are reduced to well into the submicrometer region. Furthermore, the inherent capability for in situ processing is a compelling advantage of the laser-based processes over other photolithographic techniques. Significant effort has already been invested in the design and production of prototype systems which integrate excimers into existing stepper technology. In the future, work is expected to continue on all the aspects of this emerging technology: excimer laser engineering, the design of appropriate optical components, and the development of new photochemical processes and novel resists tailored to match excimer laser radiation.

5. ACKNOWLEDGEMENTS

We thank S. Fiorillo, T. J. Pack, and D. J. Sullivan for expert technical assistance. This work was sponsored by the Department of the Air Force, in part under a specific program supported by the Air Force Office of Scientific Research, by the Defense Advanced Research Projects Agency and by the Army Research Office.

REFERENCES

1. See, for instance, H. L. Stover, Proc. SPIE 633, 2 (1986).
2. For a detailed review, see M. Rothschild and D. J. Ehrlich, to be submitted to J. Vac. Sci. Technol. B.
3. V. Pol, J. H. Bennewitz, G. C. Escher, M. Feldman, V. A. Firtion, T. E. Jewell, B. E. Wilcomb, and J. T. Clemens, Proc. SPIE 633, 6 (1986).
4. M. Nakase, T. Sato, M. Nonaka, I. Higashikawa, and Y. Horiike, Proc. SPIE (to be published).
5. M. Endo, M. Sasagu, Y. Hirai, K. Ogawa, and T. Ishihara, Proc. SPIE (to be published).
6. M. Kameyama and K. Ushida, Proc. SPIE (to be published).
7. R. T. Kerth, K. Jain, and M. R. Latta, IEEE Electron Device Lett. EDL-7, 299 (1986).
8. D. J. Ehrlich, J. Y. Tsao, and C. O. Bozler, J. Vac. Sci. Technol. B 3, 1 (1985).
9. M. Rothschild, C. Arnone, and D. J. Ehrlich, J. Vac. Sci. Technol. B 4, 310 (1986).
10. M. Rothschild and D. J. Ehrlich, J. Vac. Sci. Technol. B 5, 389 (1987).
11. Y. Horiike, R. Yoshikawa, M. Nakase, H. Okano, H. Komano, and T. Takigawa, paper B&C1.2, presented at the 1986 Fall Meeting of the Materials Research Society (Boston, Dec. 1-6, 1986).
12. J. G. Black, D. J. Ehrlich, M. Rothschild, S. P. Doran, and J. H. C. Sedlacek, J. Vac. Sci. Technol. B 5, 419 (1987).
13. P. D. Brewer, D. McClure, and R. M. Osgood, Jr., Appl. Phys. Lett. 49, 803 (1986).
14. M. Rothschild and D. J. Ehrlich (unpublished).
15. T. H. Baum, E. E. Marinero, and C. R. Jones, Appl. Phys. Lett. 49, 1213 (1986).

EXCIMER LASER PATTERNING AND ETCHING OF METALS

P. H. KEY, P. E. DYER, R. D. GREENOUGH
Department of Applied Physics
University of Hull.   U.K.

INTRODUCTION

Metal films are extensively used in electronics applications and their patterning to form conductor tracks is often achieved by conventional wet etching; these techniques involve a number of processing steps each of which is time consuming and a potential source of error. Whilst laser techniques of metal deposition for the direct writing of these tracks have been developed [1,2] and are gaining in importance, it may well be convenient and cost effective, where larger areas of metal are involved, to employ traditional surface metalisation techniques and to effect patterning by laser micro machining.

The small signal absorption of metals is generally much greater at UV wavelengths than in the IR or visible regions of the spectrum. This has potential advantage for efficient processing using UV lasers in addition to the greatly improved resolution offered by the shorter wavelength. These factors together with the availability of the Rare Gas Halide (RGH) family of excimer lasers as a convenient source of pulsed, high power radiation makes the UV and VUV processing of metal films an attractive proposition.

In this paper we identify two regimes of metal processing based on film thickness and present results of studies of their etching and patterning using excimer lasers.

EXPERIMENTAL

We have used a projection etching technique in which the demagnified image of a RGH laser illuminated mask is projected onto target materials mounted in an evacuable cell. This scheme is favoured because (i) relatively large scale, and therefore structurally less demanding, master masks may be used (ii) complex beam steering equipment needed for raster or vector scan pattern generation is unnecessary, and (iii) the mask damage and fouling associated with contact or proximity print techniques is eliminated.

Excimer lasers by virtue of their low spatial coherence, which virtually eliminates speckle in the image plane, are particularly suitable as an illumination source and their short wavelengths allow, in principle, sub-micron resolution to be achieved. In the present arrangement, using a 15 cm focal length lens with a relative aperture of $>$ f/10 and demagnifications of 0.25 to 0.1, an edge resolution of $\simeq$ 5µm is estimated from the spherical aberration and diffraction contributions for the single lens projection system.

Whilst some of our studies have been carried out at 193 nm (ArF) and 248 nm (KrF), most of the work has centred on the discharge-pumped XeCl (308 nm) laser since this has long gas lifetime, reliable output and high (1 KHz) pulse repetition frequency (prf) capability [3].

D. J. Ehrlich and V. T. Nguyen (eds.), Emerging Technologies for In Situ Processing, 105–112.
© *1988 by Martinus Nijhoff Publishers.*

THIN FILMS
    We have shown [4] that thin metal films deposited on substrates may
be ablated with a single excimer laser pulse, leading to high quality
patterning with good edge resolution (fig 1).

FIGURE 1.
SEM of single excimer
laser pulse patterning
of 500 Å gold film on
Kapton substrate.

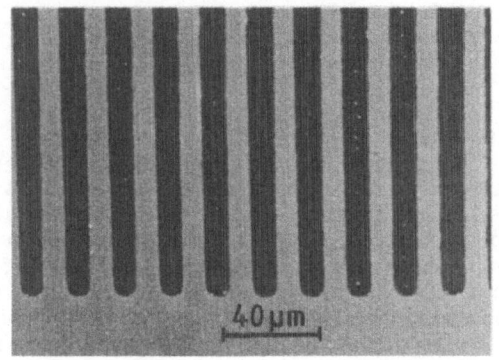

    Ablative removal of the irradiated area of the film occurs at a well
defined threshold fluence which is dependent upon the metal thickness and
its optical and thermal properties. The threshold fluences were found to
be approximately an order of magnitude lower than predicted by an
evaporation limited removal mechanism and closer to the fluence required
to melt the film. To illustrte this, a typical plot of threshold fluence
as a function of film thickness is shown in fig 2 together with the
theoretical curves for attaining melt and evaporation conditions derived
from calculations using the optical and thermal properties of the metal.

FIGURE 2.
Threshold fluence at
193 nm as a function of
film thickness for Ni
films deposited on Mylar.

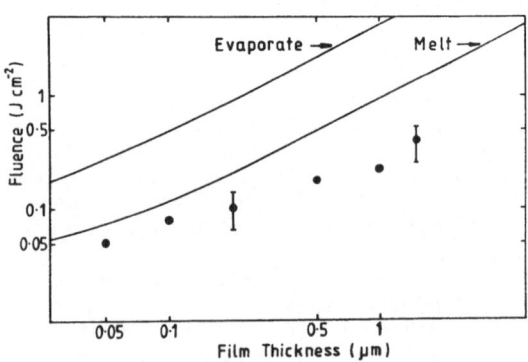

    Measurements of a variety of metal films (Cu, Ni, Au, Ag, Cr, Aℓ)
show the ablative thresholds are similarly low and are largely independent
of substrate material (polymers, glass, quartz, aluminium) and the
operating environment (vacuum, air, inert gas).
    In the case of films which show little wavelength dependence in their
optical absorption in the UV spectral region such as copper, no
significant difference in the threshold fluence for ablation was found for
the ArF, KrF or XeCl lasers. This suggests that the removal process is

independent of the photon energy and is thermal rather than photonic.

The low threshold for removal is thought to be due to the high thermal conductivity of metals which allows the heat generated at the surface by the absorption and thermalisation of photons to be communicated to the film-substrate interface during the laser pulse. The rapid interfacial temperature rise leads to degassing or decomposition of the substrate and the explosive removal of the film in its molten or near molten state [4, 5].

The upper limit to film thickness, ℓ, that may be 'explosively' removed by a single pulse may thus be approximated by: $\ell = (4D\tau)^{\frac{1}{2}}$ where D is the thermal diffusivity of the film and τ the laser pulse length [6]. For typical discharge pumped excimer lasers and high thermal conductivity metals, such as copper, ℓ is $\leq$ 2.5 μm. This limit to the maximum thickness that can be patterned on a single shot basis is borne out experimentally.

THICK FILMS AND FOILS

For thermally thick metal target with thickness > ℓ, the rapid diffusion of heat from the irradiated zone necessitates high peak powers to induce thermal etching of the surface and, for an evaporative mechanism, threshold values of at least several Joules cm^{-2} can be estimated depending on the wavelength. For copper, several values of damage fluence have been reported in the literature, ranging typically from 3 J cm^{-2} in the UV to 13.7 J cm^{-2} in the IR [7, 8, 9] and showing strong dependence on pre-irradiation surface preparation. Processing in the high fluence evaporative regime is of limited value for micro machining applications because it can lead to the loss of feature resolution, damage to the substrate material and problems arising from the onset of laser induced breakdown and detonation waves in the local gas environment.

In preliminary experiments on bulk samples the ablation of copper at 193 nm in a poor vacuum (p $\approx$ 0.01 torr) was observed to occur at fluences as low as 1 J cm^{-2} and with material removal rates of 10 Å per pulse but with poor etch feature resolution. In ambient air the etch rate increased to 75 Å per pulse with a slight improvement in pattern resolution. In this case, the improved etch rate is thought to be due to the laser initiated/enhanced oxidation of the irradiated surface in air with the subsequent laser ablation of the oxide layer. These results serve to illustrate the feasibility of patterning via the formation and ablation of metal-gas surface compounds in a multipulse, low fluence regime. With a suitable reactive gas, high etch rates can be achieved as has been demonstrated by Koren et al [10] for aluminium in a chlorine atmosphere.

We describe here the copper-chlorine - RGH laser system and show that at low fluences (< 1 J cm^{-2}) and a range of pressures (0.05 - 50 torr) high etch rate, well resolved patterning can be achieved in free standing, thermally thick foils.

Detailed studies of the $Cu-Cl_2$ system have been carried out by Sesselmann and Chuang [11] and they report that the chloride layer, formed by the dissociative chemisorption of the gas at a clean copper surface, is generally not stoichiometric $Cu.Cl$ but rather $Cu.Cl_x$ with x varying from 0 to almost 2 depending upon the exposure conditions.

We have measured the etch rates at 308 nm in Cl_2 using rolled copper foils of thickness 25, 50 and 75 μm with quality ranging from 'engineering' to 99.9% purity. Apart from degreasing in alcohol, the foils were used as received i.e., with a native oxide layer approximately

50 Å thick. The oxide layer was found to act as an effective etch resist in that, at room temperature, no reaction between the gas and surface was observed even after several hours exposure at a pressure of 10 torr. Upon irradiation the oxide layer was removed in the area defined by the projected image and subsequent laser pulses resulted in the ablation of part of the chloride layer formed in the interpulse period. An example of a deep etched (25μm thick) copper foil is shown in fig 3 and the etch rate per pulse as a function of laser fluence in fig 4.

FIGURE 3.
Etched pattern in 25μm
free standing Cu foil.

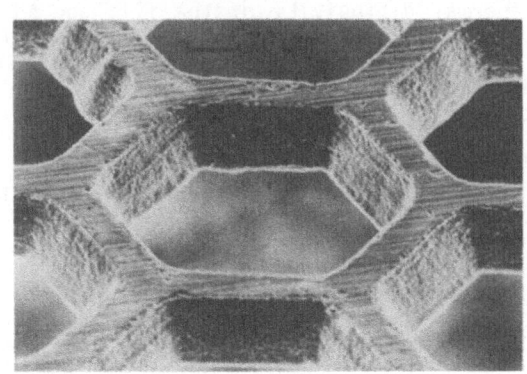

FIGURE 4.
Etch Rate vs. Fluence
for Cu in 1 torr Cl_2,
308 nm, 1Hz.

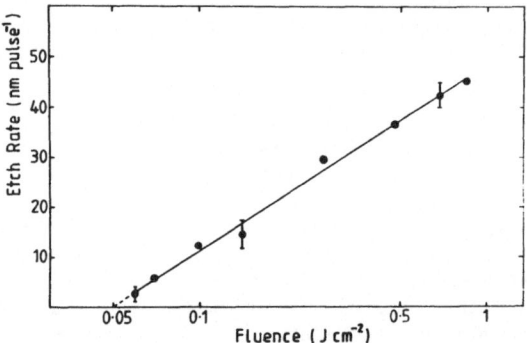

The etch rate data shows an apparent threshold, $F_T \simeq 0.05$ Jcm^{-2}, for removal and that above this the fluence dependence of the etch depth d, follows a behaviour expected from Beers law. In this respect the interaction shows similarities with the laser induced etching of organic polymers [12] and may be expressed by:

$$d = {}^1/_\alpha \ \ell n \ (F/F_T)$$

From the data the constant α is found to be $6.2 \pm 0.4 \times 10^5$cm^{-1} and the equivalent absorption coefficient for the chloride is deduced to be $1.8 \pm 0.2 \times 10^5$cm^{-1}, assuming a chloride to copper thickness ratio of 3.45 calculated from the relative molecular weights and densities. The absorption coefficient of CuCl was also obtained from direct measurements of the sub-threshold laser transmission of various film thicknesses prepared by exposing copper films deposited on transmitting quartz substrates, to Cl_2 at $> 10^9$L so as to completely transform the Cu to CuCl. From this data the absorption coefficient was found to be 6.3 ± 0.6 x 10^4cm^{-1}, significantly lower than from the bulk material etch rate data and probably due to the excess Cu in the non-stoichiometric chloride formed as a surface layer on the copper foil.

From the foregoing, we might expect the absorption coefficient to increase with prf since a reduction in the interpulse period will limit the growth of the chloride and result in a larger Cu gradient across the thinner layer. Support for this behaviour was obtained by observing that the etch rate per pulse decreased with increasing prf, the absorption coefficient deduced from the data being shown as a function of prf in fig 5.

FIGURE 5.
Variation of absorption coefficient of the chloride layer at 308 nm as a function of pulse repetition frequency.
● - Cu + 2.5 torr Cl_2
■ - from CuCl transmission

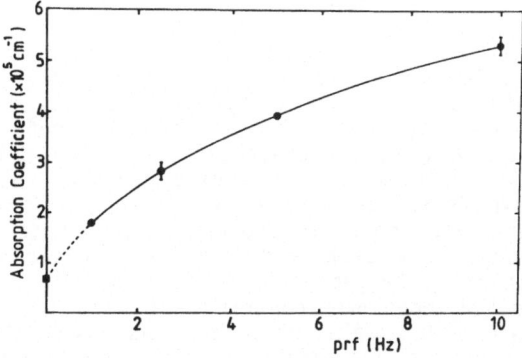

The exact form of this curve will depend upon the growth characteristics of the chloride which may be expected to be a function of both Cl_2 pressure and sample temperature.

Time resolved interferometry studies of the chloride growth rate as a function of chlorine pressure and substrate temperature indicate that the initial growth is linear in time, t, but rapidly develops into a diffusion limited regime such that the thickness varies as $t^{\frac{1}{2}}$. A comparison of the chloride growth and etch rate data reveals that, at moderate fluences and a given prf, there will be a low pressure regime in which all of the chloride formed in the interpulse period is removed and hence the etch rate will exhibit a strong dependence on pressure. At higher pressures, where the initial growth exceeds the removal rate, a quasi-equilibrium thickness of chloride will develop after a few pulses and will be maintained throughout the etch process, receding into the sample until all the copper is consumed with an etch rate which is independent of pressure. The measured etch rate, with a fluence of 0.4 J cm^{-2} at 5Hz, as a function of Cl_2 pressure is shown in fig 6.

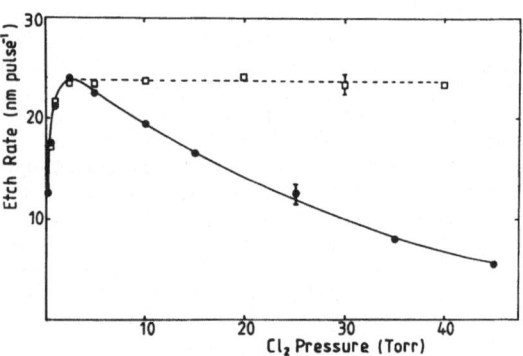

FIGURE 6.
Etch Rate of Copper
vs Cl_2 pressure with
XeCl at 5Hz.
● - constant cell
Fluence = 0.4 Jcm^{-2}
□ - constant target
Fluence = 0.4 Jcm^{-2}.

The high pressure decline in etch rate is consistent with the absorption loss at 308 nm in the Cl_2 gas in the cell, no decrease in etch rate with high Cl_2 pressure being observed when the fluence incident upon the target was kept constant as shown in the data in Fig 6.

At 308 nm photo-dissociation of the chlorine can occur leading to possible enhancement of the etch process. To test this, chloride layers of known thickness were grown on copper foils and irradiated with the cell evacuated so as to prevent further interpulse growth; the etch rates were found to be unchanged from those measured in the presence of chlorine with the same target fluence. We may thus deduce that photon excitation of the gas does not contribute to the process. The effect of a high pressure (< 1.5 atm) Ar buffer gas was to depress the etch rate due, it is thought, to inhibition of the ablation process by ambient pressures higher than the vapour pressure of the products.

In summary our present understanding of the $Cu-Cl_2$ etching process using the XeCl laser is that of laser ablation of a chloride layer formed by a dark reaction in the interpulse period. The etch rate is controlled by the laser fluence, prf and gas pressure; the latter two define regimes which may be distinguished as: (i) A low exposure regime (<5 x 10^5L) in which all of the chloride layer grown in the interpulse period may be removed at fluences < 1 Jcm^{-2} and in which the etch rate exhibits a fluence dependency over only a limited range above threshold (ii) A high exposure regime (> 5 x 10^5L) in which a quasi-equilibrium thickness of chloride develops, and the etch rate exhibits a strong fluence dependence. The prf is also influential in this case as the properties of the film depend upon the interpulse period.

PATTERN RESOLUTION
High spatial frequency components of the optical information are lost through the finite aperture projection system resulting in a finite gradient rather than a step cut-off of fluence at the edge of the image features; this will affect both the reproduction fidelity of the pattern and the edge resolution of individual features.

In the case of single pulse ablative removal of thin metal films the effect of a removal threshold is to improve the feature edge resolution over that expected from the optical system; however resolution better than the thermal diffusion length has not been observed. The fidelity of pattern reproduction, observed as variations in the mark-space ratio of the image of a line-pair test mask, depends critically on the fluence distribution in relation to the threshold.

In the case of deep multipulse etching, the fluence variation and concomitant differential etch rate over the region of a feature edge produces a sloping wall as seen fig 7.

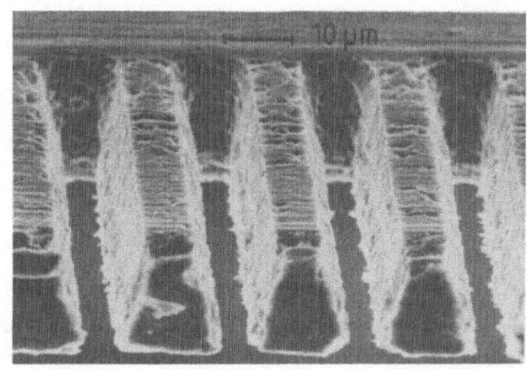

FIGURE 7.
SEM of deep etched 25μm thick copper foil with $F_T/F \approx 0.125$.

As the wall develops the local irradiation is no longer at normal incidence and the effective fluence is reduced. This continues, with the wall angle decreasing, until the effective fluence falls to the threshold value and etching ceases. It can be shown that the side wall angle is determined by the ratio of the threshold to initial incident fluence [13].

The side wall angle can be minimised by working at high fluence; however, this leads to a reduction in the fidelity of image reproduction with mark-space ratios < 1. Thus, in the case of deep etching, the attainment of steep walls and faithful pattern reproduction may be incompatible and a compromise based on particular objectives may have to be made.

CONCLUSIONS

We have identified two distinct projection etching regimes for metal films. In the single pulse explosive removal of thin films, effective patterning can be achieved at a threshold fluence which is lower than expected for an evaporative removal mechanism and offers an attractive means for fast, high resolution, large area pattern generation.

In the case of multiple pulse etching in a reactive gaseous environment and specifically for copper in low pressure chlorine, the etch rate is limited by the surface chemistry and hence dependent upon the interpulse exposure conditions.

In both cases the existence of a threshold fluence at which the removal process occurs is advantageous in that the etching may be controlled so as to improve upon the resolution of the optical system used.

REFERENCES

1 Baüerle D: Laser Induced Chemical Vapour Deposition in 'Laser processing and diagnostics' ed D Baüerle, Springer-Verlag 1984.
2 Ehrlich D J, Tsao J Y: Laser direct write applications. ibid.
3 Bishop G J, Dyer P E, et al: Performance and characterisation of a high pulse repetition frequency (> 1kHz) UV-preionised XeCl laser, presented at VII National Quantum Electronics conference, Malvern, UK, 1985.
4 Andrew J E, Dyer P E, Greenough R D, Key P H: Metal film removal and patterning using a XeCl laser. Appl Phys Lett 43, 1076 (1983).
5 Zalekas V J, Koo J C: Thin film machining by laser induced explosion. Appl Phys Lett 31, 615 (1977).
6 Ready J F: The effects of high power laser radiation. Academic Press 1971.
7 Agev V P, Garbunov A A, et al: Heating of metals by nanosecond XeCl laser radiation pulses generating a surface plasma, Sov J Quantum Electron 13 (7), 954 (1983).
8 Hussla I, Viswanathan R: Excimer laser induced ablation of clean and CO-covered polycrystalline copper surfaces. Surf Sci 145 L488 (1984).
9 Ursu I, Apostol I, et al: On the influence of surface condition on air plasma formation near metals irradiated by microsecond TEA CO_2 laser pulses. J Phys D: Appl Phys 17, 709 (1984).
10 Koren G, Ho F, Ritsko J J: Excimer laser etching of Al metal films in chlorine environments. Appl Phys lett 46 1006 (1985).
11 Sesselmann W, Chuang T J: The interaction of chlorine with copper. Surf Sci 176, 32 and 67 (1986).
12 Andrew J E, Dyer P E, Forster D, Key P H: Direct etching of polymeric materials using a XeCl laser. Appl Phys Lett 43, 717 (1983).
13 Dyer P E, Jenkins S D, Sidhu J, Development and origin of conical structures on XeCl laser ablated polyimide, Appl Phys Lett 49 (A), 453 (1986).

ION PROJECTION LITHOGRAPHY

Gerhard Stengl, Hans Löschner, Ernst Hammel,
IMS - Ion Microfabrication Systems GmbH,
A-1020 Vienna, Austria,

Edward D. Wolf,
National Nanofabrication Facility, Cornell University, Ithaca,
NY 14853, USA,

Julius J. Muray,
SRI International, Menlo Park, CA 94025, USA.

1. INTRODUCTION
    Ion Projection Lithography (IPL) uses demagnifying ion-
optics to project a reduced ion image of an open stencil mask
onto a substrate. The basic principles of this lithographic
technique and experimental results obtained with a test bench
Ion Projection Lithography Machine (IPLM-01) have already been
presented [1-3]. This paper adresses new IPLM-01 results and a
discussion of the potential of IPL techniques for in situ
processing.

2. IPL PATTERN TRANSFER INTO ORGANIC RESIST LAYERS
    In an IPL machine the open stencil mask is exposed to 5 - 10
keV ions with power densities up to 50 mW/cm^2. Because of the
subsequent acceleration of the ions passing through the mask
openings to final energies between 50 keV and 100 keV (or
higher, depending on system design) and with 10x reduction at
the wafer substrate power densities of several 10 W/cm^2 are
realized. This corresponds to ion current densities of several
100 µA/cm^2. Thus the dose rate can exceed 10^{18} ions/cm^2/sec, so
that insensitive high resolution organic resist layers like
PMMA can be exposed with $\leq$ $^1/_{10}$ sec chip exposure times.
Although lateral straggling using light ion species (H$^+$, He$^+$)
is extremely small [4], with overexposure and overdevelopment,
negative slopes of the patterned PMMA layer can be achieved
[5,6]. Thus, with lift off techniques (Fig. 1) the Au-lines of
Fig. 2 could be obtained (using a simple sputter coater).
    With heavy (at least 10 times) overexposure crosslinking
occurs in the exposed sites leading to "negative" developed
patterns in the PMMA resist. Fig. 3 shows an example obtained
with an exposure of $2x10^{14}$ ions/cm^2 and a 45 sec development in
a 1:3 solution of MIBK/IPA. The 5000 lines/mm (0.2 µm pitch)
correspond to a 10x reduction of an open stencil mask with
smallest openings of 1 µm width [2]. For the IPLM-01 the chip
field at 10x reduction is 2 mm x 2 mm.
    The resist surrounding the exposed sites can be removed by a
blanket DUV or ion beam exposure prior to development (Fig. 4).
The result in Fig. 5 is due to a 5x reduction (4 mm x 4 mm
field size) exposure.

D. J. Ehrlich and V. T. Nguyen (eds.), Emerging Technologies for In Situ Processing, 113–120.
© 1988 by Martinus Nijhoff Publishers.

114

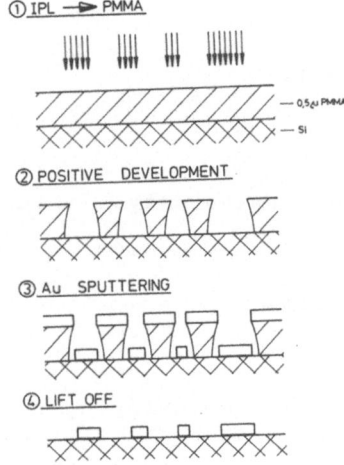

FIGURE 1. Lift off techniques using standard sputtering and resist stripping.

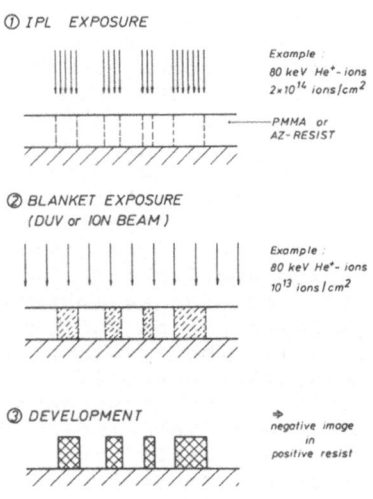

FIGURE 4. Image reversal techniques with additional blanket exposure to remove residual resist.

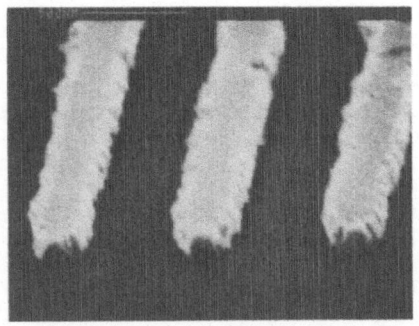

FIGURE 2. Au-lines after resist stripping, SEM bar = 2 μm

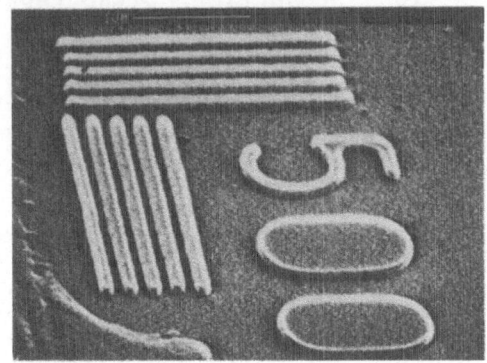

FIGURE 3. PMMA overexposure with 2×10^{14} He$^+$-ions/cm^2 resolving 5000 lines/mm (with 10x reduction), SEM bar = 1μm.

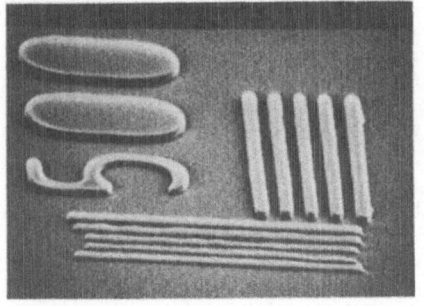

FIGURE 5. PMMA exposure according to procedure of Fig. 4 resolving 2500 l/mm (5x reduction), bar = 2 μm.

A blanket ion beam exposure (using the IPLM-01 with removed test mask) of about 10^{13} ions/cm^2 was used in order to clean the areas between the image reversal patterns.

Although the PMMA structures due to the image reversal techniques have a thickness of only about 0.2 μm they can be used very effectively as a conformal masking material for subsequent reactive ion etching. Furthermore, as the image reversal PMMA is a carbonized material [7] there are tremendous changes in electrical conductance which may find applications in the future [8]. The ion bombarded PMMA also blocks UV and DUV optical transmission which may permit its use for photomasks [6].

3. IPL PATTERN TRANSFER INTO INORGANIC RESIST LAYERS

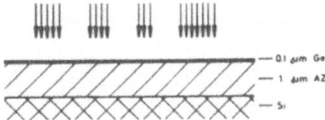

Fig. 6 shows the principles of IPL pattern transfer into an organic/inorganic double layer resist with dry development. An inorganic 0.1 μm thick evaporated Ge layer is used as the top lithographic resist with negative development in a CF$_4$-plasma and subsequent O$_2$-RIE pattern transfer into a 1 μm thick AZ-1450 organic base layer. SEM photos of the structures thus obtained are shown in Fig. 7. The edge roughness is due to a large surface roughness in the Ge layer. Similar results were obtained with evaporated SiO$_x$ top layers.

FIGURE 6. IPL exposure and dry development of Ge/AZ double layer resist.

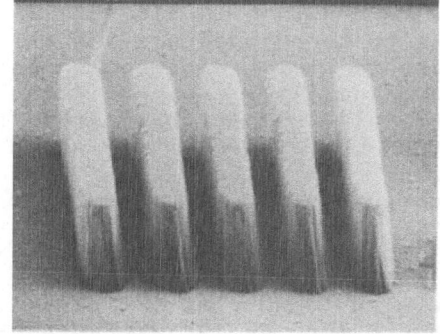

FIGURE 7. Ge/AZ-resist after dry development: a) SEM bar = 5μm, b) SEM bar = 1 μm, SEM < = 70°.

An important advantage of ion projection lithography is the possibility to use conventional well proven materials like Si, Ge, SiO_2 and Si_3N_4 as inorganic resist layers. The most simple development is wet chemical ion-bombardment enhanced etching [3]. Figure 8 shows the possibility of pattern transfer into a 0.3 µm thick SiO_2 layer conformal to a silicon substrate where 3 µm deep grooves have been etched prior to thermal oxidation. This result demonstrates also the excellent depth of focus [2] of the IPL techniques.

a) 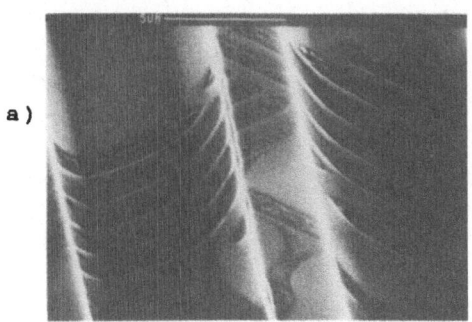   b)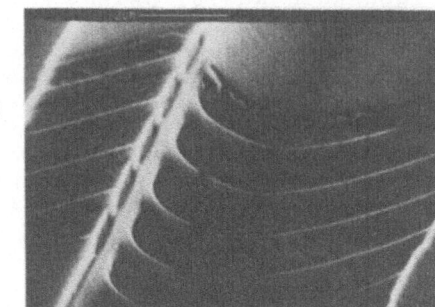

FIGURE 8. IPLM-01 exposure into 3 µm deep trenches of oxidized Si with subsequent ion-bombardment enhanced wet chemical etching in 5% HF: a) SEM bar = 5 µm, b) bar = 2 µm. Due to a contaminated surface, wet chemical etching is partly inhibited in the IPL bombarded sites.

As the resolution capabilities and the stability of organic structures will not meet the requirements for reliable pattern transfer, inorganic resist materials will gain in importance in the sub-0.5-µm regime.

4. PROSPECTS OF RESISTLESS IN SITU IPL PATTERN TRANSFER

IPL techniques provide an effective use of self-developing organic (e.g. nitrocellulose [9,10]) or inorganic (e.g. fluoride [11]) resists. The latter have been demonstrated so far for E-beam lithography [12]. The inorganic materials are often quite insensitive which is coupled to their extremely high resolution [13]. Because an IPLM can deliver high ion current densities, these insensitive materials may be exposed with subsecond chip exposure times.

However self-development may impose a contamination problem to IPLM equipment. Therefore it seems to be of advantage to develop techniques of "sensitizing" layers and/or surfaces with the help of an IPL exposure and then to transfer the sample to another vacuum chamber where a subsequent processing step leads to the required selective transformation of the ion beam exposed sites.

This principle may, as an example, be the formation of an etch resistant layer at the exposed sites in a plasma chamber with a proper composition of the etch gas - as has been demonstrated in the case of UV or electron beam exposure of Ge-Se-inorganic resists with subsequent dry development [14] - or epitaxial growth selective to the ion implanted sites [15].

Because of the large depth of focus, with IPL techniques, materials conformal to the wafer topography can be used. Thus imaging, development and pattern transfer can be implemented directly without the necessity to introduce planarization (as is required for submicron photolithographic techniques).

5. INTRAFIELD DISTORTION OF THE ION-OPTICAL COLUMN

Compared to the use of a single focused ion beam, ion projection lithography promises an advantage in exposure throughput of 3 to 6 orders of magnitude [1]. This advantage requires the ability to control the distortion of the ion image projected to the substrate in order to maintain the required positioning accuracy and overlay.

Using the IPLM-01 setup a first evaluation of the intrafield distortion of the ion-optical column was performed using a 5x reduction projective lens. The distortion level of the test mask pattern under permanent ion beam exposure was evaluated with the help of a proximity exposure [Fig. 9] using a Ge/AZ double-layer resist and dry development subsequent to the 5 keV He+-ion beam exposure. Fig. 10 shows the test mask and Fig. 11 the shadow exposure of a 20 mm x 20 mm grid field with 13 resolution test patterns within this grid region.

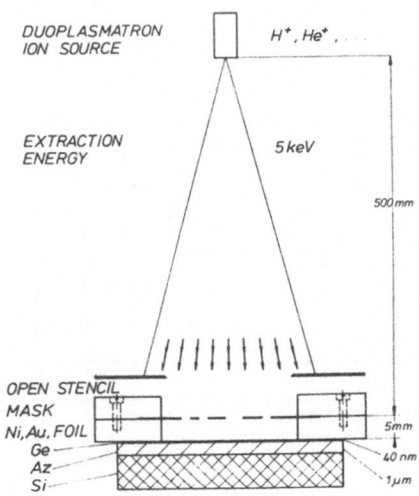

FIGURE 9. Proximity exposure with 5 keV ions to investigate mask distortion.

FIGURE 10. Top view of the 2 µm thick Ni mask with test grid (20 x 20 mm²) clamped between invar steel rings.

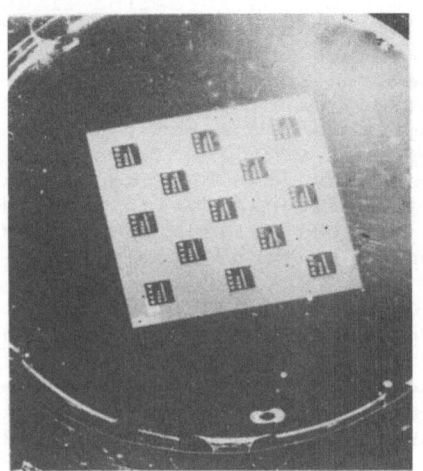

FIGURE 11. Proximity (shadow) exposure of the test grid in a Ge/AZ resist with 13 resolution test patterns.

FIGURE 12. Resolution in one of the test patterns. 7 mm gap replication of 500 lines/ mm, SEM bar = 20 μm.

The smallest structures of the test mask (1 μm lines and spaces, Fig. 12) could be transferred even though a 7 mm gap between mask and wafer was used during shadow printing. (This result is due to a duoplasmatron source [16] with an improved virtual source size of ≤ 10 μm.)

A Varian scanning E-beam system [17] was used as a metrology tool in order to evaluate the distortion level of the exposed grid. The results in Fig. 13 show a shear distortion of about 8 μm which is due to the crude method to clamp the 2 μm thick nickel open stencil test mask [18] to an invar steel frame (Fig. 11).

Using E-beam metrology the distortion level of chip fields exposed with the IPLM-01 at 5x reduction was evaluated. Fig. 14 shows a result for a specific machine set up with the mask influence already subtracted.

A quite large pincushion distortion of about 5 μm is obvious.

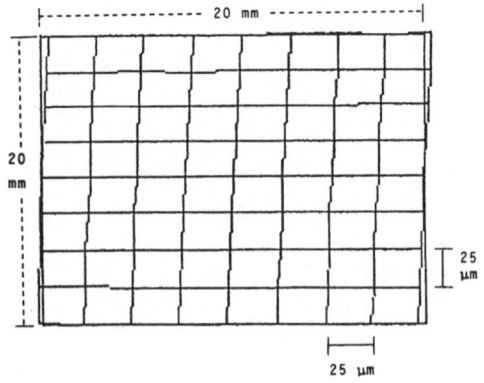

FIGURE 13. Mask grid distortion map as extracted from proximity exposure.

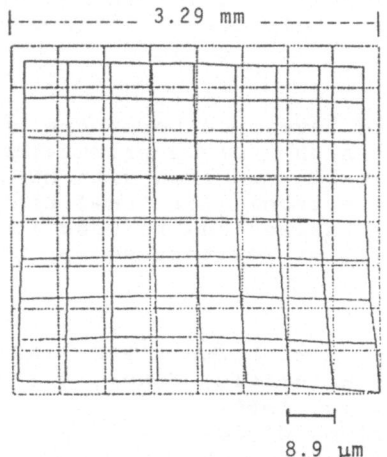

FIGURE 14. Demonstration of a pincushion distortion at a selected IPLM-01 setup.

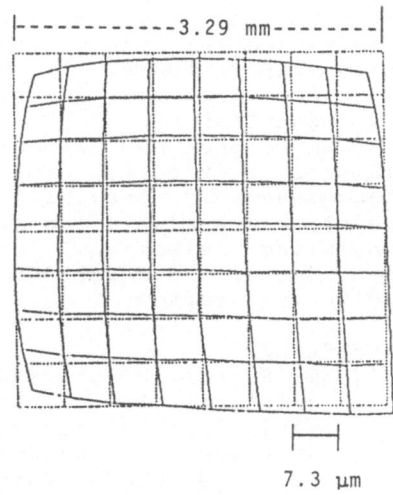

FIGURE 15. Demonstration of a barrel distortion at another selected IPLM-01 setup.

This distortion level can be changed dramatically just by changing machine parameters as is demonstrated in Fig. 15 which shows a barrel distortion pattern. (To obtain the result of Fig. 15 the voltage at the immersion lens was changed from 80 kV to 78 kV and the wafer Z-position was altered by 2 mm with respect to the pincushion setup.) The unsymmetrical (shear and trapezoid) contributions to the distortion evident in Figs. 14 and 15 are mainly due to substantial stray magnetic fields in the test bench setup of the IPLM-01.

The change from pincushion to barrel cushion distortion gives the possibility of an optimized machine set up. Furthermore, with the help of ion-optical correction elements (e.g., electric multipoles, Fig. 16) an electronic fine tuning of the intrafield distortion level can be developed [1,10].

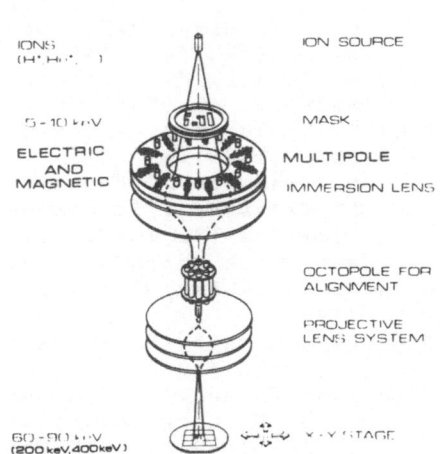

FIGURE 16. Fine tuning of distortion with ion-optical correction elements.

Thus, mix & match techniques may lead to a combined use of ion projection lithography with photo-, electron-, or x-ray lithographic tools.

6. CONCLUSION

Ion Projection Lithography is being developed to become an integral part of the future in situ high volume production techniques. The possibility of combining IPL machines with MBE, reactive (chemical) ion beam etching, electron- (or ion-) beam annealing, electron- (or ion-) beam inspection, focused ion beam repair and selective (metal) deposition /epitaxy equipment is very attractive.

REFERENCES
[1] G. Stengl, H. Löschner, J.J. Muray, Ext. Abstr. 18th (1986
     Int) Conf. Solid State Devices and Materials, Tokyo, p.29
     (The Japan Society of Applied Physics, 1986).
[2] R. Fischer, E. Hammel, H. Löschner, G. Stengl, P. Wolf,
     Microelectronic Engineering 5, p.193 (North-Holland,1986).
[3] G. Stengl, H. Löschner, P. Wolf, Nucl. Instr. Meth.
     B19/20, 987, (1987).
[4] A.J. Muray, J.J. Muray, Vacuum 35(10), 467 (1985).
[5] Y. Wada, M. Migitaki, K. Mochiji, H. Obayashi, J. Electro-
     chem. Soc. 130(5), 355 (1983).
[6] G. Stangl, F.G. Rüdenauer, G. Stengl, H. Löschner, W.
     Maurer, P. Wolf, Appl. Phys. Lett. 47(12), 1358 (1985).
[7] W.L. Brown, Radiation Effects 98, 115 (1986).
[8] T. Venkatesan, M. Feldman, B.J. Wilkens, W.E. Willenbrock,
     J. Appl.Phys.55(4), 1212 (1984).
[9] M.W. Geis, J.N. Randall, R.W. Mountain, J.D. Woodhouse,
     E.I. Bromley, D.K. Astolfi, N.P. Economou, J.Vac. Sci.
     Technol.B 3(1), 343 (1985)
[10] G. Stengl, H. Löschner, J.J. Muray, Techn.Proc. SEMICON/
     West, p. 42 (SEMI Inc., 1986).
[11] E. Kratschmer, M. Isaacson, J. Vac. Sci. Technol. B4(1),
     361, (1986).
[12] M. Isaacson, A. Muray, M. Scheinfein, I. Adesida,
     E. Kratschmer, in Nanometer Structure Electronics, Y.
     Yamamura, T. Fujisawa, S. Namba Eds., p. 58 (Ohmsha Ltd.
     and North Holland, 1985).
[13] A. Macrander, D. Barr, A. Wagner, Proc. SPIE 333, 142
     (1982).
[14] P.G. Hugett, K. Frick, H.W. Lehmann, Appl. Phys. Lett. 42
     (7), 592 (1983).
[15] P.N. Favennec, H. L'Haridon, M. Salvi, L. Henry, A. Le
     Core, D. Lecrosnier, A. Regreny, M.A. Diforte Poisson,
     J.P. Duchemin, NATO Workshop on Emerging Technologies
     for in Situ Processing, Cargese, Corsica, May 4-8, 1987,
     this volume.
[16] G. Stengl, H. Löschner, W. Maurer, P. Wolf, J. Vac. Sci.
     Techn. 4(1), 194 (1986).
[17] Varian VLS E-beam system installed at the company AMI,
     Unterpremstätten, Austria
[18] Electroplated nickel mask fabricated by H. Kraus, company
     Heidenhain, Traunreut, FRG.

UV LIGHT-ASSISTED DEPOSITION OF Al ON Si FROM TMA

J.E. BOUREE* AND   J. FLICSTEIN**
* CNRS, Laboratoire de Physique des Solides, 92195 MEUDON, France
**CNET, Laboratoire de Bagneux, 92220 BAGNEUX, France

1. INTRODUCTION

Recently, in the field of microelectronic devices, a lot of investiga-tions have been concerned with laser-induced deposition of metal films on semiconductor or insulating substrates (1-3). More precisely the localized deposition (micron scale) of aluminum on silicon or on silica from laser-assisted CVD is of great interest for maskless device fabrication techni-ques (4) as well as for circuit mask repair (5). Up to now, several authors have demonstrated that a two-step process consisting of surface nucleation followed by growth was needed to deposit Al thin film on Si and SiO2 sub-strates (see Table 1). Among them, Tsao et al. (6) used a focused cw ultra-violet (frequency-doubled Ar ion laser) irradiation followed by a cw infrared CO2 laser irradiation, triisobutylaluminum (TIBA) being the gaseous source. Following the same idea, Bourée et al. (7) first illuminated a flow of trimethylaluminum (TMA) molecules with UV light (mercury lamp) and then submitted the sample to a focused cw visible Ar ion laser irradiation.In these two examples, the two-step process was performed by using two diffe-rent light sources. Higashi et al. (8) used a similar approach with only one light source, namely a pulsed UV (KrF excimer) laser while heating the substrate up to 200°C. Roughly speaking, for each of these experiments, UV

TABLE 1. Al light-assisted deposition on Si and SiO2 from metalorganic precursors.

Growth mechanism \ Decomposition mechanism	one light source		two light sources	
	photolysis	pyrolytic autocatalysis	photolysis	pyrolytic autocatalysis
Surface nucleation + growth step (connected steps)	cw UV laser 257 nm (this work)			
sequential steps — surface nucleation	pulsed UV laser 248 nm (8)	heated substrate (8)	cw UV laser 257 nm (6) UV lamp 254 nm (7)	
sequential steps — growth	pulsed UV laser 248 nm (8)	heated substrate (8)		cw IR laser 10.6 µm (6) cw visible laser 514 nm (7)

121

D. J. Ehrlich and V. T. Nguyen (eds.), Emerging Technologies for In Situ Processing, 121–129.
© 1988 by Martinus Nijhoff Publishers.

exposure is used to photolytically decompose metalorganic surface adsorbates, leaving Al sites which then serve to activate the subsequent thermal
decomposition of TIBA or TMA to grow Al films. However, if the first two
techniques are well suited for repairing or customizing circuits, the third
one, which is a projection patterned technique, is compatible with conventional photolithographic processing. This paper aims to show the possibility of Al direct writing on Si in a single-step process (surface nucleation not disconnected with growth stage) by using a single light source :
high power UV laser.

2. EXPERIMENTAL
2.1. Sample preparation
The substrates are (100) oriented p-type silicon. The sample surface is
chemically polished prior to the cleaning procedure. A careful degreasing
in organic solvents followed by an etch in buffered HF, a rinse in deioni-
zed water and a drying with N_2 are conducted before introduction into the
cell.

2.2. Experimental apparatus
The experimental light CVD apparatus is schematically shown in Fig. 1. It
is composed of a gas feeder connected to the cell and two light sources :
a UV lamp (Fig. 1.a) and a UV laser (Fig. 1.b).

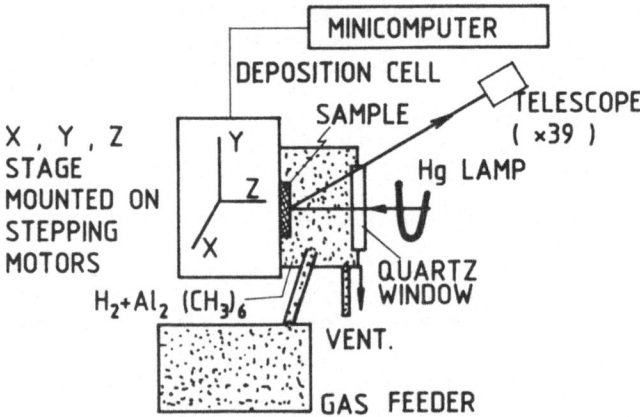

FIGURE 1.a. Experimental set-up of the UV light CVD system using a mercury
lamp for Si substrate pre-nucleation.

The stainless steel gas feeder was described in detail in a previous
work (7).
The low pressure mercury lamp used in this work irradiated preferentially
at 254 nm. The measured power density on the Si surface was about 5 mW/cm².
The cw UV laser beam employed was generated by frequency doubling an Ar
ion laser beam (514.5 nm) in intracavity mode. The maximum available output
power at 257 nm was 50 mW, as measured with a calorimeter. The beam was focused to about 15 μm spot diameter through a lens assembly onto the Si
surface (see Fig. 1.b).
Experimental set-up concerning the cw visible laser (Ar ion laser emitting at 514.5 nm) was described elsewhere (9).

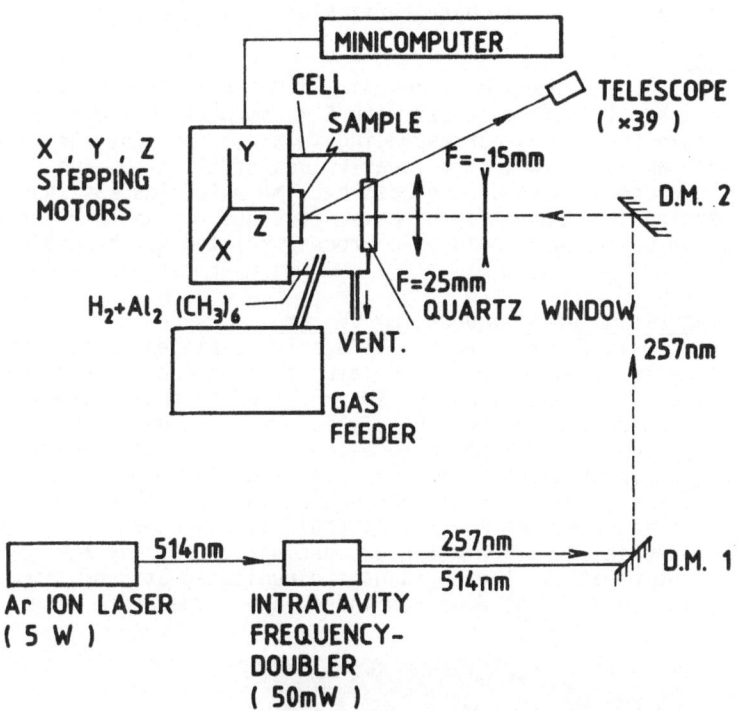

FIGURE 1.b. Experimental set-up of the UV light CVD system dedicated to Al direct writing via UV laser. (D.M. = DICHROIC MIRROR).

2.3. Experimental procedure

In order to remove residual gas impurities and passivate the different pipes before beginning each experiment, the cell and the pipes are repeatedly pumped out using a sorption pump to 10^{-3} Torr and flushed with pure H_2. The system is then degassed at about 100°C for 3 hours in flowing H_2 (D_{H_2} = 1 l/min). Typical experimental conditions are listed in Table 2.

TABLE 2. Experimental conditions

Total cell pressure	: 1 atm
Partial pressure of TMA	: 9.5 Torr
Hydrogen flow in TMA container	: 0.6 l/h
Substrate temperature	: 293 K
Water vapor concentration	: 0.7 ppm vol

It should be mentioned here that, mainly due to photolytic deposition, the average scanning speed along the X and Y axes was selected between 5 to 100 µm/s.

2.4. Characterization techniques

The deposited layers are visualized in situ with a telescope and extensively analysed ex situ by the following techniques : dark field optical microscopy, scanning electron microscopy (SEM), energy dispersive X-ray analysis (EDAX) dedicated to light elements, Auger electron microspectro-

124

scopy (AES) as well as electrical characterization techniques.

3. RESULTS AND DISCUSSION
   In previous work (9), a Si sample placed in a flow of TMA was irradiated
with a cw visible laser (514.5 nm). SEM with EDAX as well as Auger micro-
analysis indicated, in the region corresponding to the laser spot, the mixed
presence of a large amount of C and of lower values of Al, Al_xO_y and SiO_z.
Thus the pyrolytic decomposition of the adsorbed TMA molecules was shown
to lead to an uncontrolled deposition process. Consequently another approach
was attempted, which was a pure photolytic process related to the use of a
UV light source.

3.1. Photolytic deposition of Al induced by UV lamp
   It was shown previously (9) that when a Si substrate placed in flowing TMA
was exposed for a few minutes to the UV lamp, its surface was subject
to a minute change which,   analysed by SEM, resulted in the nucleation
of an Al fine grain structure. As the photolysis was going on (UV exposure
of 30 minutes), these grains were shown to be made up of Al domed crystal-
lites and some spheroids of about 50 nm size (see Fig. 2). Dome type grains
are most probably the sign of heterogeneous nucleation resulting from
photodissociation of adsorbed TMA whereas spheroid type grains
are most probably associated with homogeneous nucleation (10) in the gas
phase. It must be noted that all these grains are nucleated at room temperature because low fluence UV light does not induce any temperature rise.

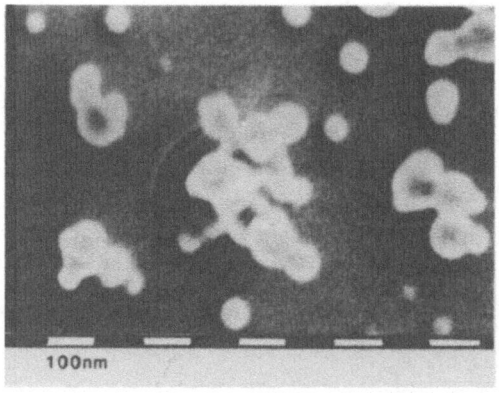

FIGURE 2. SEM micrograph of a UV lamp pre-nucleated Si substrate after an
exposure of 30 minutes.

   When UV exposure time increases, the grains coalesce and
finally after three hours a continuous thin film is observed (9). An ex situ
Auger microanalysis performed on the photodeposited film indicates the
presence of aluminum, oxygen and traces of carbon (see Fig. 3). Al
metal and Al_xO_y peaks can be distinguished from their energy position (64
and 52 eV respectively). Sputtered erosion profiles combined with point and
scanning analysis suggest that the chemical nature arises from initial
oxygen contamination of the Si surface and subsequent Al atmospheric oxidation during the sample transfer to the Auger spectrometer.
   Thus it may be said that the photolytic deposition of Al induced by low
fluence UV lamp is the key for the understanding of the nucleation preceding

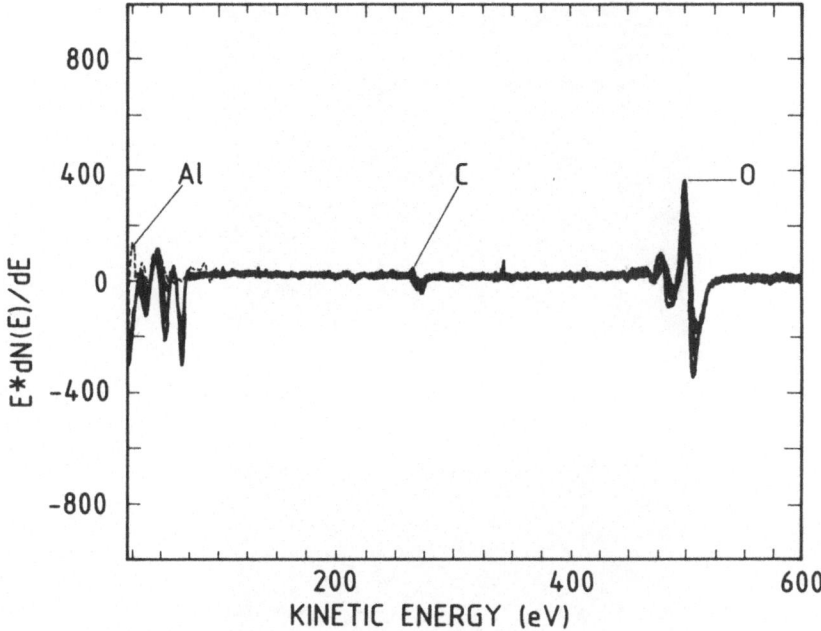

Figure 3. Sputtered Auger microspectroscopy profile of an Al thin film deposited by UV lamp.

the growth. However the deposition rate obtained with this UV source was too slow for a possible technological application (9).

3.2. Photolytic deposition by UV lamp followed by pyrolytic deposition by cw visible laser

In order to increase the deposition rate, a combination of UV lamp (prenucleation step) and visible laser (continuous growth step) was used. It was demonstrated (7) by fitting the experimental parameters (UV pre-exposure time, visible laser power and TMA dilution in hydrogen), that in this two-step two.source process (see Table 1), Al deposition was observed far below the threshold temperature of pyrolytic decomposition of the TMA molecule (350°C) : e.g. the calculated temperature at the spot center was 50°C corresponding to 1W visible laser power. Consequently, the UV exposure was shown to subsequently induce an enhanced surface pyrolysis of TMA (pyrolytic autocatalysis). Moreover three main features are remarkable. The Al deposit width was completely determined by the laser spot size impinging on the Si surface. Deposits contained traces of oxygen, but they were carbon contamination free, as determined from SEM + EDAX analysis (see Fig. 4).

3.3. UV lamp + visible laser-assisted direct writing of Al

The two-step two.source process was used for direct writing of Al on Si. One additional parameter was needed to obtain uniform lines : the average scanning speed along the X axis. Optimized results were obtained (7) for a scanning speed of 41 µm/s, an incident visible laser power of 2W and with TMA diluted in hydrogen : in order to do this, 1.2 l/h hydrogen was sent in a TMA container and further mixed with 1.8 l/h hydrogen flow. A few millimeter-long Al line drawn in these conditions is shown in the SEM micrograph of

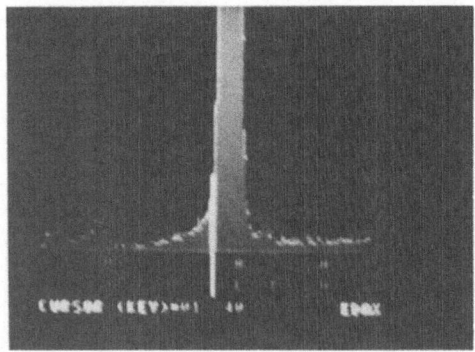

Figure 4. SEM + EDAX analysis of Al deposit using UV lamp + visible laser.

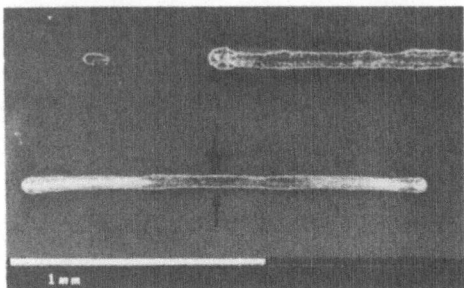

Figure 5. SEM micrograph of a direct writing Al line using UV lamp + visible laser (P = 2 W, v = 41 μm/s).

Fig. 5. This line has a small variation (typically 4%) of width definition. Big round-shaped grains up to 10 μm diameter are formed between nodular edges. The thickness and the shape of the lines obtained with this procedure were determined by using a mechanically moving stylus (Tencor). Fig. 6 shows that a crater-shaped line was obtained for a laser power of 2.3 W while for 2 W the shape of the line followed the laser intensity distribution.

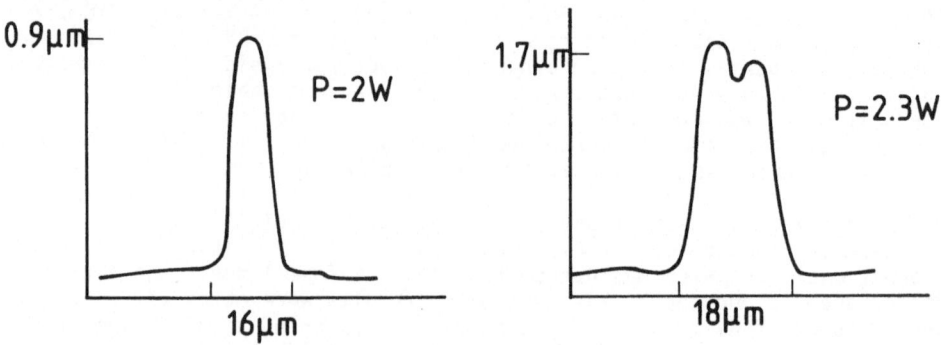

Figure 6. Kinetical parameters of direct writing Al lines.

Taking into account in the latter example the effective laser beam dwell time, an optimized growth rate of 10 µm/s was determined. This confirms indeed that the UV lamp gives the basis for the further enhanced pyrolytic decomposition of TMA molecules by visible laser. The electrical measurements were performed on as-grown Al lines connecting Ti/Au/Ti plots deposited on the Si substrate. Two point probe measurements give ohmic resistivity around 5 mΩ.cm (see Fig. 7), as compared with 2.6 µΩ.cm for the bulk. Surface contamination as related to grain morphologies (8) and deposition rate appear to play a leading part in the increase of resistivity. So, more work is necessary to improve the electrical characteristics.

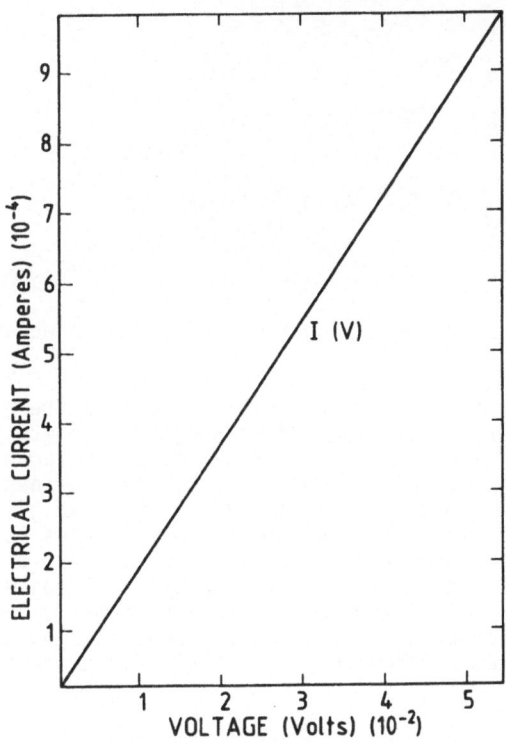

Figure 7. Electrical characteristics of a direct writing Al line using UV lamp + visible laser.

3.4. UV Laser-assisted direct writing of Al

It has been possible to get direct writing of Al on Si substrate by using the two-step two-source process. The low level of C contamination obtained after dilution of TMA seemed to indicate that the outdiffusion time of the generated carbon species is the limiting step. So a different approach arises naturally which makes use of a single source, namely a high power UV laser to photolyse the adsorbed TMA molecules. The direct writing was obtained in a one-step process at a low scanning speed of 5 µm/s for an incident laser power of 8 mW and a spot diameter of 16 µm. Surprisingly enough, for such a line a high growth rate of about 700 Å/s was registered. Furthermore, the optical micrograph (Fig. 8) typically displays a remarkable increase in size along the scanning direction of the deposited Al grains (up to 30 μm length). Both of these growth effects are in contrast with the variable line definition shown up. In each case, a very small grain nsures the electrical continuity between two big grains. To optimize this unfavourable microstructure, experiments are under completion to improve the line definition.

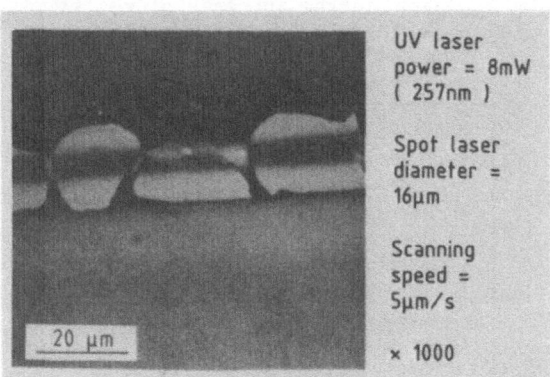

FIGURE 8. Optical micrograph of a direct writing Al line using UV laser.

4. CONCLUSIONS
It has been shown that it was possible to overcome the nucleation barrier of Al on Si single crystal, using TMA heterogeneous photolysis at very low UV fluence and room temperature. This pre-nucleation technique can be generalized to other CVD systems having large physical nucleation barriers. Moreover the surface photoreaction induced by the lamp gives the basis for the further enhanced surface pyrolysis of TMA molecules (pyrolytic auto-catalysis) observed at near room temperature and due to an Ar ion laser. The laser direct writing of Al was obtained after TMA dilution in hydrogen. Electrical measurements show ohmic resistivity higher than in the bulk.

Finally using a high power UV source, the results show the interest of UV continuous cold direct writing in a one-step rapid process.

ACKNOWLEDGEMENTS
This work has been supported by a contract from CNET-Meylan. Part of this work was jointly done with Dr Y.I. NISSIM (CNET-Bagneux). The authors would like to acknowledge Dr. M. RODOT (CNRS-Meudon), Dr. G. AUVERT (CNET-Meylan) for their interest in this work and Mr. R. DRUILHE (CNRS-Meudon) for his support on the experimental part. They would like to thank Mr. R. MELLET, Dr. J.F. BRESSE, Dr. A.C. PAPADOPOULO, Mr. B. FENAILLE and Dr. J.F. PALMIER for conducting respectively the metalorganic compound purification, Auger, SEM, SEM + EDAX and electrical characteristic analyses. These persons are with CNET-Bagneux except Mr. FENAILLE who is from USINOR.

REFERENCES

1. Ehrlich D.J. and Tsao J.Y., J. Vac. Sci. Technol. B1 (1983) 969.
2. Bäuerle D., in Laser Processing and Diagnostics, Springer Series in Chemical Phys. 39 (Springer, Berlin, 1984) p. 166.

3. Osgood R.M. and Gilgen H.H., Ann. Rev. Mater. Sci. 15 (1985) 549.
4. Tsao J.Y., Ehrlich D.J., Silversmith D.J. and Mountain R.W., IEEE Electron Dev. Lett. EDL-3 (1982) 164.
5. Randall J.N., Ehrlich D.J. and Tsao J.Y., J. Vac. Sci. Technol. B3 (1985) 262.
6. Tsao J.Y. and Ehrlich D.J., Appl. Phys. Lett. 45 (1984) 617 ; in Laser Chemical Processing of Semiconductor Devices, MRS Symposium Extended Abstracts (1984), p. 84.
7. Bourée J.E., Flicstein J. and Nissim Y.I., in Photon, Beam and Plasma Stimulated Chemical Processes at Surfaces, MRS Symposium (1986).
8. Higashi G.S. and Fleming C.G., Appl. Phys. Lett. 48 (1986) 1051 ; Higashi G.S., Blonder G.E. and Fleming C.G., in Photon, Beam and Plasma Stimulated Chemical Processes at Surfaces, MRS Symposium (1986).
9. Bourée J.E., Nissim Y.I., Flicstein J., Licoppe C. and Druilhe R., in Energy Beam-Solid Interactions and Transient Thermal Processing, MRS-Europe Symp. Proc. (1985), p. 119 ; Bourée J.E., Flicstein J., Nissim Y.I and Licoppe C., in Beam Induced Chemical Processes, MRS Symposium Extended Abstracts (1985), p. 71.
10.Chakraverty B.K. in Crystal Growth : an introduction (North Holland, Amsterdam, 1973) p. 50.

E-BEAM INDUCED DECOMPOSITION OF INORGANIC SOLIDS

Mino GREEN, Constantine AIDINIS and Olasebikan FAKOLUJO
Electrical Engineering Department, Imperial College of Science & Technology, Exhibition Road, London SW7 2BT, U.K.

ABSTRACT
    A general model for the e-beam induced decomposition of covalent and ionic inorganic solids is outlined. The primary electron beam produces energetic electron/hole pairs; these decay to the band edges, and either recombine or participate in the chemical decomposition of the solid. The kinetics and mechanism of several systems is discussed, including $CdC\ell_2$ and PbI_2. The importance of temperature in decomposition is remarked and it is suggested that induced decomposition mostly starts at 0.6 melting point (K). The usefulness of various inorganic systems, as resists, for x-ray masks, for local indiffusion sources and for metal inter-connects is noted.

1. INTRODUCTION

We have, over the last fifteen years, been studying the photon and electron induced decomposition of various inorganic solids (1-11) in thin film form. The purpose of this work has been to measure the kinetics, and deduce the mechanism of the decomposition processes; also to demonstrate the potential usefulness of these decomposition processes in various solid state micro-lithographic procedures.

It is the intention in this paper to present a summary of our work with thin film inorganic materials, noting some of the unifying features of such systems, to compare and contrast some aspects of organic and inorganic systems, and to speculate about various processing applications.

2. FUNDAMENTAL CONCEPTS

Our concern is with polycrystalline thin films ranging in thickness from about 0.04 to 1 μm. These materials are non-metallic, having band gaps from ca. 2 to 8eV, and are variously described as covalent or ionic solids. In all cases the first thing, which is required for decomposition, is the creation of electron/hole pairs by the inducing radiation. Thus for photolysis we need light of energy equal to or greater than the band gap energy. For radiolysis by e-beams, the general tendency is to work with electrons in the 1-50 KeV energy range; the primary, high energy electron, creates a shower of electron/hole pairs along its path, these latter usually have energies about three times that of the band gap (12). The energetic electron/hole pairs created near the peaks in the density of states will rapidly lose energy to the lattice, falling to band gap energy. We are next concerned with the fate of these excess electron/hole pairs which are energetically at the band edges, and spatially distributed throughout the grain in which they were created.

D. J. Ehrlich and V. T. Nguyen (eds.), Emerging Technologies for In Situ Processing, 131–136.
© *1988 by Martinus Nijhoff Publishers.*

The excess electron/hole pairs may recombine, and/or may participate in the chemical decomposition of the solid. Recombination might occur in the bulk of a grain at various singularities, at the grain boundary, or at the free boundary of the film. Chemical decomposition will occur by reactions such as:

$$M^{y+} + ye^- \rightarrow M \quad \text{(metallic element)}$$

and

$$yX^- + yh^+ \rightarrow yX \quad \text{(non-metallic element)}$$

for substances like MX_y.

There will be the usual sequence of chemical steps, one being rate determining. The site of the reactions may be in the bulk or at the surface. There may well be a rate determining transport step and the build-up of reaction products will cause vast difficulties in interpretation.

3. EXAMPLES OF INDUCED DECOMPOSITION

The conclusions of a number of studies of induced decomposition are given below, and it is suggested that these may form the beginnings of a useful classification.

3.1 Cadmium Chloride, $CdCl_2$ (7,8). This material is readily evaporated in a vacuum system to give polycrystalline thin films whose mean grain size is about equal to the film thickness. The vapour pressure of $CdCl_2$ is such that at 467K (194°C) the evaporation rate of a deposited film would be $10^{-3} nms^{-1}$, while the rate is about hundred times higher at 523K (250°C); this means that $CdCl_2$ can be used as an e-beam resist material up to about 473K (200°C) without serious loss by evaporation. Electron beam irradiation results in the reaction,

$$CdCl_2 \xrightarrow{\text{e-beam}} Cd + 2Cl$$

(The chlorine atoms mostly combine to desorb as Cl_2 molecules). Above about 363K (90°C) cadmium (as well as the Cl_2) is volatile, so the physical result of e-beam irradiation is progressive thinning of the exposed area. The total removal of reaction products makes mechanistic analysis very considerably easier.

It has been found in all the chemical systems we have studied that the rate of decomposition at fixed flux increases with temperature, typical of an activated process. We have, in addition to temperature, investigated the effect on reaction rate of electron energy, beam current density, film thickness, film doping and mean grain size. A typical decomposition rate curve is shown in Figure 1.

The mechanism proposed for $CdCl_2$ decomposition is: holes are trapped at surface chloride ions which have become special sites because of some local spatial singularity, the chloride ion plus trapped hole moves to a new surface site - this is the rate determining step - from which chlorine either desorbs directly or combines with another (mobile) surface chlorine atom and then desorbs. The cadmium ion, which is made singular by the chloride ion removal, rapidly traps the necessary two electrons and, the system being warm enough, the atom desorbs into the vacuum.

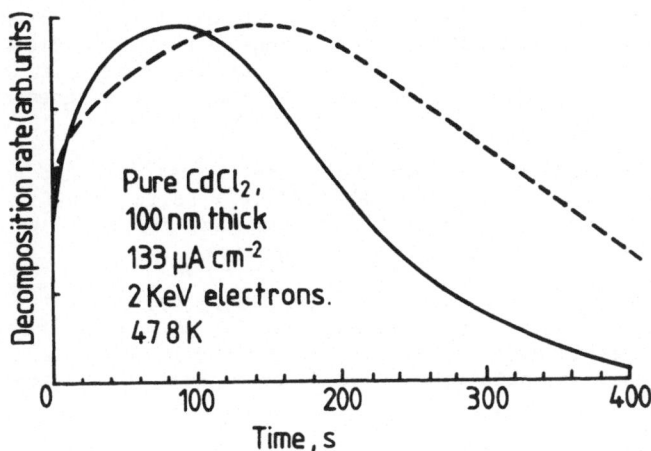

FIGURE 1. Chlorine evolution rate versus time for e-beam irradiated
CdCℓ₂ (PbCℓ₂ decomposition is shown as the dashed line)

The curve shown in Figure 1 is explained, in a manner consistent with
the proposed mechanism, by suggesting that the initial rising portion is
a result of increasing the surface concentration of chloride ion
singularities as the decomposition of the film progresses. The decaying
portion of the curve results from a decreasing concentration of
electron/hole pairs per unit area because the film is thinner than the
Grun range (see later). Finally the exponential tail must correspond to
the film becoming discontinuous. The chlorine evolution rate may be
written

$$J = \ell C [1 - p \exp\{(\ell - L)/\ell_o\}]$$

The initial film thickness is L; the film thickness at any time is ℓ and
ℓ_o is a constant proportional to the initial rate of increase in the
density of special surface chloride ion sites. The constant C is
proportional to the reaction rate constant and the flux density, and p
is related to the factor by which the initial density of special sites
increases over the full course of the decomposition.

The above type of decomposition should be classified as, <u>surface
decomposition</u> system, with only <u>one solid phase</u>.

Lead chloride (10) and iodide (1-3): Lead is not volatile, so that
the decomposition of these lead halides results in the evolution of
halogen and the formation of lead clusters throughout the lead halide
matrix. Study of the initial stages of photo-decomposition of PbI₂ at
low decomposition rates (1) showed that rate was proportional to the
area concentration of excess electron/hole pairs. Analysis of all the
variables, plus a comparison with the electrical transport properties of
the PbI₂, led to the proposal that the rate limiting step was the
transport of negative ion vacancies - generated by halide ion discharge
- from the free surface to the interior, where they were annihilated
together with positive ion vacancies.

E-beam decomposition kinetics of PbCℓ₂ (10) is not the same as that
for PbI₂, it is closer in apparent behaviour to CdCℓ₂ while retaining
some essential differences. The dashed line on Figure 1 is the

normalized (to peak maximum) rate of $PbCl_2$ (10) decomposition for 2KeV electrons; flux 140 μAcm^{-2}, film 90 nm thick, temperature 483K (210°C). The mechanism has not been worked out as yet, but it seems that the growth of lead clusters plays a role in reducing the chemical decomposition rate, possibly by decreasing the diffusion length of the excess carriers.

The above decomposition should, for PbI_2, be classified as: bulk transport, two solid phases, while for ($PbCl_2$/$CdCl_2$) solid solution it should be surface decomposition: two solid phases, and for $PbCl_2$ we do not yet know the mechanism, but it is two solid phases.

3.2 <u>Miscellaneous systems</u>. SnI_2 (4-6), Al_3 (9), GaI_3 (9) and AsI_3 (9) all completely photo-decompose, and should readily be decomposed by energetic e-beams. The disadvantage of the tri-iodides of Al, Ga and As are that they are highly volatile (the induced decomposition was carried out at 250K).

4. A ROUGH GUIDE TO STIMULATED DECOMPOSITION

We wish to be able to know under what conditions a solid might be photo- or e-beam decomposed. It is obvious that we need to produce free carriers at the band edges. It is also obvious that we need to make the chemical pathway as favourable as possible as compared with the physical (recombination) pathway. Clearly this can be achieved by increasing the temperature, which has little effect on recombination, but which exponentially increases chemical reaction rates. There would seem to be a correlation between chemical rate and the binding energy of the solid, and it is therefore proposed that there should also be a correlation with melting point, for example approximately 5% usage of electron/hole pairs at 0.6 x melting point (K); this fits our data fairly well.

The absorption of radiation is straightforward with photons, but there is a greater lack of precision where electrons are concerned. A useful guide for electron absorption is to use the Grun range, R, which is defined as the value of x obtained upon extrapolation to zero of the straight line part of the dE/dx vs x curve: dE/dx is the energy loss with distance and x is the distance into the solid. An empirical expression (13) which applies well in the 1 to 50 KeV energy range, in nm, is

$$R = 100E^{1.44}/\rho$$

where ρ is the density of the thin film, in g/cc; and E is the primary electron energy in KeV.

It would seem that mean grain size does not have a noticeable effect upon recombination rate, i.e. internal grain boundaries do not offer efficient recombination centres. However, mean grain size is important in determining resolution when these inorganic materials are used as resists, since the electron/hole pairs produced by the radiation pervade the grain and are not a cloud along the particle track, as might be nearer the case for organic resists. One final point about grain boundaries, is that we take it that electron/hole pairs do not diffuse from grain-to-grain, because the grain boundary is a barrier (double-Schottky) to carrier transport.

5. APPLICATIONS

A partial listing of potential applications is given below, together with some comments.

5.1 <u>Resists</u>: Inorganic resists are significantly different in behaviour and usage from organic resists. They are thin films (~ 50nm thick) and so might well function without need for planarization. Resolution is determined by mean grain size, and is not sensitive to development procedures. Inorganic resists do not suffer significantly from the proximity effects that organic resists have to endure. Since inorganic resists are evaporated, it is possible to deposit over a limited area (using mechanical masking), so making local processing possible. Removal of resist is usually a matter of aqueous solution wash at room temperature.

Organic resists are very much more sensitive than inorganic resists (5 μCcm^{-2} compared with $2mCcm^{-2}$). There are possible inorganic over layer schemes which might be employed in order to improve sensitivity, but these have yet to be tested.

Inorganic resists would appear to be considerably more chemically resistant in reactive ion (or plasma) etching, giving the probability of selectivities of 50 and more.

5.2 <u>Metal masks and patterns</u>: When substances such as $PbCl_2$ are decomposed the lead left behind tends to be globular in texture. The texture can be made continuous and less globular when a solid solution of 0.8 $PbCl_2$: 0.2 $CdCl_2$ is decomposed. Such a system should be useful in the fabrication of X-ray masks for VLSI.

The possibility exists to deposit thin films of metal compound and to decompose these using an e-beam system. Such a procedure might be useful in making metal inter-connects. Thus MoO_3 (m.pt. 1068K) might be a candidate for Mo inter-connects (decomposition temperature estimated to be 640K).

5.3 <u>Local sources</u>: Another possibility inherent in e-beam decomposition of inorganic thin films is the production of local indiffusion sources. Thus e.g., the decomposition of zinc iodide on GaAs to give a local indiffusion zinc source exists. It may, of course, be the case that the local source material is too volatile at the indiffusion temperature, in which case it would be necessary to coat the sources in a chemically inert, removable, low-volatility material.

6. CONCLUSIONS

The e-beam decomposition of semiconductors and insulators follows along much the same lines as photolysis, namely the production of electron/hole pairs by the stimulating radiation followed by the use of a fraction of these excess carriers in the decomposition process.

E-beam stimulated decomposition of various inorganic solids can be used in a variety of solid state fabrication processes.

REFERENCES

1. Albrecht MG & Green M: J Phys Chem Solids, <u>38</u> (1977) 297.
2. Albrecht MG: Photolysis of Lead Iodide. PhD Thesis, University of London (1975).
3. Green M & Albrecht MG: Thin Solid Films, <u>37</u> (1976) L57.
4. Kuku TA: Photolysis of Tin Di-iodide, PhD Thesis, University of London.
5. Green M & Kuku TA: J Phys Chem Solids, <u>44</u> (1983) 999.
6. Kuku TA & Green M: Thin Solid Films, <u>144</u> (1986) L119.
7. Green M, Aidinis CJ & Fakolujo OA: J App Physics, <u>57</u> (1985) 631.

8. Aidinis CJ: Cadmium Chloride as an Electron Beam Resist. PhD Thesis, University of London (1987).
9. Green M & Lane SJ: Photolysis of Iodides of Ga, Aℓ and As. Submitted for publication.
10. Green M & Fakolujo OA: E-beam induced decomposition of $PbCℓ_2$ and related systems, to be published.
11. Green M & Khaleque F: E-beam induced decomposition of CsCℓ, submitted for publication.
12. Klein AC: J Appl Phys, <u>39</u> (1967) 2029.
13. Brodie I & Muray JJ: The Physics of Micro Fabrication, Plenum, New York (1982).

ELECTRONIC CONNECTION THROUGH SILICON WAFERS

T.DUPEUX, Thomson-LCC
P.DEROUX-DAUPHIN, G.NICOLAS, D.LETI/MEM/IRDI/CEA

1. INTRODUCTION

Connecting both sides of a silicon substrate has direct applications in the fabrication of multilevel packaging, interconnection networks and many microelectronic components and devices.
To solve this problem, different solutions have been proposed. In LETI, we have defined a new process used for I/O interconnections of thin film recording heads.
A YAG laser have been chosen to drill holes, the two steps of polysilicon and tungsten CVD ensure the electrical conduction.

Requirements for interconnections in integrated magnetic thin film recording heads	
Total number of holes per wafer	= 3 000 (Ø 4")
	= 6 000 (Ø 6")
Drilling speed required	= 3 to 5 holes/s
Diameter hole	= less than 50 μm
Axe to axe distance between holes	= 200 μm

2. FORMING THE VIA

Different methods have been proposed : anisotropic wet etching, reactive plasma etching, laser enhanced photochemical etching, laser drilling.

2.1. Anisotropic wet etching

This method has given good results in the development of mechanical components on silicon (integrated sensors, microinjectors, nozzles for ink-jet printing [1] [2] [3] [4] [5].
The three main parameters to consider are :
. crystallographic orientation of silicon
. choice of the etchant
. etch rate dependence on dopant.
PETERSEN [1] has given in Table 1 a review of wet etching capabilities. The main limitations are low effective aspect ratio of via holes 750 μm x 25 μm (fig.1) and low etch rate (several hours needed for 520 μm)

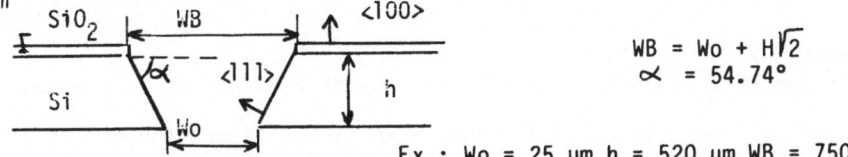

$$WB = W_0 + H\sqrt{2}$$
$$\alpha = 54.74°$$

Ex : W_0 = 25 μm h = 520 μm WB = 750 μm

137

D. J. Ehrlich and V. T. Nguyen (eds.), Emerging Technologies for In Situ Processing, 137–143.
© 1988 by Martinus Nijhoff Publishers.

Etchant (Diluent)	Typical Compos.	Temp °C	Etch rate (um/min)	Anisotropic (100)/(111) E.rate ratio	Dopant Depend.	Masking films (etch rate of mask)
HF HNO$_3$ (water, CH$_3$COOH)	10 ml 30 ml 80 ml	22	0.7:3.0	1:1	<10^{17}cm^{-3} n or p reduces etch rate by about 150	SiO$_2$(300A/min)
	25 ml 50 ml 25 ml	22	40	1:1	no dependence	Si$_3$N$_4$
	9 ml 75 ml 30 ml	22	7.0	1:1	/	SiO$_2$(700A/mn)
Ethylene diamine Pyraoca-thecol (water)	750 ml 120 ml 100 ml	115	0.75	35:1	>7 x10^{19} cm^{-3} boron reduces etch rate by about 50	SiO$_2$(2A/min) Si$_3$N$_4$(1A/min) Au,Cr,Ag, Cu, Ta.
	750 ml 120 ml 240 ml	115	1.25	35:1		
KOH (water, isoprop.)	44 ml 100 ml	85	1.4	400:1	> 10^{20} cm^{-3} boron reduces etch rate by about 20	Si$_3$N$_4$ SiO$_2$(14A/min)
	50 ml 100 ml	50	1.0	400:1		
H$_2$N$_4$ (water, isopropyl)	100 ml 100 ml	100	2.0	/	no dependence	SiO$_2$ Al
NaOH (water)	10 gr 100 ml	65	0.25-1.0	/	> 3 x10^{20} cm^{-3} boron reduces etch rate by about 10	Si$_3$N$_4$ SiO$_2$(7A/min)

TABLE I

2.2. Reactive plasma etching
With wellknown RF plasmas(13,56 MHz) etch ratesof 2 μm/mn can be reached in silicon.
Using a microwave excitation (2,45 GHz) in an RIE configuration, it is possible in LETI to obtain near 20 μm/min etch rate [13]. However, the aspect ratio remains low.

2.3. Laser photochemical processing

T.J.CHUANG [6] and D.EHRLICH [7),[8] have given large reviews of etching performances versus material   gases and laser sources.
EHRLICH proposed to use a 3W Argon laser focused on Si wafers enclosed in a 400 Torr Chlorine atmosphere. He obtained 40 μm x 5 μm conical holes through 250 μm thick   wafers.

```
┌─────────────────────────────────────────────────────┐
│        Laser photochemical etching from D.EHRLICH    │
│                                                      │
│   Laser  : 3 W Argon                                 │
│             6 μm diameter spot                       │
│                                                      │
│   Wafer  : Si <100>   250 μm thick                   │
│                                                      │
│   Gas    : Cl₂  400 Torr                             │
│                                                      │
│   Holes  : 40 x 50 μm conical via                    │
└─────────────────────────────────────────────────────┘
```

Despite this high etch rate (10 μm/s) our application would require nearly 1 minute per hole, that is to say, more than 1 hour per wafer.
However, this method might be used with new powerful lasers to drill several holes at the same time with the use of a protective mask.

Columbia University researchers (OSGOOD, PRUCNAL) use a laser assisted etching technique to drill very small holes < 10 μm in silicon [14] [15] [16] [17] into which optical fibers are inserted and glued.
Good technical results seem to have been obtained, with no damage to the chips surface, but the efficiency of the solution in volume is also a problem for our application.

2.4. Laser drilling

Processes based on focused laser beam and microdrilling are principally used by micromechanic technologies   like, for example, watch making.
Our investigations began by testing laser equipment on the market today. The first results showed that pulsed and Q switched YAG lasers could be used.

Drilling speeds of 50/100 holes per second were obtained with I.R. λ = 1.06 μm. YAG lasers through 500 μm thick silicon wafers. We have obtained several thousand   50 μm diameter holes per wafer.
However, after these first experiments, many difficulties remain in controlling parameters. We observed, like ANTHONY [10] from General Electric Schenectady, that high power pulsed YAG laser caused spalling and cracking of the silicon around the entrance and exit holes. Also we noticed a lot of material projections around holes. With a YAG in the continuous mode Q switched, we obtained smaller holes. ANTHONY has obtained comparable results [10] in 330 μm thick single-crystal sapphire wafers, and has explained that the pulse sequence used to drill a hole had a particular importance. If the integrated power level chosen is too low, it

is necessary to use a high number of successive pulses and then the hole has a better aspect. However, if the number of pulses is too high, then randomly geometric displacements of the bottom of the drill hole can occur.
To drill very small diameter holes with a laser it is necessary that the power level be chosen such that the number of pulses remains low (in the range of 100 to 200 successive pulses), then the geometric displacement becomes negligible. Another consequence is the decrease in cracks and spalling in the silicon around the hole.

The inner surface quality of the hole is a function of some important parameters. For a given material and a given laser source it is necessary to optimize the duration of pulses and the separation time between successive pulses. Good results are obtained with a separation time higher than 10^{-3} sec. in air atmosphere. If one uses a blow of oxygen during the drilling sequence, the gas accelerates the reaction and the results are quite different , the inner hole surface seems cleaner but the diameter of the hole increases for the same power of the laser beam. With Nitrogen on the contrary, it is possible to obtain smaller holes.
The inner surface hole quality and the possibility of obtaining high effective aspect ratios through silicon wafer holes are principally determined by a good compromise of these parameters.
In Table II, we resume our drilling parameters and hole characteristics.

Drilling parameters
Laser Microcontrole 904 DT
TEM 00 YAG = 1,06 µm Q switched
P = 3,5 W
Number of pulses = 100
Pulse width = 200 10^{-9}s
Fast air bearing XY stage (FAB 200)
(laser interferometer)
Stage speed = 200 mm/s
Travel range = 200 x 200 mm
Total accuracy 2,5 µm
Wafer supported on the sides (clear area below the holes)
Hole characteristics (silicon wafer 550 µm thick)
Entrance diameter = 35 µm
Exit diameter = 10 µm
Conicity = 1.5°
Drilling speed = 3-5 holes/sec

TABLE II

Technology	Characteristics	Results
Capillarity wetting	Silver particles 4 μ Ø in a carrier, epoxy liquid or polyimide liquid.	Limitation to holes Ø>100um Hole size = 25 Particle Ø
Wedge extrusion	A moving wedge extrudes the liquid into the laser drilled holes.	Idem Ø >100 μ
Wire insertion	W wires 13 μm introduced in holes Copper electroplating which locks the wire.	Ø < 100 μ possible but impracticable for a large number of holes
Electroless plating	P.C.B. method electroless copper	Ø<75 μ possible, erratic results. Yield problem due to inner surface state of holes (etching pb.)
Electro-forming	Metal film deposited on the back-face. Metal electroplating with periodic reverse plating.	Entrapped bubbles non uniform growth rate Low electrical resistance
Double side sputtering and electro-plating	Necessity to maintain a fluid velocity to keep the holes open during pla-ting.	Limitation to Ø holes > 62,5 μm

TABLE III

CVD process	
1st step	Polysilicon deposition LPCVD on SiO_2 thermal thickness : 1.2 μm $SiH_4 \xrightarrow{H_2} Si + 2 H_2$ t = 640°C P = 0.5 Torr Polysilicon acts as a catalyst for Tungsten deposition Equipment used : Applied Materials
2nd step	Tungsten deposition CVD thickness 0.4 μm Reduction of WF_6 by H2 $2 WF_6 + 3 Si \rightarrow 3 Si F_4 + 2 W$ Si acts as a catalyst for W deposition (first atoms) $WF_6 + 3 H_2 \rightarrow W + 6HF$ at = 400°C P = 0.4 Torr Equipment used : ANICON (hemispheral chamber)

These results are in good agreement with T.R.ANTHONY papers from General
Electric Schenectady [11] who obtains 25 micron diameter holes through
300 μm thick silicon wafers.

The protection of wafer surface from the debris and melted projections
ejected during the drilling operation is done by using a several μm thick
coating of colloïdal silica, Emulsitone, or Kryolin spray-on film [11] on
the whole surface of the wafer· These protections are easy to apply
and   remove.

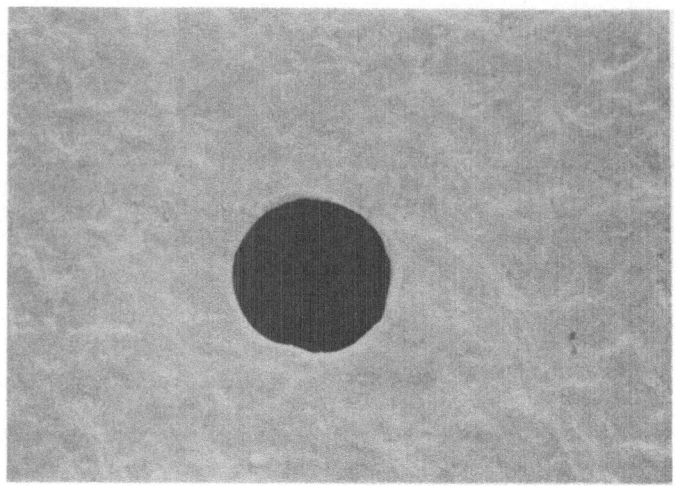

PHOTO : 10 μm Ø hole laser drilled with protective coating

3. FORMING ELECTRICAL CONNECTION

ANTHONY experimented with different technologies to form conductors in laser
drilled holes [12]. Table III compares the solutions he has tried : capi-
larity wetting, wedge extrusion, wire insertion, electroless plating ,
electroforming, double side sputtering and then through hole plating.
The last method showed the best results but depth/diameter ratio is limited to 6. In LETI, we are working to obtain 20 to 50 depth/diameter ratios
and we use a CVD process in an ANICON isothermal chamber.
We get good results with 50 μm diameter 650 μm deep holes. We are still
trying to achieve processing of smaller holes and have not yet found any
limitation.

4. CONCLUSION
The following processes can be used to form through wafer vias :
- wafer face organic protection
- laser drilling
- ultrasonic cleaning
- wet etching
- thermal SiO_2
- CVD polysilicon
         tungsten
- electroplating Au, Cu, Ni, etc...

    Having described our "today process" on the fabrication of through-wafer via conductors in silicon, we conclude that hole diameters of 10 to 35 microns are possible and can establish electrical connections through both sides of a wafer with a resistance less than $1\,\Omega$. The problem of drilling less than 10 µm diameter holes is now being   studied in LETI but it is necessary to modify the   optics and laser beam. "Hybrids on Silicon" and 3 D Hybrid devices are the main applications of these new technologies.

REFERENCES
1. K.Peterson : Proceedings on the IEEE Vo.70 No.5 May 82
2. J.B.Angell, S.C.Terry, P.W.Barth : "Pour la Science" June 83
3. L.Csepregi : Microelectronic Engineering 3, 1985, 221-234
4. E.Bassous : IEEE Trans.on Elect.Devices N.10 Oct.78
5. E.Bassous, E.F.Baran : "The fabrication of high precision nozzles by the anisotropic etching of <100> silicon, J.Electrochem.Soc. 125 p.1321, 1978
6. T.J.Chang : J.Vac.Science Tech. B5 1985
7. D.J.Ehrlich, R.M.Osgood, T.F.Deutsch : "Laser chemical technique for rapid writing on surface relief in silicon" Appl.Phys.Letters Vo.38 p.1081 1981
8. D.J.Ehrlich, J.Y.Tsao : A review of laser-microchemical processing, J.Vac.Sci.Techno. B1(4) Oct/Dec.1983
9. D.J.Ehrlich, D.J.Silversmith, R.W.Mountain, J.Tsao : IEEE Trans.on Comp.Hybrids and Man.Tech. Vo.5 No4 Dec. 1982
10.T.R.Anthony : "The random walk of a drilling laser beam", J.Appl.Phys. 51(2) Feb.1980
11.T.R.Anthony : "Diodes formed by laser drilling and diffusion", J.Appl.Phys. 53(12) Dec.1982
12.T.R.Anthony : "Forming electrical interconnexions through semiconductor wafers" J.A.P. 52(8) August 81
13.B.Charlet, T.Dupeux, L.Peccoud : "6th Intern.Colloquium on Plasma and Sputtering" June 1/5th 1987.
14.Podlesnik, Gilgen, Osgood : "Deep UV induced wet etching of AsGa" Appl.Phys. Lett. 45(5) sept.84
15.Osgood, Sanchez, Rubio, Ehrlich, Daneu : Localyzed laser etching of compound S.C. in aqueous solution. Appl.Phys. Lett. 40(5) March 82
16.Brewer, McClure, Osgood : "Dry laser assisted rapid HBr etching of GaAs". Appl.Phys. Lett. 47(3) August 85
17.Maegele : "A better way to connect optical fiber and chips". Electronics January 6th 86.

THE DEVELOPMENT AND USE OF NOVEL PRECURSORS FOR PHOTOLYTIC DEPOSITION OF
DIELECTRIC FILMS

C.J. BRIERLEY, F.W. AINGER, C. TRONDLE
Plessey Research Caswell Limited Towcester
Northampton, NN1Z 8JU, U.K.

ABSTRACT

The use of ultra-violet light in the deposition of electronic
materials from the gas phase is a rapidly emerging technology. Many of
the conventional MOCVD precursors that are available however do not
absorb light strongly at wavelengths above 200nm and it is, therefore,
often necessary to resort to wave lengths of less than 200nm, where most
organic materials have significant absorptions. An alternative approach
is to synthesize precursors containing chromophores which specifically
absorb at wavelengths that are more easily accessible. We have used
this approach to produce a number of precursors, for the deposition of
Al_2O_3, with photo-absorption in the 240nm to 300nm range. Basing our
work on aluminium tri-isoproxide it has been possible to substitute one
of the isopropoxide groups with a β-diketonate group, which forms a
six-membered conjugated ring with the aluminium atom. Depending on the
nature of the β-diketonate, these components have absorptions in the
230-320mm range, with molar extinction coefficients in excess of 10,000
at λmax. The low temperature deposition of Al_2O_3 for some of these
compounds in the presence of UV light from a high pressure Mercury Xenon
arc lamp (with a large spectral output between 260 and 320nm) and the
frequency doubled line of an argon ion laser (257nm) is discussed.

Introduction

Interest in the use of UV light as a means of promoting chemical
reactions at material surfaces has developed rapidly in recent years.
Photo-induced reactions involving gas phase molecules at solid surfaces
have been of particular interest. For a photolytic reaction to take
place it is important that a photon of energy is absorbed by a molecule
either in the gas phase or when absorbed at a surface [1]. In some
circumstances absorption of UV light by substrate molecules may allow
reactions to occur between the substrate and non-absorbing gas phase
molecules [2].

A cursory glance at the literature reveals that the choice of
precursors for photolytic deposition has been almost entirely made on
the basis of what precursors are available at the time; thus for most
metal and semiconductor deposition systems the alkyls which have been
developed for pyrolytic MOCVD processes have been used. These materials
usually absorb very poorly at wavelengths beyond 220mm and often only at
wavelengths less than 200nm (Vacuum UV). Table I gives the peak
absorptions of some common precursors.

145

D. J. Ehrlich and V. T. Nguyen (eds.), Emerging Technologies for In Situ Processing, 145–152.
© *1988 by Martinus Nijhoff Publishers.*

Table I   PEAK ABSORPTIONS (λmax) FOR A RANGE OF COMMON PRECURSORS

PRECURSOR	λmax
$Cd(CH_3)_2$	220
$Hg(CH_3)_2$	203
$Zn(CH_3)_2$	200
$Te(C_2H_5)_2$	246
$In(CH_3)_3$	210
$Ga(CH_3)_3$	195
$(Al(CH_3)_3)_2$	180
SiH_4	160
Si_2H_6	200

There are, however, several fundamental disadvantages to the use of light at wave lengths less than 220nm. These include:

1. The lack of readily available light sources.
2. The need to exclude oxygen from the beam path (to prevent absorption losses and ozone formation).
3. The high cost of optics.
4. The greater tendency for homogeneous gas phase reactions to occur.

The last point may ultimately prove the most telling and occurs as a result of the more general absorption that occurs in organic bonds at shorter wavelengths, in this way molecules can physically 'fall apart' in the gas phase, leading to gas phase nucleation (dust formation) [3], a loss of spatial resolution [4] and increased carbon contamination [5].

It is apparent from the lack of literature on the subject that very little consideration has been given to the design and synthesis of precursor molecules that have absorptions at longer wavelengths in the UV, although this is potentially a far more useful avenue to investigate than that of using light sources at shorter wavelengths.

This paper describes the way in which such a line of research can prove extremely fruitful.

Background

The aim of this work was to develop a low temperature photolytic deposition technique for Al_2O_3 as an insulator/passivant for GaAs device processing. Al_2O_3 was chosen rather than Si_3N_4 on the basis that it had greater potential for epitaxial deposition onto GaAs for some longer term objectives.

Two UV light sources were available: the first was a 1KW input power XeHg arc lamp; such a lamp produces 35% of its total radiation between 230 and 400mm, but very little below 200nm, as shown in Fig. 1. The second UV source consisted of an argon ion laser with an intracavity frequency doubler producing in excess of 100mW @ 257nm. The precursors aluminium tri-isopropoxide and aluminium tri-butoxide have been used successfully for pyrolytic deposition of Al_2O_3 in this laboratory.

Metal alkoxides are particularly suitable precursors for metal oxide deposition because most spontaneously break down to the oxide without the need for an oxidising gas. However, metal alkoxides in general have virtually no photo-absorption at wavelengths above 200nm and experiments with the XeHg lamp confirmed that deposition could not be initiated or sustained at less than the pyrolytic breakdown temperature (350-400°C) of these precursors. Attempts were therefore made to synthesize a suitable precursor.

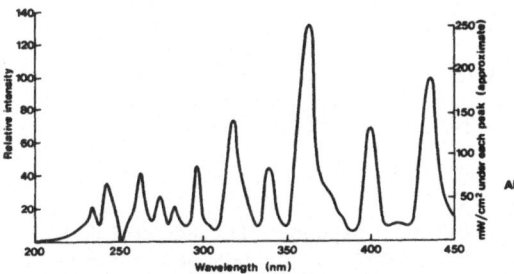

Fig. 1 **Spectral Output of Xe Hg Lamp (1kw)**

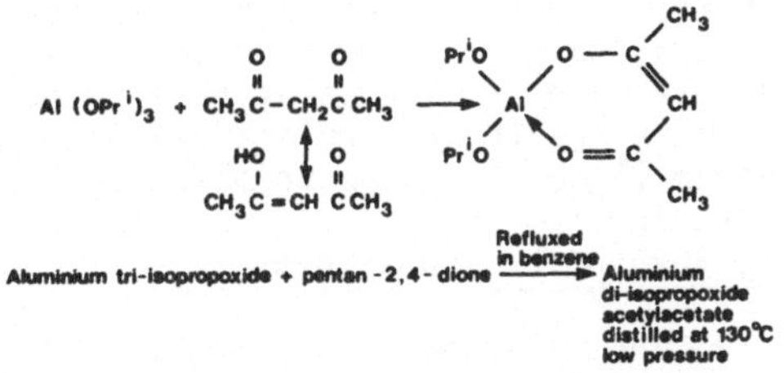

Fig. 2 **Synthesis Route for Aluminium Di-isopropoxide Acetylacetonate**

Synthesis

The metal acetylacetonate compounds are known to possess a UV absorption band centred at 290nm[6] therefore the possibility of replacing one alkoxy group of an aluminium trialkoxide with a substituted β-diketonate group has been explored with the intention of producing a volatile oxide precursor with a significant absorption above 250nm.

The preparation of the β-diketonates was achieved as shown in Fig. 2 by the reaction of equimolar amounts of aluminium alkoxide and β-diketone in refluxing benezene. The reaction was forced to the right by removal of isopropanol as the benzene azeotrope. After evaporation of the benzene the di-isopropoxy aluminium-β-diketonate was distilled at reduced pressure.

The acetylacetonate, which forms a 6-membered resonance stabilized ring with the aluminium, has an absorption peak above 250nm. Substitution of the ring with electron withdrawing or donating groups destabilises or stabilises the contributing structures resulting in hypsochromic (blue) or bathochromic (red) shifts in the absorption band. The effect of the substitutions carried out is shown in Table II.

Table II **Properties of Aluminium Precursors of General Formula**

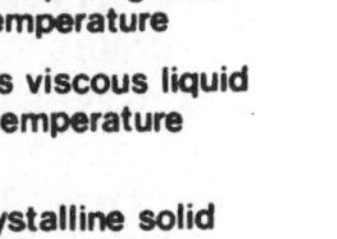

Substituents for di-isopropoxy aluminium diketonate	Physical characteristics	Vapour pressure @ 120°C (mbars)	λ max in CH$_3$Cl	λ max (vapour)
1 R = H, R$_1$ = OC$_2$H$_5$, R$_2$ = OC$_2$H$_5$	Extremely viscous colourless liquid (glass) at room temperature	0.042	257 nm	256 nm
2 R = H, R$_1$ = OC$_2$H$_5$, R$_2$ = CH$_3$	Colourless viscous liquid at room temperature	0.140	267 nm	266 nm
3 R = H, R$_1$ = OCH$_3$, R$_2$ = CH$_3$	Yellow crystalline solid at room temperature MP ~60°C	0.092	269 nm	261 nm
4 R = H, R$_1$ = CH$_3$, R$_2$ = CH$_3$	Yellow crystalline solid at room temperature MP ~105°C	0.226	287 nm	284 nm
5 R = H, R$_1$ = C(CH$_3$)$_3$, R$_2$ = C(CH$_3$)$_3$	Colourless liquid, thermally unstable	–	293 nm	–
6 R = H, R$_1$ = C(CH$_3$)$_3$, R$_2$ = CF$_3$	Brown liquid, thermally unstable	–	297 nm	289 nm

148

It should be noted that compounds 3 and 4 are listed as crystalline solids. In fact these materials crystallize only after a period of several days after melting and cooling, it is therefore possible to handle them as liquids at room temperature. This phenomenon is believed to be a result of a relatively slow conversion that takes place from the dimer (liquid) to the trimer (solid). This is explained in detail in Ref. 7.

UV Absorption Charactersitics

The UV absorption characteristics of the compounds listed in Table II were measured both in a solution of CH_3Cl and in the gas phase. By using a 5×10^{-4} molar solution, the absorbance in a 1mm path length could be measured and from this the molar extinction coefficient (absorbance $mole^{-1}cm^{-1}$) calculated; this was found to be similar for each compound at approximately 14,000.

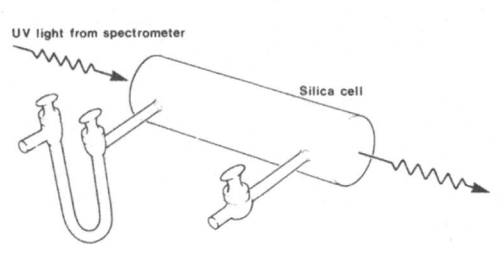

Fig. 3   **UV Gas Cell**

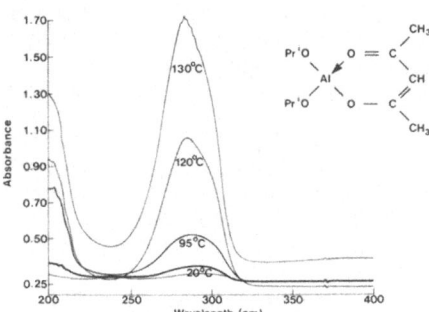

Fig. 4   **Gas Phase UV Spectra of Al(OPr^i)$_2$Acac**

Gas phase measurements were made by means of the gas cell shown in Fig. 3. A small amount of the material was loaded into the U tube and the whole vessel evacuated to better than 10^{-1}mbar. Then with the tap open between the gas cell and U tube the vessel was heated to a uniform temperature in an oven. A UV spectrum was then taken using a multiple photodiode spectraphotometer; such a measurement can be completed in less than a second during which no appreciable cooling of the gas cell occurs. The gas phase absorption for aluminium di-isopropoxy actylacetonate is shown at various temperatures in Fig. 4. Given the molar extinction coefficient for the material in solution it is possible to ascertain the vapour pressure at a given temperature from the gas phase absorbance at λmax. Table II gives the vapour pressure for some of the materials at 120 °C and Fig. 5 gives an Arrhenius plot of vapour pressure against temperature.

Some important points are noted from Table II.

1) By modifying the nature of the organic group at positon R_1 and R_2, it is possible to substantially shift the absorption band.

2) A range of values for λmax between 256 and 289nm has been achieved. Larger alkyl groups cause a bathochromic shift while alkoxy groups cause a hypsochromic shift.

3) A small shift occurs between the λmax measurement for solution and vapour. This shift is greatest in compound 3 and means that for this precursor λmax (261nm) is very close to that of the laser line of 257nm, along with compound 1 with λmax of 256nm.

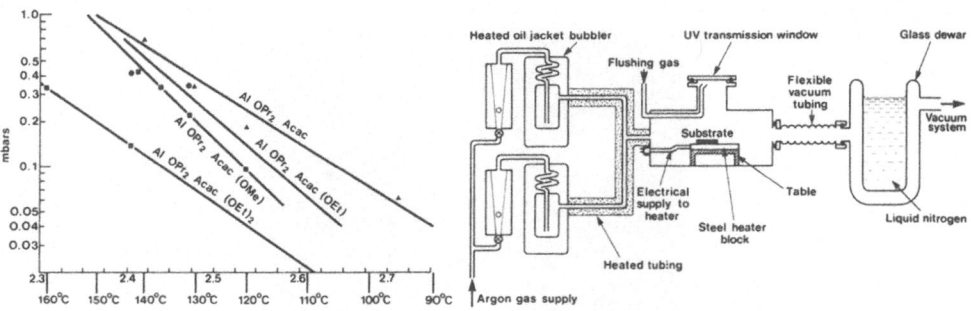

Fig. 5   Vapour Pressures of Aluminium Precursors
(Gas phase UV measurements)

Fig. 6   Photolytic Deposition Chamber

Deposition Experiments

UV activated deposition of Al_2O_3 from these materials at temperatures below $400\,^0C$ has been achieved. It will be noted from Fig. 1 that the XeHg arc lamp has a fairly uniform distribution of output in the 250nm-300nm range albeit consisting of discrete peaks. It is therefore likely to be effective in stimulating the deposition of any of the precursors developed. Two of the precursors shown in Table II have been used in deposition experiments, compounds 3 and 4. The latter has the highest volatility, while Compound 3 has a λmax at 261nm close to the available laser line and also has a reasonable volatility. It is also easy to handle (being a meta-stable liquid at room temperature) and therefore chosen in preference to compound 1.

Deposition experiments have been carried out in the apparatus shown schematically in Fig. 6. The precursor is loaded into one of the bubblers under dry nitrogen and heated by means of recirculated oil to between $120\,^0C$ and $130\,^0C$. The argon carrier gas was passed through the bubbler at low pressure and admitted to the deposition cell which is pumped down to a pressure of approximately 0.5 mbar. A flushing gas is used to keep the UV transmission window clean, which in our experience has never clouded. In most experiments the substrate is maintained at $200\,^0C$ which is some $300\,^0C$ lower than the pyrolytic deposition temperature for these components while being sufficiently high to ensure the precursor cannot condense on the substrate.

Both precursors deposited Al_2O_3 at a similar rate when the XeHg radiation is incident upon the substrate. Figure 7 shows a GaAs substrate coated with Al_2O_3 (approx. 1000Å) from a collimated beam of XeHg radiation having a power intensity at 250-300nm of approx. $40mW/cm^2$. The stripe across the centre is the shadow of a wire suspended above the substrate where deposition has been reduced.

Fig. 7   **Al₂O₃ Deposited by XeHg Arc Lamp**   (Shadow of wire across centre)

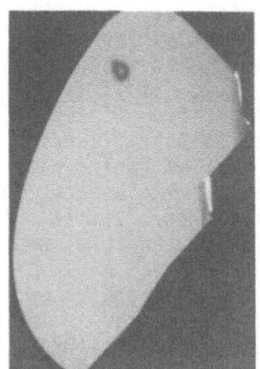 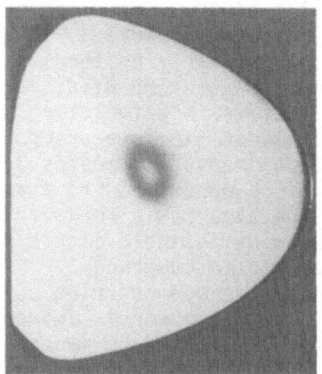 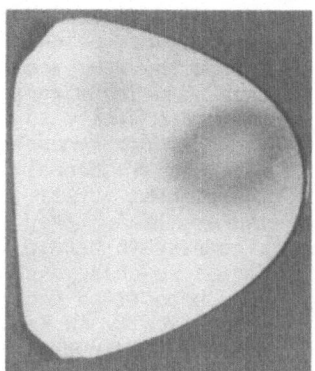

Fig. 8   **Examples of Al₂O₃ Deposition Under Expanded Elliptical Laser Spot**

Figure 8 shows the effect of the elliptical laser spot (257nm), expanded to three different diameters, on the deposition of Al_2O_3 from compound 3. The deposition rate at the centre of the beam seems fairly independent of light intensity suggesting that the rate is limited by the precursor concentration in the gas phase. The partial pressure of the precursor in the gas phase in these experiments is estimated to be about 2×10^{-3}mbar when it has been diluted by the window flushing gas.

Deposition rates of up to 1μm/hour have been achieved using the focussed XeHg arc lamp and a precursor partial pressure of approximately 10^{-2}mbar.

The thickness of the layers has been measured using an ellipsometer with a HeNe laser, the refractive index being between 1.65 and 1.70. The layers are amorphous showing no sign of crystallinity and no features under SEM analysis.

Conclusions and Discussion

A group of related precursors with photo-absorption maxima in the range 250-300nm have been developed which can be used for the photolytic deposition of Al_2O_3 from a XeHg arc lamp and the 257nm line of a frequency doubled argon ion laser. Initial experiments suggest that the deposition process is extremely efficient, being more often limited by the vapour concentration of the precursor rather than the light intensity. Thin amorphous films of Al_2O_3 are deposited. No homogeneous gas phase reaction is observed suggesting that the precursor controllably decomposes on the substrate surface. One possible pathway would involve absorption of a photon by the conjugated acetyl acetonate group which breaks away to yield an aluminium di-isopropoxy radical. The alkoxy groups are lost when the radical becomes absorbed by the substrate surface. It is planned to investigate the process of radical formation and surface decomposition using a laser spectroscopic technique.

The vapour pressure of these precursors has been calculated by their absorbance in the vapour at λmax once the molar extinction coefficient had been established in solution. In addition the absorbance cross section of the gas can be calculated and compared with other precursors at present available. Table III shows the absorbance cross section of various precursors at their λmax, not only do the β-diketonates have a more practical wavelength they also possess a higher absorption cross section. The chief disadvantage of these materials is their relatively low vapour pressure compared with alkyls, but sufficient volatility can be achieved by heating to an appropriate temperature.

Table III

PRECURSOR	λmax	Cross Section $Å^2$
Cd $(CH_3)_2$	220nm	0.4
Hg $(CH_3)_2$	203nm	0.4
Zn $(CH_3)_2$	200nm	0.38
Te $(C_2H_5)_2$	246nm	0.12
$(Al(CH_3)_3)_2$	180nm	0.3
Al(OPri)β-dik(4)	284nm	0.55
Al(OPri)β-dik(3)	261nm	0.55

In many photolytic deposition processes that have been reported there has been a very poor match between the precursor absorption and UV light source particularly for wavelengths above 200nm. By far the most common light source that has been used below 200nm is the ArF line (193nm) of the excimer laser. While most organic materials have a reasonable absorbance cross section at this wavelength, there are significant disadvantages in using vacuum UV wavelengths (<200nm). It is therefore clear that the synthetic approach of including a chromophore in the precursor molecule should have enormous advantages when applied to other deposition systems.

Acknowledgements

This work has been carried out with the support of the Procurement Executive, Ministry of Defence.

References

1. See for Example, S. J. S. Irvine, J. B. Mullin, D. J. Robbins and J. L. Glasper, J. Electro. Chem. Soc. 132, (1985), p968

2. K. A. Bertness, C. E. McCants, T. Chiang, P. H. Mahowald, A. K. Wahi, 'Comparison of Low Intensity Laser Enhancement of Oxygen Chemisorption on GaAs using O_2 and N_2O', Proc. MRS Fall Meeting Dec. 1986, Session B 6:43, Boston, Mass.

3. T. R. Jervis, S. K. Menon and D. W. Carroll, 'Mechanisms for the Deposition of Thin Metallic Films by Laser Driven Gas Phase Reactions', Proc. MRS, Fall Meeting, Dec. 1986, Session B.1.3., Boston, Mass.

4. G. S. Higashi, G. E. Blonder, C. G. Fleming, 'Wavelength Dependent Activation Selectivitiy in Aluminium Chemical Vapour Deposition', Proceedings of "Microphysics of Surfaces, Beams and Absorbates Topical Meeting", February 1987, Santa Fe, New Mexico

5. V. M. Donnelly, D. Brasen, A. Applebaum, M. Geva, "Excimer Laser Induced Deposition of InP', J. Vac. Sci. & Tech., Vol. 4 No 3, 1986, p716-721

6. D. C. Bradley, R. C. Mekrotra and D. P. Gaur, 'Metal Alkoxides', Published by Academic Press, London (1978), Chapter 4 p215

7. J. H. Wengrovius, M. F. Garbauskas, E. A. Williams, R. C. Going, 'Aluminium Alkoxide Chemistry Revisited', J. Am. Chem. Soc., 108, 1986, p982-989

FOCUSED ION BEAM INDUCED DEPOSITION

J. MELNGAILIS, A.D. DUBNER, J.S. RO, G.M. SHEDD*, H. LEZEC, and C.V. THOMPSON

RESEARCH LABORATORY OF ELECTRONICS, MASSACHUSETTS INSTITUTE OF TECHNO-LOGY, CAMBRIDGE, MA 02139;

ABSTRACT
Ion induced deposition is a novel method of thin film growth in which a local gas ambient is created near an ion bombarded surface. The ion bombardment causes the gas molecules to break up and some of the gas constituents to deposit on the surface. If a focused ion beam is used, then this becomes a technique for maskless, resistless, patterned deposition. Depositions of films from gases of $Al(CH_3)_3$, WF_6 and $Ta(OC_2H_5)_2$ have been reported. The films for the most part have contained high (approaching 50%) concentrations of impurities such as O or C, presumably due to the lack of ultrahigh vacuum conditions. Gold deposition has been observed from dimethyl gold hexafluoroacetylacetonate ($C_7H_7F_6O_2Au$), with both focused ion beams and broad beams. In many cases, the gold films are much purer (less than 5% C or O) and have exhibited resistivities from 20 to $1000\mu\Omega cm$ (Bulk gold resistivity is $2.5\mu\Omega cm$.) Deposition yields (atoms deposited per incident ion) of 4 to 100 have been observed. But the higher yields correlate with higher resistivity and higher impurity content. Preliminary transmission electron microscope examination shows the gold films to start out as unconnected islands of 40 to 60nm dimensions. The mechanisms for the deposition is at present not well understood. Some hypotheses will be discussed. Ion-induced deposition appears to be a promising technique for in-situ deposition of metals or insulators with submicrometer resolution.

1. INTRODUCTION

An in-situ processing machine that can start from a bare wafer of a semiconductor and yield chips with a variety of devices must possess a number of capabilities, such as: the addition and removal of material, alteration of material, patterning, and analysis. Of special interest are processes which combine two steps into one, for example patterned deposition of material. The focused ion beam, because of its flexibility, is likely to be a key tool in future in-situ processing machines. The initial steps in this direction have already been taken at the Optoelectronics Joint Research Laboratory, where molecular beam epitaxy, focused ion beams and other processes have been combined in one machine.[1]

The focused ion beam can implant, mill away material, and deposit material in patterns with submicrometer resolution.[2] For implantation the dopants of Si and GaAs can be focused on the sample with submicrometer resolution and energies from a few tens of keV to 300 keV. The beam can be accurately deflected to achieve the desired pattern.

---

*Present address: Department of Materials Engineering, North Carolina State University, Raleigh, NC 27695.

D. J. Ehrlich and V. T. Nguyen (eds.), Emerging Technologies for In Situ Processing, 153–161.
© *1988 by Martinus Nijhoff Publishers.*

The focused ion beam can also be used in more classical, two step, patterning, e.g. lithography. Numerous inorganic and organic films
(including most resists) can, by focused ion beam implantation, be
rendered sensitive or insensitive to subsequent material removal treatments, such as plasma etches.

The ion beam, if it is left to dose an area heavily, will also mill
away material (sputtering). By feeding a gas such as Cl_2 to the area
being milled and permitting chemical action to play a role the material
removal rate can be enhanced by factors of 5 to 10. Conversely if other
gases such WF_6 or $Al(CH_3)_3$ are fed to the surface, material deposition
can be induced by the ion beam. In addition, the focused ion beam when
combined with mass analysis of the sputtered-off material forms a high
resolution secondary ion mass spectrometer (SIMS). Alternatively, if the
secondary electrons are detected a scanning ion microscope (much like the
scanning electron microscope) is formed. This latter capability is very
important in aligning a desired focused ion beam fabrication step to
existing features.

This paper will discuss the most recent and least developed of the
focused ion beam processes, namely ion induced deposition. We will
review the existing results in the field, discuss mechanisms, and mention
some of our own results.

2. ION INDUCED DEPOSITION*
2.1 Creating the Local Gas Ambient

FIGURE 1. (From ref. 3)
Schematic of apparatus for
ion induced deposition. The
sample is in a box in which
a gas ambient (10-30mTorr) is
maintained. The pressure in
the vacuum chamber surrounding
the box remains ~10^{-6} torr.
Ions enter through a small hole
at the top.

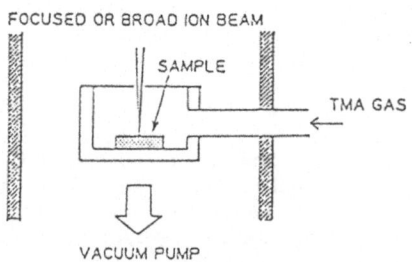

FOCUSED OR BROAD ION BEAM

SAMPLE

TMA GAS

VACUUM PUMP

The earlier experiments in ion induced deposition used a box constructed around the sample, as shown in Fig. 1. A small 120 μm hole

---

*The term "ion induced deposition" is much more appropriate for the
phenomenon under discussion than "ion assisted deposition". The deposition is not "assisted"; it will not take place without the ion beam.
Furthermore, the term ion assisted deposition is used (aptly) when a film
deposited by, say, evaporation is simultaneously bombarded by ions to
enhance its properties.

located in the top lid permitted the ion beam to be incident on the sample, and, within limits of the hole to be scanned [3] [4] [5]. The pressure inside the vessel could be maintained at 10-30 millitorr, while outside the pressure could be 3 or more orders of magnitude lower as required for ion beam operation. Both a broad beam from an implanter or a pencil beam from a focused ion beam machine has been used. [3] [4] [5]

An alternative scheme is to use a capillary gas feed aimed at the surface in close proximity (≤1 mm) to produce a local gas ambient[6] [7]. (See Fig. 2) This has the advantage that the sample can easily be moved

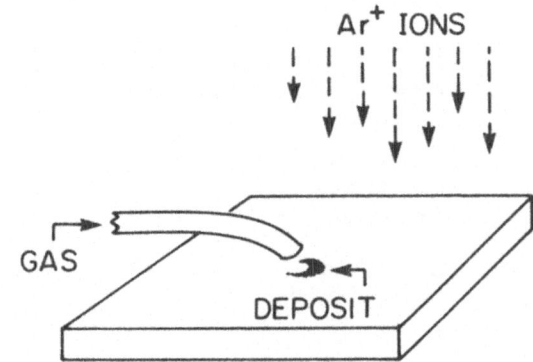

FIGURE 2. (top) An alternate apparatus with a capillary gas feed aimed at the surface where the ions are incident. (Bottom) Examples of deposition with this apparatus. Deposits are on glass slides photographed with back lighting. The rounded indentation in the bottom of each deposit is due to the shadow cast by the tubing.

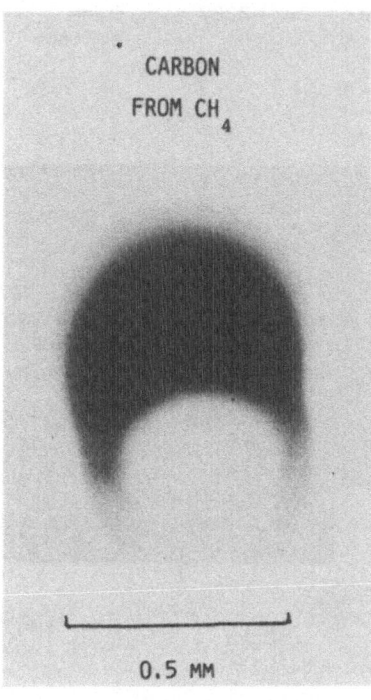

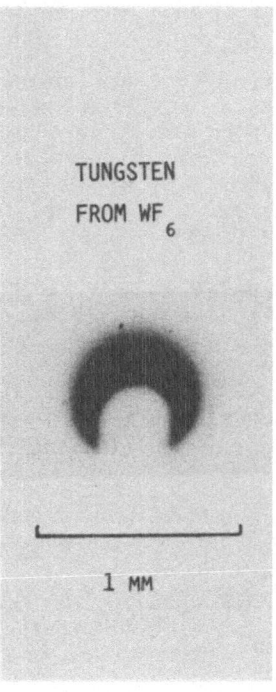

for exposure of large areas, and the tubing itself can be used as a flow controller by choosing its length and the upstream gas pressure.[8] The disadvantage, at least for scientific studies, is that the pressure is more difficult to measure. The capillary gas feed is used in the commercial photo mask repair process based on focused ion beams [9] [10] [11], where carbon films are deposited to repair clear defects (absence of chrome).

The flow rate can be calibrated and modeled [12], and the pressure as a function of distance from the nozzle can be calculated.[13] In some

TABLE 1. Ion Induced Deposition

Gas (Ref.)	Ion, Energy	Current Density (A/cm^2)	Yield (atoms /ion)	Deposit Composition	Resistivity $\mu\Omega$cm
$Al(CH_3)_3$ (4),(5)	Ar+ 50 KeV Au+ 50 KeV (Implanter)		13	Al: 20-58% O: 16-39% C: 64-3%	
$Ta(OC_2H_5)_5$ (7)	Ar+ 50 KeV		27	Ta: 56% O: 27% C: 17%	
WF_6 (RT) (7)	Ar+ 50 KeV		31	W: 75% O: 25%	
WF_6 (80K) (7)	Ar+ 50 KeV		3800	high O content	
WF_6	Ar+ 750 eV (Ion Miller)	4×10^{-4}	5	W: 90% O: 5% C: 5%	350 (Bulk W = 4.6)
Hydro- carbon (9,10,11)	Ga+ 20-30 KeV (FIB)		10		
$C_7H_7F_6O_2$ A (6)	Ga+ 15 KeV (FIB)	0.05 scanned	4-5	Au: 75% Ga: 20% O: <5% C: <5%	500-1300 (Bulk Au = 2.4)
$C_7H_7F_6O_2$ Au	Ar+ 70 KeV (Implanter)	7×10^{-6}	100	Au: 60% O: 5% C: 35%	10,000
$C_7H_7F_6O_2$ Au	Ar+ 750 eV (Ion Miller)	4×10^{-4}	~1	Au: 95% other <4%	20
$C_7H_7F_6O_2$ Au	Si+ 50 KeV (Implanter)	7×10^{-6}	8-16	Au: 75% O: 10% C: 10%	130

instances, the flow rate with N_2 gas has been measured, and then scaled to gases of different molecular weight.[12]

For practical applications the capillary gas feed clearly has many advantages. In addition, the uniformity of the pressure as a function of position can be increased by having, for example, four gas feeds aimed at the area where the beam is incident.

In either type of apparatus pressures from 10^{-4} to 10^{-2} torr can readily be created without degrading the vacuum in the region where the beam is produced. This is sufficient to produce films by ion induced deposition.

2.2 Deposition of Films

Numerous gases have been used to deposit films. Unfortunately, attempts to deposit, for example, Al and W have so far yielded films laden with impurities such as oxygen and carbon. The results of most of the depositions so far are summarized in Table 1.

In fact, all species of ions that have been tried have worked. Heavier ions are more effective (higher yield) than lighter ones.[5] [7] The oxygen present in films such as W, Al, or Ta is thought to be due to the background pressure in the "ordinary" vacuum chambers (not ultrahigh vacuum (UHV)) used. If the base pressure before gas is introduced is 3×10^{-7} torr, then a monolayer of oxygen is likely to form in bare Al or W in 3 to 10 sec. Some of the films were grown with an ion current density of $1 \mu A/cm^2$ (Ref. 7) or (6×10^{12} ions/cm^2 sec.). If the yield is 15, say, then a monolayer of metal would be grown in 17 sec. Thus there is ample

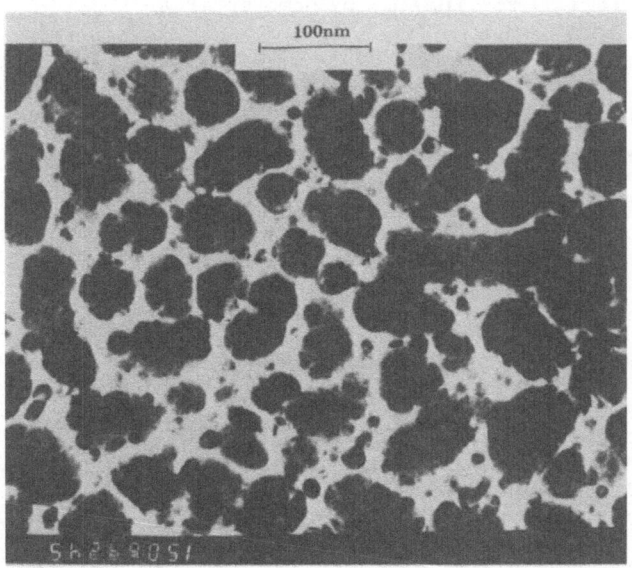

FIGURE 3. A transmission electron micrograph of a gold film on SiO_2 nominally 60nm thick. Deposited using 75 keV Si^+ ions from an implanter.

time for the film to incorporate oxygen as it is grown. Thus either a better vacuum or a higher growth rate should improve film purity.

Resistivity of the film is another property of interest. As shown in the table, the resistivity of the gold films deposited is from 8 to 4000 times higher than the handbook value for pure gold. Some of this increase in resistivity is seen to correlate with increased impurity content. An additional factor appears to be the granularity of the film. Preliminary transmission electron microscope studies [14] show that some of the gold films grown at high yields (Table 1) are largely made up of unconnected islands even at thicknesses of 60nm. (See Fig. 3) One tungsten film grown at low yield that was examined showed voids but otherwise a much more continuous structure.[14]

The reasons for the microstructure of the films is not at present well understood, nor for that matter, is the mechanism of film growth, i.e. how is the energy of the incident ion coupled into breaking up tens or thousands of gas molecules. There is considerable circumstantial evidence that the break up occurs in gas molecules adsorbed in the surface: a) Deposition has resolution equal to the beam diameter.[3][4][6] If the break up occured in the gas above the sample, this would not be true. b) At the gas pressures present in the vicinity of the sample surface ($^{\sim}10^{-2}$ torr), the mean free path is about 10mm. Thus the probability of an ion colliding with a molecule is less than 1. Yet yields of 5-4000 have been reported. c) The deposition yield is a strong function of substrate temperature.[7] This would not be expected unless adsorption played a role.

TABLE 2. Mechanisms of Ion Induced Deposition (Film deposited from WF_8 at 20 mTorr, Ions 50 keV $1\mu\text{Å/cm}^2$ by Gamo et al.)

Ion	Deposition Yield (a)	Secondary Electron Yield of Mo (b)	Rate of Energy Loss to Electrons (c) keV/μm	Rate of Energy Loss to Nuclei (c) keV/μm	Ion Range in W R_p (μm)	Range Straggle R_p (μm)
He	5.9	2.3	225	15	.110	.090
Ar	14.1	2.0	770	690	.015	.014
Xe	27.9	1.8	830	2900	.008	.005

Notes on table:

a) Calculated from data of K. Gamo et al. Microcircuit Engineering 5, 163 (1986) assuming pure tungsten is deposited.
b) J. Ferrón et al. J. of Phys. D 14, 1707 (1981). We have been unable to find corresponding data for tungsten, but we expect molybdenum to be qualitatively similar.
c) Calculated using the TRIM Program, original developed by J.P. Biersack and J.F. Ziegler, see for example, Ion Implantation Techniques, Eds. H. Ryssel and H. Glawischnig, Springer-Verlag, (1982), p. 122.

The lowering of the substrate temperature to 80° K has been shown to increase the yield from WF_6 by a factor of 5000 over the yield at room temperature.[7] However, the film is reported to have a high density of oxygen.[7] Presumably at the lower temperatures more gas molecules condense on the surface. Our own experience with gold deposition (see Table 1) is that high yield correlates with high carbon concentration in the gold film. We speculate that this may be due to the inability of the decomposition products (mostly hydrocarbons) to escape from the surface at high growth rates.

If we believe for the time being that the break up of gas molecules occurs in an adsorbed monolayer, the coupling mechanism is still not clear. Because of the high yields, (Table 1) direct ion/adsorbed-molecule collisions appear to be insufficient as the only mechanism.

Table 2 is meant as an aid in examining the possibilities. Secondary electrons are thought to play a role in _electron_ beam induced deposition[15], and may need to be considered also here. However, the results in Table 2 indicate that secondary electron emission is likely to be very low, and it does not increase with the mass of the ion.

Energetic ions incident on a surface lose energy by coupling to the electrons in a solid and by colliding with the nuclei of a solid. Note that in Table 2 the energy loss to electrons does not correlate well with ion induced deposition yield, while the energy loss to nuclei has a stronger dependence on ion mass than the yield. We note that (and this may have no physical meaning) the total energy loss rate (electrons + nuclei) taken to the power 1/2 correlates reasonably well with the yield.

Imagine a monolayer of adsorbed gas molecules of thickness, say, 5Å. The 50 keV He ion loses energy (in tungsten) at an initial rate of 120eV/5Å. Thus there should be enough energy available to produce the decomposition of the gas adsorbed on the surface. The heavier ions lose energy at an even greater rate. (Table 2). However, the detailed mechanism by which this occurs is hard to visualize. Direct ion/molecule collision appears unlikely since it could not account for the high yields. However, the ion as it enters the solid excites a small area which could even be visualized as molten.[16] This has also been described by the term "thermal spike."[17] [18] [19] If we assume (somewhat arbitrarily) that the diameter of the spike is equal to 100Å while its depth is equal to the range of the ion in tungsten then the deposition of 50 keV of energy in this cylinder would cause its temperature to rise to 360° C for He ions, 2,500° C for Ar ions, and 5000° C for Xe ions. This thermal spike can be estimated to decay in 10^{-11} sec. The use of macroscopic concepts like temperature, specific heat, and thermal conductivity in this extreme situation is highly suspect. However, we seek no more than an order of magnitude, qualitative picture. The decomposition of the adsorbed gas molecules might thus be imagined as caused by the highly excited area of the surface which surrounds the point of entry of the ion.

This thermal spike hypothesis of ion induced deposition may also play a role in explaining the columnar or granular growth we have observed. See Fig. 3. A surface protrusion of lateral dimensions on the order of a few hundred angstroms or less which is hit by an ion would provide a larger excited area than a flat surface. In addition, the protrusion would stay excited (hot) longer. Thus these structures would grow faster than (and perhaps at the expense of) the flat surface. Note that an isolated gold sphere of 10 nm radius would have its temperature raised by 1000° C due to the impact of a 50 keV ion. TEM examination of the earlier stages of growth should permit clarification of this thermal spike picture. In addition, the growth over prepatterned structures, say gratings, is expected to proceed faster on the corners and vertical edges than on a flat surface, and may shed light on the mechanism.

3. SUMMARY AND POTENTIAL IMPACT IN IN SITU PROCESSING

Ion induced deposition by dissociation of ambient gas molecules appears to take place with any number of ion species and with either broad beams or focused beams. With focused beams, depositions occur only where the beam is incident and have resolution equal to the beam diameter. Conductors have been deposited but due to granularity and impurities the resistivity is often more like that of polysilicon i.e. about 2 orders of magnitude higher than pure metals. The mechanism of deposition is not well understood. Some alternatives have been considered including thermal spikes.

A number of near term applications can be identified: x-ray-lithography-mask repair, circuit repair, or circuit prototyping.

If we allow imagination some freedom focused ion beam induced deposition promises exciting capabilities for in-situ processing:
  a) deposition with 30 nm resolution (minimum beam diameter reported but not yet applied to deposition),
  b) high aspect ratio and vertical step coverage,
  c) deposition of conductors, insulators, and semiconductors (perhaps single crystal)
  d) compatibility with MBE, i.e. patterned contacts and conductors
Some of the key issues that need to be addressed to bring these possibilities closer to reality are:
  - understanding of the process,
  - control of impurities (presumably UHV deposition is the first step)
  - increased growth rate
  - control of columnar growth
  - availability of low energy ions

4. ACKNOWLEDGEMENTS

This work was supported by Draper Laboratory, Hitachi and by DARPA/ONR. (Contract No. N00014-84-K-0073.)

REFERENCES

1. E. Miyauchi and H. Hashimoto, J. Vac. Sci. Technol. **A4**, 933 (1986), E. Miyauchi and H. Hashimoto, Nuclear Instr. and Methods in Physics Research **B21**, 104 (1987), see also I. Hayashi, this volume.
2. For a recent review of focused ion beam applications and references, see J. Melngailis, J. Vac. Sci Technology **B5**, 469 (1987)
3. K. Gamo, N. Takakura, N. Samoto, R. Shimizu, and S. Namba Japan. J. Appl. Phys. **23** L293 (1984)
4. K. Gamo, N. Takakura, D. Takehara, and S. Namba Extended Abstracts, 16th International Conference on Solid State Devices and Materials (Kobe Japan 1984) p. 31
5. K. Gamo and S. Namba in Proceedings of Symp. on Reduced Temperature Processing for VLSI.(Electrochem. Svc. Pennington, NJ 1986) Vol. 86-5.
6. G. M. Shedd, A. D. Dubner, H. Lezec, and J. Melngailis, Appl. Phys. Lett. **49**, 1584 (1986)

7   K. Gamo, D. Takehara, Y. Hamamura, M. Tomita and S. Namba, Microelectronic Engineering 5, 163 (1986)

8. C. M. Horwitz Rev. Sci. Instruments 50, 652 (1979)

9. J. R. A. Cleaver, H. Ahmed, P. J. Heard, P. D. Prewett, G. J. Dunn, H. Kaufman, Microelectronic Engineering 3, 253 (1985)

10. N. Economou, D. Shaver, B. Ward, SPIE March 1987 Santa Clara

11. M. Yamamoto, M. Sato, H. Kyogoku, K. Aita, Y. Nakagawa, A. Yasaka, R. Takasawa, O. Hattori, SPIE Vol. 632 (1986)

12. G.M. Shedd Focused Ion Beam Assisted Deposition of Gold, S. M. Thesis Massachusetts Institute of Technology (1986) unpublished

13. H. Lezec (unpublished result)

14. J.S. Ro et al. in preparation

15. R.R. Kunz and T.M. Mayer, Appl. Phys. Lett. 50, 962 (1987).

16. N. Winograd, Prog. in Solid State Chem. 13, 285 (1982)

17. F. Seitz and J.S. Koehler, in Solid State Physics Vol. 2 (Academic Press 1956) p. 351, F. Seitz and D. Turnbull Ed.

18. W.L. Brown in Beam Solid Interactions and Phase Transformations, edited by H. Kurz, G.L. Olson, and J.M. Poate (North-Holland, Amsterdam, 1986), Vol. 51, p. 53.

19. R.P. Webb and D.E. Harrison, Jr., Vacuum 34, 847 (1984).

LASER-INDUCED METAL DEPOSITION FOR CLEAR DEFECT REPAIR WORK ON X-RAY MASKS

R. Putzar, H.-C. Petzold, U. Weigmann
Fraunhofer-Institut für Mikrostrukturtechnik
D-1000 Berlin 33, Dillenburger Strabe 53

1. INTRODUCTION

X-ray lithography is a one-to-one shadow projection of an absorber structure on an X-ray transparent membrane onto a resist coated wafer. The typical arrangement is shown in Fig. 1. The X-ray radiation is provided by BESSY, the Berlin Electron Storage ring for SYnchrotron radiation. Synchrotron radiation is nearly parallel and is emitted by a well-defined electron beam. A proximity gap between mask and wafer protects the mask against mechanical damage.

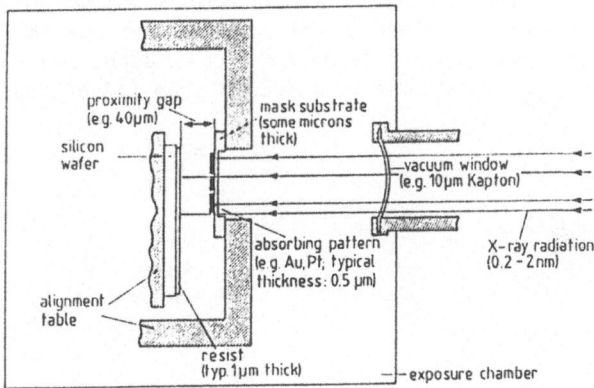

Fig. 1: Schematic exposure arrangement for X-ray lithography using synchrotron radiation

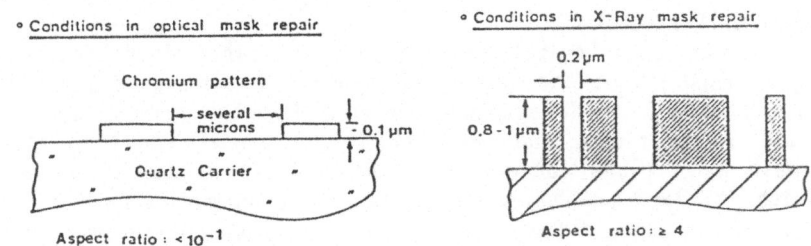

Fig. 2: Specific demands in X-ray lithography

D. J. Ehrlich and V. T. Nguyen (eds.), Emerging Technologies for In Situ Processing, 163–169.
© 1988 by Martinus Nijhoff Publishers.

An important difference between optical and X-ray lithography (see Fig. 2) is that there are no materials available for the wavelength range of soft X-rays (0.5 nm < λ < 2nm) which would be fully transparent even at large thicknesses (such as glass in the optical case) or which would fully absorb the radiation in very thin layers (such as chrome in the optical case). In optical lithography, the aspect ratio is below 0.1, whereas in X-ray lithography it can be greater than 4.

In order to obtain a sufficiently transparent mask substrate, a low-Z element with low absorption must be selected, for example silicon (Z=14). The thickness of the membrane is about 2 μm. The absorber material consists of high-Z materials like gold (Z=79) or tungsten (Z=74).

Masks can have two types of defects: opaque defects, i.e. excess absorber material and clear defects, i.e. missing absorber material. Opaque defects are repaired by a focussed ion beam. The excess absorber atoms are sputtered away by collision with the ions. The repair of sub-μm clear defects (see Fig. 3) is done by means of laser deposition of heavy metals (such as gold or tungsten) onto the defect area. Since the spatial resolution of this deposition process is insufficient, excess deposited material has to be removed by a focussed ion beam in a second step (as an opaque defect).

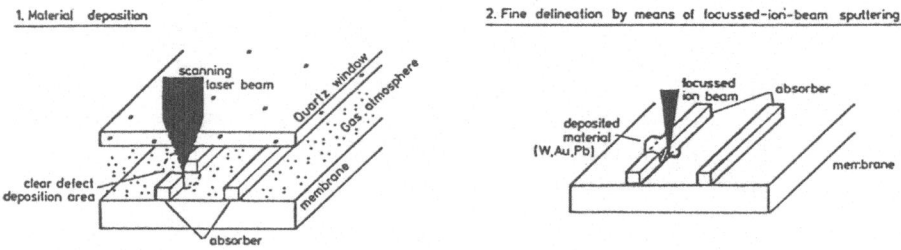

Fig. 3: Repair of sub-μm clear defects by means of laser deposition

2. EXPERIMENTAL ARRANGEMENT

Fig. 4 shows our experimental arrangement for first investigations on laser induced deposition. The Ar+ laser provides wavelengths from 360 nm to 514 nm and of 257 nm (frequency-doubled). For the production of patterns in the 10 μm range, the laser beam is expanded and then focussed onto the mask by a plano convex lens. The mask is placed in a reaction chamber which is fixed on an x-y-stage. The stepping motors have a resolution of 0.2 μm per step. A HeNe-laser beam is aligned collinearly into the optical path of the Ar+ laser by a beam splitter. This allows a visual observation of the Ar+ laser focus position relative to the defect area on the mask surface. The process can be observed through a microscope which looks through a filter and the laser mirror upon the mask surface.

The diameter of the focal spot is about 10 µm, which is sufficient for basic investigations on laser deposition. For higher spatial resolution, there will be an upgraded version in the future.

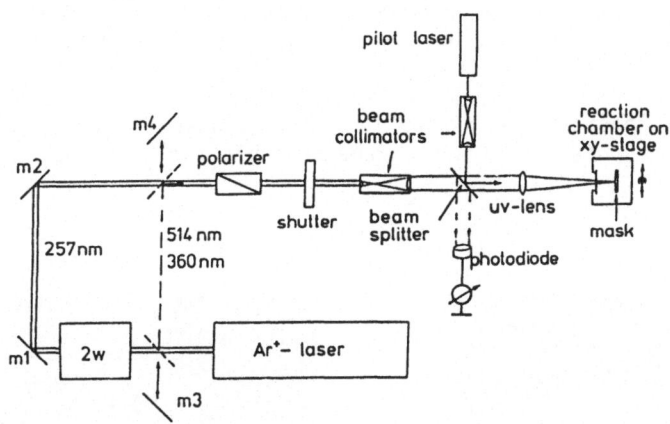

Fig. 4: Experimental arrangement for laser-induced deposition

3. PYROLYTIC AND PHOTOLYTIC DEPOSITION
Laser-induced deposition can mainly be achieved by two processes: the pyrolytic and photolytic dissociation of reactant molecules.

3.1 Pyrolytic deposition
The pyrolytic deposition is based on local substrate heating by means of laser light, which is not absorbed by the gas molecules. Due to collisions of the precursor gas molecules with the hot substrate surface, the molecules are thermally dissociated.

The molecules are typically an organometallic compound with a central metal atom and several organic ligands. The thermal dissociation can occur in one or several steps. The isolated metal atom will then deposit on the substrate surface. Temperatures in the focal spot can exceed 1000°C and the typical laser intensity is about 10^4 W/cm² or more.

For our first investigations the substrate was a 2 µm silicon membrane. The reaction gas was $W(CO)_6$, which was used at its vapor pressure of several 10^{-2} mbar. The laser wavelength was 514 nm. Fig. 5a shows a tungsten spot on a silicon membrane deposited by laser pyrolysis and Fig. 5b its image transferred into resist by X-ray lithography. There are several observations to be discussed: Firstly, the minimum diameter of the deposited spots was about 150 µm. The reason for this high value in comparison with the laser focus spot diameter of 10 µm is the high thermal conduction within the thin silicon membrane. The heat flux out of the focus spot led to deposition of material outside this region. Besides, irreversible distortions of the mask membrane were observed. Secondly, when the laser power was too low, there did not occur any deposition of material on the mask surface. Thirdly, when the laser power was above the destruction threshold of the silicon membrane , that is the peak temperature was too high, the mask was destroyed and a hole arose in the centre of the deposited spot (see Fig. 6a), which could be transferred into resist, too (Fig. 6b).

Because of the considerable distortions and because of the large diameter of the deposited spots, pyrolysis is unacceptable for X-ray mask repair.

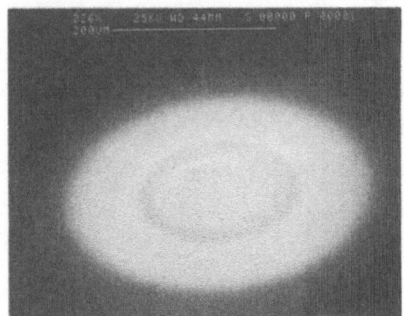

Fig. 5a: W-spot on Si membrane
deposited by laser pyrolysis

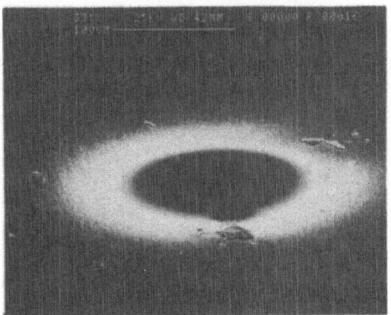

Fig. 5b: W-spot transferred from
Fig. 5a into resist by
X-ray lithography

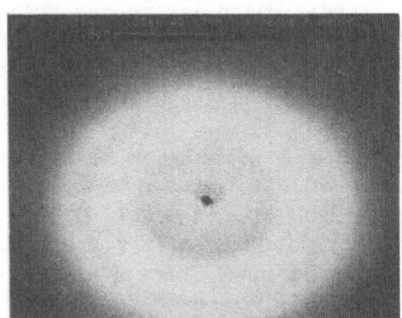

Fig. 6a: W-spot on Si membrane
deposited by laser pyrolisis
(peak temperature above destruction threshold of Si
membrane)

Fig. 6b: W-spot transferred from
Fig. 6a into resist by
X-ray lithography

3.2 Photolytic Deposition

A more promising way is laser induced photolytic deposition. In this case the laser radiation is absorbed by the gas phase molecules and breaks chemical bonds directly, that is, non-thermally. The dissociation may occur in one or several steps. After the dissociation, the isolated metal atoms will deposit on the mask surface.

Principally, the same chemical compounds are suitable for photolysis as well as for pyrolysis. In the case of photolysis, however, the temperature rise in the focal spot is negligible and typical laser intensities amount to about 10^3 W/cm^2. Since the photon energy should exceed the binding energy of the reactant molecules, laser wavelengths below 300 nm , i.e. ultraviolet light, are required.

 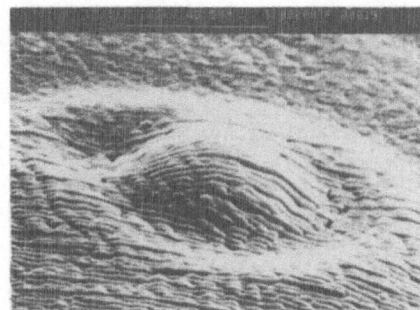

Fig. 7a   Fig. 7b

Fig. 7a: W-spot on Si membrane deposited by laser photolysis (Typically: λ=257 nm, P=1-7mW, t_{dep}= 2-6min, $p(W(CO)_6)$<2.5 Pa)

Fig. 7b: Deposition pattern center from Fig. 7a

For our investigations we used a silicon membrane with a thin plating base ($\approx$ 50 nm). The laser wavelength was 257 nm and the reaction gas was again $W(CO)_6$. Fig. 7a shows a W spot deposited from $W(CO)_6$ for typical values of the experimental parameters. The ring structure, which can be seen in Fig. 9a, derives from the diffraction effect of the aperture. When the deposition time was increased, the thickness and the lateral dimension increased, too. The maximum thickness of the deposit was about 1 μm.

Fig. 7b shows a detail of the deposition pattern center which is shown in Fig. 7a. The central diffraction spot has a typical diameter of 4 to 5 μm.

Although the thickness of the deposited spots of about 1 μm should result in a sufficient X-ray contrast, it was not possible to transfer the structure into resist by X-ray lithography. An Auger analysis showed that the photolytic tungsten deposit contained a smaller amount of oxygen which, however, cannot explain the low X-ray absorption. The carbon content was negligible. Instead, the low X-ray absorption seems to be due to the bad film morphology as can be seen in Fig. 7b.

We have performed the same experiments with WF_6, where H_2 was added as a buffer gas. Here, however, only very thin layers could be deposited, which led to insufficient X-ray absorption.

4. CONCLUSIONS
4.1 Pyrolytic Deposition
By means of laser induced pyrolytic deposition it is possible to deposit tungsten spots from a $W(CO)_6$-atmosphere onto a silicon membrane, which are sufficiently opaque for X-ray lithography. The minimum diameter of the deposited spots was about 150 μm due to the high lateral heat conduction within the thin silicon membrane. Therefore and because of the considerable distortions of the mask membrane, pyrolysis is unacceptable for X-ray mask repair.

4.2 Photolytic Deposition
By means of laser induced photolytic deposition it is possible to deposit tungsten spots with the same lateral dimension as predicted by the diffraction limit. Although the thickness of the deposited spots of about 1 μm should be sufficient for X-ray absorption, the deposited spots could not be transferred into resist by X-ray lithography. The deposited material contains a certain amount of oxygen, which cannot explain the low X-ray opacity. Nevertheless, photolysis is more suitable for X-ray mask repair than pyrolysis, because there are no distortions due to heating of the thin membrane.

5. FURTHER INVESTIGATIONS
Further investigations are planned to test several other reaction gases and the addition of buffer gases. Another major problem will be to reduce the diameter of the deposited spots. The questions to be answered are as follows:
- Which systems (gas/wavelength) are the most promising ones for photolytic deposition with our available wavelengths (257 nm, 360 nm, 514 nm)?
  The most important criterion is a sufficient X-ray absorption of the deposited spot. For a high spatial resolution and avoidance of irreversible membrane distortions, the temperature in the focal spot should not significantly exceed the room temperature.
- Which systems will provide the highest deposition rate at this temperature?
- Is there possibly a combination of pyrolytic and photolytic processes necessary for an optimum deposition process?

REFERENCES:

1. A.E. Adams, M.L. Lloyd, S.L. Morgan, N.G. Davis, in Laser Processing and Diagnostics, ed. D. Bäuerle, p. 269 (July 1984).
2. D. Bäuerle, in Surface Studies with Lasers, eds. F.R. Aussenegg, A. Leitner, M.E. Lippitsch, p. 178 (March 1983).
3. T.F. Deutsch, D.D. Rathman, Appl. Phys. Lett. 45 (6), p. 623 (1984).
4. D.J. Ehrlich, Solid State Technology, p. 81 (December 1985).
5. A. Heuberger, Solid State Technology, p. 93 (February 1986).
6. R. Solanki, P.K. Boyer, G.J. Collins, Appl. Phys. Lett. 41 (11), p. 1048 (1982).
7. R. Solanki, P.K. Boyer, J.E. Mahan, G.J. Collins, Appl. Phys. Lett. 38 (7), p. 572 (1981).

CONFIRMATION OF THE WAVELENGTH DEPENDENCE OF SILICON OXIDATION INDUCED BY VISIBLE RADIATION

IAN W BOYD & F. MICHELI

Electronic & Electrical Engineering, University College London,
Torrington Place, London WC1E 7JE, UK

ABSTRACT

We use a novel technique to amplify small increases in the oxidation rate of laser irradiated silicon to successfully isolate a photonic contribution to the reaction induced by cw argon laser radiation. This method presents clear evidence of optical enhancement over the usual thermally induced growth rate.

1. INTRODUCTION

For in situ processing, laser and arc lamp radiation sources are increasingly being used to induce a wide range of film growth and pattern definition processes in microelectronic materials [1]. The application of visible and ultraviolet radiation towards oxidizing single crystal silicon (c-Si) in dry oxygen (O_2) has in particular received much attention [2]. Indeed, in recent papers, a small non-thermal component to the reaction has been invoked to explain observed enhancements in the growth rate [3-8]. It is important to realise that so-called "photonic enhancements" have been seen to be present for a wide range of irradiation conditions for incident wavelengths from the ultraviolet through the visible to the infrared.

Figure 1 shows a hypothetical spectrum of radiation induced reactions for oxidation of silicon, where the quantum energy of the radiation is greater than the silicon bandgap [8]. In addition to this relatively high photon energy regime, the application of infrared radiation around 9.6 μm can also promote a photothermal enhancement through strong absorption by the Si-O bond stretching vibrational mode. A problem in isolating the so-called photonic contribution, however, in all these regimes has been the domination of conventional thermal reactions also induced by the incident radiation. Here, by concentrating in the visible regime, we describe a novel method of amplifying small increases in the oxide growth rate in such a way that reaction enhancements may be observed.

D. J. Ehrlich and V. T. Nguyen (eds.), Emerging Technologies for In Situ Processing, 171–178.
© *1988 by Martinus Nijhoff Publishers.*

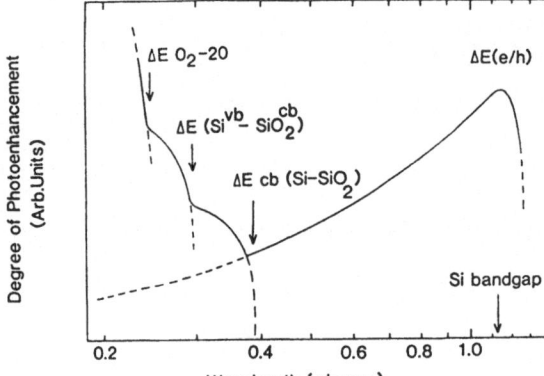

FIGURE 1. Summary of the various wavelength dependent mechanisms for photonic-induced oxidation of silicon. By assuming that each mechanism proposed in the literature is valid an estimate of the relative contribution of each is shown here for a constant beam flux [8].

2. EXPERIMENTAL

Radiation at 488 or 514 nm from a Coherent Innova 100-20 argon ion laser was directed a few degrees off normal incidence on to 400μm thick microelectronic grade c-Si of about 9mm² in area. The samples were pre-cleaned using standard production-line techniques, and held by a quartz support in thermal isolation in a stainless steel chamber containing a closed atmosphere of pure dry O_2 (BOC Electra II grade). The incident beam power was adjusted so that a constant power of 4W was absorbed at each wavelength. In this way, the minor wavelength dependence of the reflectivity did not affect the induced temperature, which was determined by an optical pyrometer to be approximately 800C. The nature of our experimental technique relaxed the requirement for the precise processing temperature to be known, as will be discussed later.

Films of different thickness were grown by varying the irradiation times. After exposure, a three dimensional contour mapping of the oxide grown was obtained by translating the sample on a computer-controlled x-y stage under an automatic ellipsometer with a sampling beam size of 26 by 50 μm. The peak thicknesses grown varied from 100 to 850A, and as is usually always necessary in ellipsometry, the refractive index of films thinner than ~450A was assumed to be constant at n = 1.46.

FIGURE 2. A three dimensional profile of the central 2mm square of a typical argon laser grown oxide on c-Si. The peak thickness is 810A, while the average thickness at the edge is approximately 450A.

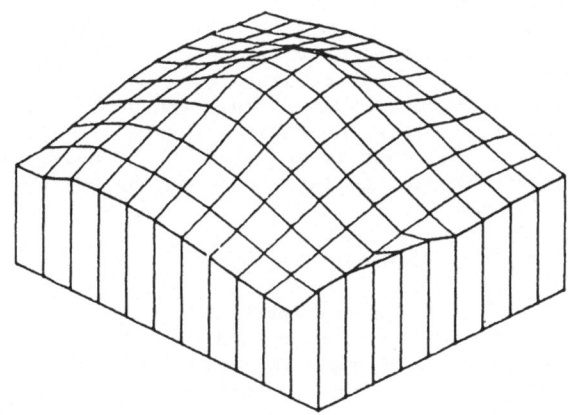

3. RESULTS

Figure 2 shows a typical mound of oxide grown in these experiments. The central peak is indicative of a non-uniform reaction across the irradiated sample, but this in itself cannot be taken as evidence of any non-thermal mechanisms, since there is no knowledge of the precise thermal profile across the c-Si. Indeed, this is a continual problem in laser processing, which, despite much theoretical and experimental effort, remains to be satisfactorily resolved. Consequently one must find qualitative grounds for isolating photonic contributions to the reaction.

The normal incidence reflectivity near a wavelength of 500nm of c-Si covered by a thin optically transparent film of thickness x can be calculated using the standard thin-film interference formula

$$R(x) = \frac{n_r^2(1-n_s)^2\cos^2\phi + (n_s-n_r^2)^2\sin^2\phi}{n_r^2(1+n_s)^2\cos^2\phi + (n_s+n_r^2)^2\sin^2\phi} \tag{1}$$

where n_r and n_s are the film and substrate refractive indices, respectively, and $\phi = 2\pi n_r x/\lambda$. Using the known optical properties of c-Si and its oxide, SiO_2, [9] it is found that R falls from 0.38 (when only the native oxide ($\approx$20A) is present) to a minimum around 0.10 after some 850A of oxide has grown [10].

Although the precise mechanisms of c-Si oxidation have yet to be satisfactorily formulated, the temperature dependence of the rate constants appropriate to the different reaction regimes observed [11] suggests that they can be fitted by simple activation-energy Arrhenius expressions of the form

$$C \exp [-E_a/kT] \qquad (2)$$

where E_a is the activation energy, k is Boltzmann's constant, and T is the induced temperature. Hence, as the oxide grows it will tend to reduce R with time, and permit more energy to enter the sample. Since the c-Si is thermally isolated, the major source of its heat loss is by "black-body" emission, and the temperature induced by the cw laser beam may be approximated, assuming $T^4-T_o^4 \approx T^4$, by

$$T \approx [P(1-R)/A\varepsilon\sigma]^{\frac{1}{4}} \qquad (3)$$

where T_o is the ambient temperature, P is the incident power, ε is the c-Si emissivity and σ is Stefan's constant. Therefore a decrease in R leads to an increase in T. Although it is known that R α T for c-Si, this is only a weak dependence up to the melting point of the material, and has therefore been neglected in our calculations. Thus, with time, the effective value of R will decrease as the oxide grows, and samples oxidized by visible radiation in this regime will undergo a temperature rise as the reaction proceeds, thereby increasing the reaction rate, which in turn further accelerates the oxide growth. Under such non-uniform temperature conditions, it is extremely difficult to extract thermal and photonic influences to the process. However, when comparing two almost identical irradiation conditions (e.g. exposure to 488 and 514 nm wavelengths) small differences due to non-thermal effects in the growth rate will be accelerated as this positive feedback amplifies any existing change in growth rate for films up to approximately 850A.

Previous work in photon-induced oxidation of c-Si [2-8] found evidence for non-thermal contributions to the reaction. In the work at Stanford [7] and at Princeton [5], a laser beam impinged upon c-Si already thermally oxidizing in a furnace. Together with further thermal contributions to the reaction from the temperature rise induced by the laser beam, these conditions were not conducive to highlighting any non-thermal effects. The experimental conditions reported here use no background heating; all the energy for the reaction is provided by the laser beam. Therefore, since thermal contributions are minimized, a higher ratio of photonic/thermal kinetics is achieved, and the photonic effect more efficiently isolated.

By providing beams of 488 and 514 nm radiation at identical absorbed power levels, thermal contributions to the reaction could be assumed to be virtually equal for each case. In fact, α, the absorption coefficient for these two wavelengths is 9070 and 6280 cm^{-1} at T_o and of the order of 10^5 and 7.5×10^4 cm^{-1} at 800C [8]. Therefore, it may be argued that if any difference was to be expected from solely thermal considerations, the shorter wavelength radiation, being absorbed more strongly, would induce higher temperatures and thus faster oxidation rates.

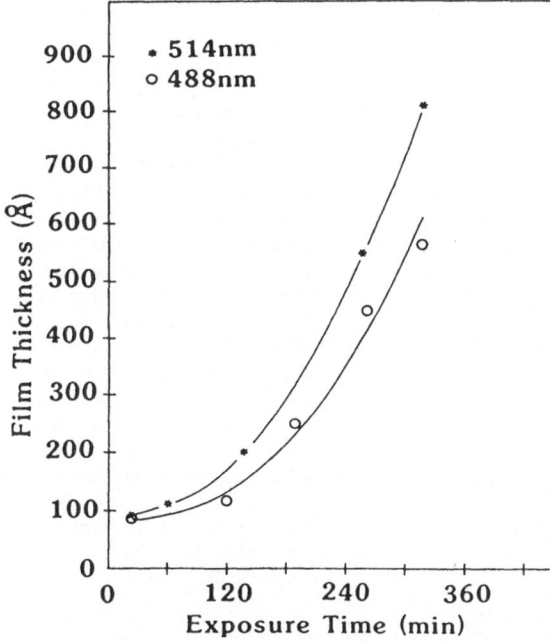

FIGURE 3. The oxide thickness grown for various irradiation times for 488nm and 514nm radiation at initially identical levels of absorbed power. The RMS error was ±5% for oxides up to 250A thick and ±10% up to 800A. During the 320 min exposure, almost 50% more oxide has been grown by the 514nm radiation. Curves have been added only for the guidance of the eye.

Figure 3 shows the oxide thicknesses grown as a function of exposure time by the two wavelengths for identical beam powers. Whereas for short irradiation times the film thicknesses produced were quite similar, at longer exposures there is a clear increase in film thicknesses grown by the longer wavelength. The 514nm radiation now evidently induces an observable increase in the reaction rate over that initiated by the 488nm photons.

4. DISCUSSION

It is well known that neither O_2, c-Si, nor SiO_2, preferentially absorb 514 nm to 488 nm radiation. However, for equal power levels, a beam of longer wavelength radiation will contain a larger photon flux, n, than a shorter wavelength beam, i.e.

$$n(514nm)/n(488nm) = 1.05 \tag{4}$$

Therefore, 514 nm radiation will create 5% more electron-hole pairs per unit area on the c-Si surface than 488 nm photons at the same power level, assuming a generation efficiency of 1. Although there are more than 8 independent models for the c-Si + O_2 reaction [2] several have highlighted the importance of

(a) Available silicon bonds for the reaction with oxygen,
(b) Conduction-band electrons in the c-Si which may be further excited into the conduction band of the silicon, and
(c) The formation of the O^- species.

Each of these criteria may be altered by changes in the population of conduction band electrons in the c-Si.

176

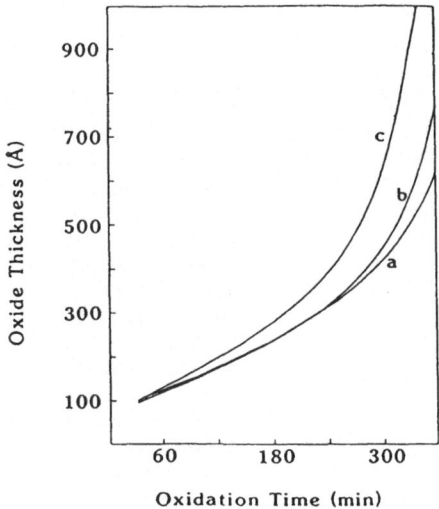

FIGURE 4. Modelling estimations of the oxidation rate of c-Si irradiated by cw Ar laser radiation, assuming only thermal contributions to the reaction. Curve (a) represents the 514nm irradiation process, while curve (b) shows the predicted 488nm case. Curve (c) shows the consequence of increasing the rate of oxidation in curve (a) by 20%, in order to reflect the trend of the data shown in figure 3.

Noting the form of the expected oxidation reaction, generally expressed by various Arrhenius-type expressions as shown in Eq (2), and realising that the photon induced reaction is clearly thermally dominated, it is possible to estimate, for illustrative purposes only, the thermal equivalence to the oxidation rate differential between the two wavelengths. We use the expression recently reported by Massoud et al [11] for thermal oxidation, to model the reaction induced by the 488 and 514nm radiation. The growth rate is given by

$$dx/dt = B(T)/[2x + A(T)]$$ (5)

where A and B account for the physical and chemical processes in the reaction. The modelling assumed only thermal influences to the reaction, and included the reflectivity and temperature changes shown in the equations above. Identical values of C and E_a were used for both wavelengths in order to simplify the calculations to induce equal growth rates in both cases at every given temperature. In other words, the initial differential in growth rate remained constant throughout the calculation.

Figure 4 (curves a & b) shows the results of the modelling. The form of the oxidation reaction is clearly similar to the results of figure 3, but we will only comment here on the qualitative behaviour. The most striking observation is that the 488nm radiation induces a faster reaction rate than the 514nm radiation, as would be expected by thermal considerations. This is opposite to the results of figure 3. In fact, in order to achieve the same trend as the data, the growth rate induced by the 514nm wavelength giving rise to curve (a) has to be increased by approximately 20% to give curve (c).

5. SUMMARY

In conclusion, we have confirmed that a photonic effect exists in the reaction of O_2 with c-Si induced by <u>visible</u> radiation, by minimizing the dominant thermal contributions to the reaction and amplifying the predictably small differences arising from non-thermal contributions. This effect appears to be related to the density of the incident photon flux rather than the absolute power level of the beam, in agreement with several recently proposed oxidation models for c-Si [2,7]. The differential growth induced by the 514 nm radiation over the 488 nm photons is estimated to be equivalent to an increase of ~ 20% in the thermal reaction rate in the latter.

It is interesting to comment at this stage on the various studies currently being carried out using rapid thermal processing (RTP) techniques [12-15]. These technologies usually involve the use of large banks of incoherent lamp sources, or specially constructed single high intensity lamps. The basic mode of operation involves some form of feedback mechanism to monitor and keep constant the temperature, most usually by means of optical pyrometry. As a consequence of the reflectivity decrease described above, it would be expected that in order to keep the processing temperature constant, the photon supply rate would have to be gradually reduced, at least during the initial stages of the reaction. Where photonic mechanisms were studied, therefore, care must be taken to account for this effect.

ACKNOWLEDGEMENTS

We acknowledge the invaluable contribution of Chris Brown of GTM Ltd in providing many of the oxide thickness measurements, Dr R. Thompson, of Hughes Microelectronics Ltd, for providing the silicon samples, and Dr J.I.B. Wilson, at Heriot-Watt University for confirming several ellipsometric data points. Part of this work was funded by SERC, and by the Nuffield Foundation.

REFERENCES

1. F. Micheli, I W Boyd, Optics & Laser Technology, 18, 313 (1986); 19, 19 (1987); 19, 75 (1987).

2. I.W. Boyd, in "Dielectric Layers in Semiconductors", ed. G.G. Bentini, E. Fogarassy, A. Golanski, Les Editions de Physique, Les Ulis (1986), and references therein.

3. R. Oren, S.K. Ghandi, Appl. Phys. Lett., 42, 752 (1971).

4. I.W. Boyd, J.I.B. Wilson, J. West, Thin Solid Films, 83, L173 (1981).

5. S.A. Schafer, S.A. Lyon, J Vac Sci & Technol, 21, 423 (1982).

6. I.W. Boyd, Appl. Phys. Lett., 42, 728 (1983).

7. E.M. Young, W.A. Tiller, Appl. Phys. Lett., 50, 46 (1987) and references therein.

8. I.W. Boyd, in "Interfaces Under Irradiation", eds., L. Laude, D. Bauerle, (Nijhoff, 1987)

9. G.E. Jellison, in "Pulsed Laser Processing of Semiconductors" Vol 23 of Semiconductors and Semimetals, ed., R.F. Wood, C.W. White, R.T. Young, p95.

10. F. Micheli, I.W. Boyd, Appl. Phys. Lett., (submitted).

11. H.Z. Massoud, J.D. Plummer, E.A. Irene, J. Electrochem. Soc., 132, 1745 (1985).

12. J. Nulman, J.P. Krusius, A. Gat, IEEE Electron Device Letters EDL-6, 205 (1985).

13. M.M. Moslehi, K.C. Saraswat, S.C. Shatas, Appl. Phys. Lett., 47, 1113 (1985).

14. A. M. Hodge, C. Pickering, A. J. Pidduck, R. W. Hardeman, in "Rapid Thermal Processing", ed., T.O. Sedgewick, T.E. Seidel, B-Y. Tsaur, MRS, Pittsburg (1986) p313.

15. J.P. Ponpon, J.J.Grob, A. Grob, R. Stuck, J. Appl. Phys., 59, 3921 (1986).

FOCUSED ION BEAM TECHNOLOGY AND APPLICATIONS

A.J. Steckl, J.C. Corelli, J.F. McDonald
H.S. Jin and R. Higuichi-Rusli, Rensselaer Polytechnic Institute
Center for Integrated Electronics, Troy, NY, 12181 USA

I. Introduction

Energetic particle beams are now widely used in the fabrication of semiconductor devices and integrated circuits [1]. In this broad arena of novel processes for semiconductor devices, the development of focused ion beam technology has been a major new development. By being able to produce a sub-micron ion beam a host of new devices and applications can be pursued . As shown in Table 1, these applications range from sub-micron scale analysis to localized ion milling to high resolution lithography to ion-assisted chemical processing and last, but by no means least, to maskless and resistless ion implantation, which for the first time can be varied in the lateral direction across the wafer surface.

In this paper we concentrate on present and potential applications of FIB technology, rather than the details of ion beam optics and system design. For additional information on the latter the recent review article by Melngailis [2] and a book by Brodie and Muray [3] are recommended.

In the next section of this paper, we briefly review the basic operation of focused ion beam systems. In Sec.III, a number of applications of FIB technology are discussed. Sec.IV deals with the liquid metal ion sources required for FIB operation, while Sec.V concentrates on computer control and related applications.

II. Focused Ion Beam Systems - Basic Operation

A focused ion beam system is composed of three major sections:

* Ion Source

* Ion Beam Column

* Sample Stage

An FIB system is shown schematically in Fig. 1. The ion source normally consists of a high brightness liquid metal ion source which generates an ion beam by field emission. To produce certain desired ion species which have high melting points it is necessary to alloy the material in order to reduce the melting temperature and hence the power input. Consequently, the alloy source generates a mixed ion beam composed of all constituent species. To isolate the desired species a mass filter, consisting of crossed electric and magnetic fields, is included in the

179

D. J. Ehrlich and V. T. Nguyen (eds.), Emerging Technologies for In Situ Processing, 179–199.
© 1988 by Martinus Nijhoff Publishers.

TABLE 1

FOCUSED ION BEAM APPLICATIONS

A. MICROANALYSIS

    O Scanning Ion Microscopy
    O Secondary Ion Mass Spectroscopy

B. MICROMILLING

    O Mask Repair
    O Circuit Restructuring

C. MICROLITHOGRAPHY

    O Organic Resists
    O Inorganic Resists

D. ION-ASSISTED CHEMICAL MICROPROCESSING

    O Ion-induced Chemical Reaction Etching
    O Ion-induced Chemical Vapor Deposition

E. MASKLESS/RESISTLESS ION IMPLANTATION

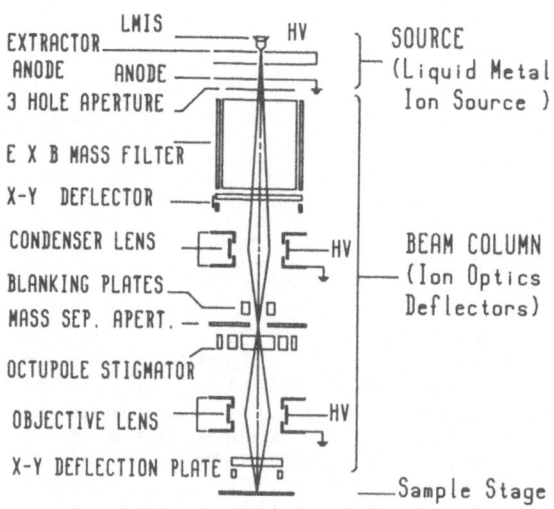

Fig.1. Focused Ion Beam system and components

TABLE 2

RPI FOCUSED ION BEAM SYSTEM (1)

○ Manufacturer　　　　　　　: VG Semicon Ltd
○ Maximum Energy　　　　　　: 100KeV
○ Minimum Beam Diameter　　: 1000Å
○ Range of Current on Target: 10pA - 10nA

○ Minimum Beam Position Step: 0.03um
○ Stage Movement Accuracy　 : ±0.05um

○ Liquid Metal Ion Sources　: Ga, Au/Si, Cu/P

　　　　　　　　　　　　　　　　Pd/B, Pd/B/As

○ UHV System Vacuum　　　　 : 1.0E-10 Torr

○ In-situ Submicron SIMS　　: 1.0E19 cm^{-3} (current)
　　　　　　　　　　　　　　　　1.0E17 cm^{-3} (future)
○ Mass Separation　　　　　　: E X B Filter
○ Computer Control　　　　　 : DEC VAX 11/750
○ Wafer Handling　　　　　　 : 10 Wafer Cassette
　　　　　　　　　　　　　　　　(4 Inch Diameter)

Future Capabilities Built into Design

　SEM for Imaging Alignment
　Laser for Diagnostics
　 & Laser-induced Chemistry

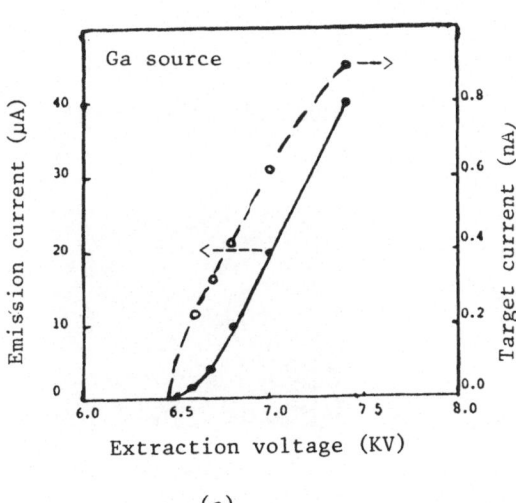

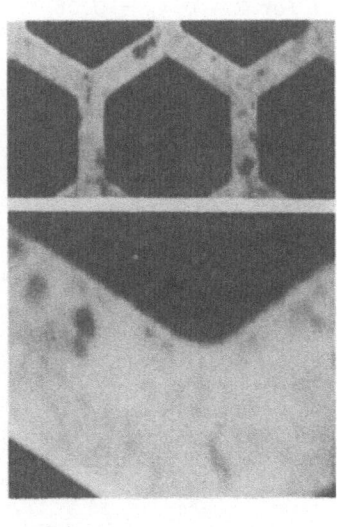

(a)　　　　　　　　　　　　　　　　(b)

Fig.2. (a) Relationship between Ga ion emission and the current
on target. (b) SIM micrograph of a calibration Cu grid

182

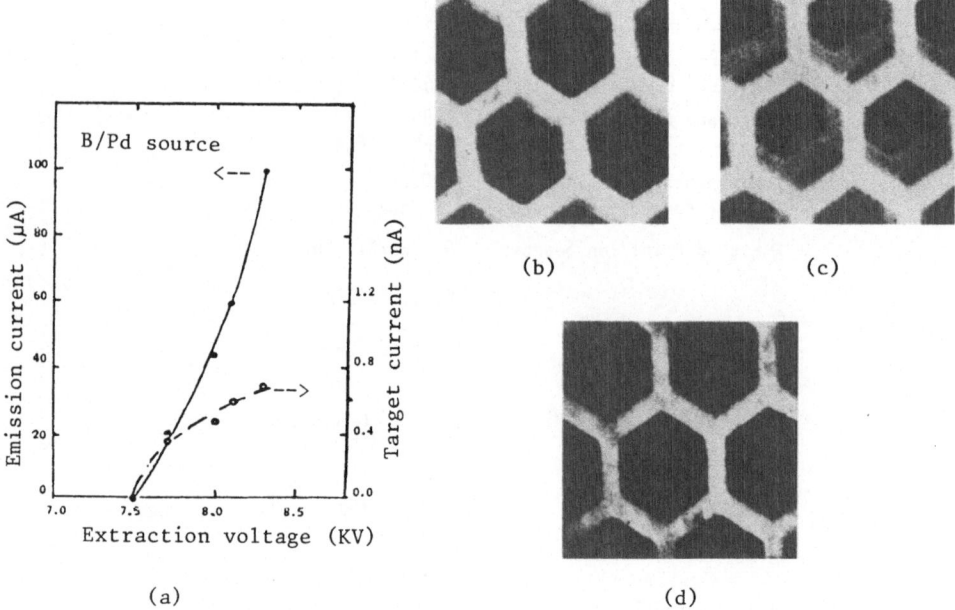

(b)　　　　　　　　(c)

(a)　　　　　　　　　　　　(d)

Fig.3. (a) Relationship between B/Pd ion emission and the current on target. SIM micrograph of a Cu grid (b) for E = 0,B=0; all species are included. Target current = 0.22 nA ; (c) for E = 0, I_B = 0.04 A.B^+ are deflected out. Only Pd^+, Pd^{++} are included. Target current = 0.15nA ; (d) for E = 0,I_B = 0.08 A, Pd^{++} are deflected out. Only Pd^+ are included. Target current = 0.14nA.

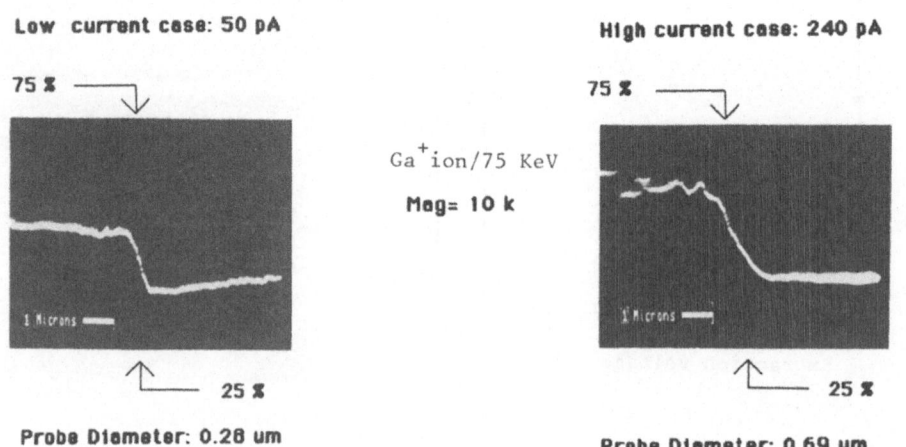

Fig.4. F.I.B. probe diameter : Al/Vacuum interface.

column. Frequently, an alloy source will have more than one usefull component (cf. PdAsB) and the ExB filter will allow for the sequential selection of each species.

The ion beam column has two electrostatic lenses. The condenser lens focuses the beam onto a plate which contains a series of fine apertures used for selection of individual species isotopes. The objective lens then focuses the ion beam onto the plane of the sample surface. The beam is deflected to form various patterns on the sample by providing the appropriate signal voltages to the X and Y deflection plates.

As an example of FIB capability we briefly describe the RPI focused ion beam system purchased from VG Semicon Ltd. of England and delivered in March of 1986. The main characteristics of the system are summarized in Table 2. The maximum accelerating energy for singly - ionized species is 100KeV. The minimum ion beam diameter is nominally 1000Å, with 1500Å being the smallest value observed thus far. The range of ion current incident on the sample is from a low of 10 pA (with the smallest beam defining aperture) to a high of 10 nA (for high total source emission and large aperture).

The RPI FIB sytem has been operated with Ga, Au/Si and B/Pd liquid metal ion sources.Fig.2(a)indicates the relationship between the the total Ga ion emission and the current on the target. Also in Fig.2(b) SIM microphotographs of a calibration Cu grid (7.6um features) are shown. Fig. 3 (a) indicates the relationship between the total emission from the B/Pd source and the corresponding current in the sample plane. The B/Pd source being an alloy source requires the operation of the ExB mass filter to separate out the desired species. An example is shown in Fig. 3 using the calibration Cu grid. Fig. 3 (b) shows the SIM image present with the electric and magnetic fields set to zero. Multiple grid images due to all the various species present in the beam (Pd+,Pd++,B[10]+,B[11]+,etc.). Because of different deflection through system, images due to different species appear slightly shifted. By applying a 40mA current to the magnetic coils, one can filter out the boron ion component of the beam, resulting in a double image, shown in Fig. 3 (c) due to the singly- and doubly-ionized Pd atoms. Finally, a single image of the grid, due to Pd+, is obtained when Pd++ is deflected out by the ExB, as shown in Fig.3 (d).

To characterize the beam profile we have investigated two types of abrupt interfaces as vehicles for profiling in the line scan mode. The first interface was obtained from an aluminum-covered Si wafer which was cleaved to provide a very sharp break in the Al film. Since the secondary electron yield of Al is high, the edge of the film is a region of very high contrast. The result of the scanning with a 75KeV Ga ion beam across this edge is shown in Fig.4. It is interesting to note the significant effect of beam current on beam diameter. The lower beam current of 50pA results in a beam diameter of approximately 1/4um, whereas for a current of 240pA we measured a diameter of close to 0.7um. This effect is due to a combination of factors, including the need to refocus the system as the beam current changes and a more fundamental effect of increased coulombic repulsion with beam current resulting in a larger beam diameter.

Ga⁺ ion/ 75 KeV, 240 pA

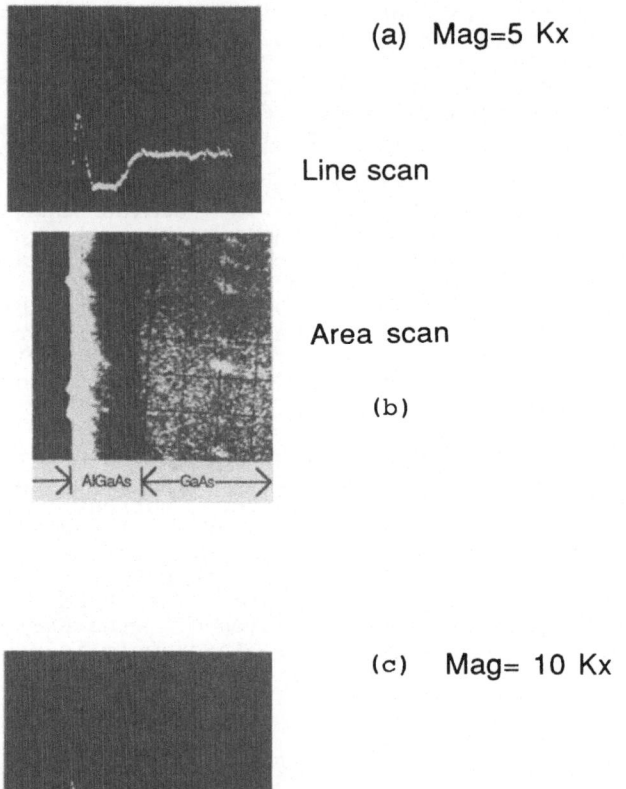

(a) Mag=5 Kx

Line scan

Area scan

(b)

(c)   Mag= 10 Kx

Line scan

Fig.5. F.I.B. probe diameter : AlGaAs/GaAs interface

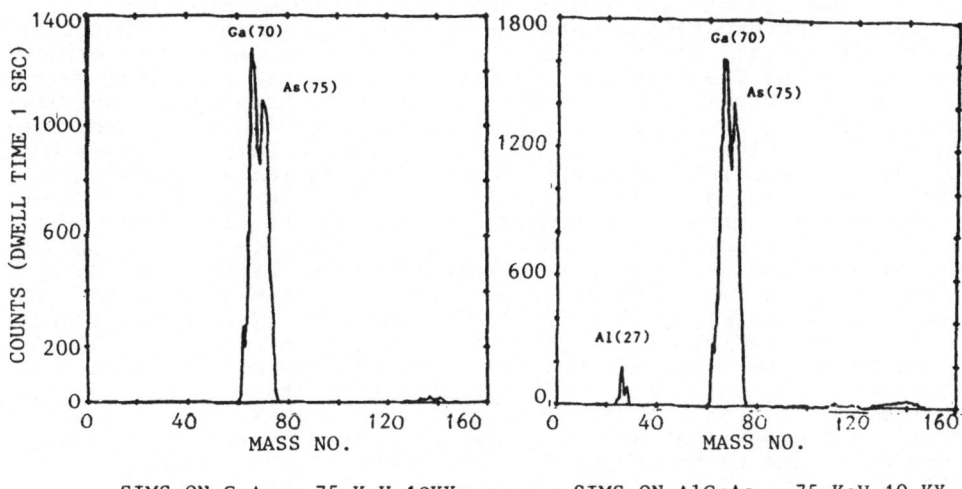

SIMS ON GaAs – 75 KeV 1OKX   SIMS ON AlGaAs – 75 KeV 10 KX

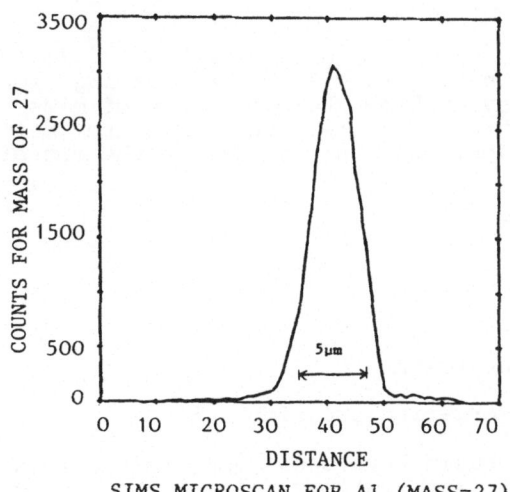

SIMS MICROSCAN FOR Al (MASS=27)

Fig.6. SIMS unit in the FIB was used to analyze
the AlGaAs/GaAs heterojunction.

The second interface employed for beam profiling is an extremely abrupt AlGaAs-GaAs heterojunction grown by molecular beam epitaxy [4]. After growth the samples were cleaved to expose the interface. The thickness of the epilayer had a nominal value of 7um. The preliminary results using this heterojunction for beam profiling are shown in Fig. 5 . The scanning ion microscope image of the cleaved sample, revealing the AlGaAs epilayer and the AlGaAs-GaAs interface, is shown in Fig. 5(a) at a magnification of 5000X. In general, the AlGaAs epilayer which is more conductive than the GaAs appears darker, except for the outer edge which presents an angle more favorable for secondary electron collection and, hence, appears very bright. A Ga ion line scan performed orthogonally to the interface results in the signal shown in Fig. 5a. One can observe a very sharp signal spike corresponding to the bright outer edge of the AlGaAs followed by the signal due to the epilayer proper and the transition to the GaAs. The scan across the heterojunction interface is seen at higher magnification, 10KX, in Fig. 5 (b). From this experiment we estimate the beam diameter in the high current regime to be approximately 0.45um. The SIMS unit in our FIB was used to provide confirmation of the heterojunction components as shown in the upper half of Fig. 6. In the lower half of Fig. 6 an Al SIMS line scan confirms that the AlGaAs layer is approximately 7um thick.

III. FIB Applications

In the domain of particle beam processing, ions are most versatile because they combine the properties of mass, charge and energy [1]. Indeed, broad-diameter ion beams are being used in various systems and applications for most of the microfabrication processes :

* Lithography

* Etching

* Doping

* Deposition/Growth

* Material Formation/Transformation

Focused ion beams have been expected to lend themselves to a wide variety of applications [5]. A list of current and future potential applications is shown in Table 1.

The incident ion beam interacts with the sample resulting in ion implantation, ion backscattering, secondary electron generation and secondary ion emission. These last two processes are illustrated in Fig. 7. The secondary electrons can be detected by various methods, including the scintillator and photomultiplier combination shown in Fig. 7. This is, in essence, the scanning ion microscopy (SIM) method routinely employed for FIB sample imaging. The advantages of SIM as a microscopy technique (when compared to electron microscopes) include a large depth of focus, higher sensitivity to shallow surface topography, lower sensitivity to charging of insulating surfaces. Of course, the SIM technique also has some serious disadvantages, namely a fundamentally lower resolution than SEM

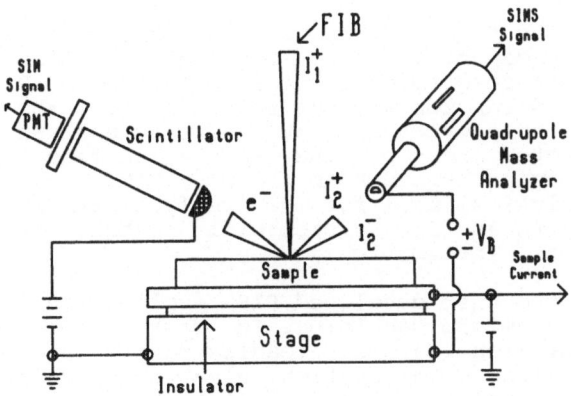

Fig.7. FIB system for Detection and Imaging

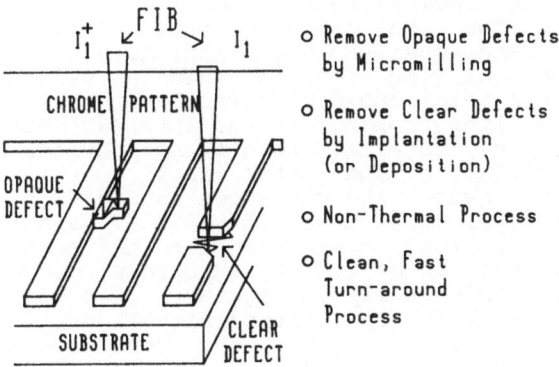

o Remove Opaque Defects
by Micromilling

o Remove Clear Defects
by Implantation
(or Deposition)

o Non-Thermal Process

o Clean, Fast
Turn-around
Process

Fig.8. FIB system for Mask Repair

due to a larger diameter beam and a constant modification of the surface during specimen observation.

Using the secondary ions for detection and imaging brings unique new capabilities of localized, sub-micron analysis. A quadrupole mass spectrometer located in the FIB system can detect the various ionized species sputtered from the substrate. This capability can provide elemental surface analysis and depth profiles of selected areas of the sample, surface elemental or composite ion imaging. Broad beam (~ 1mm) SIMS is, of course, a widely used analytical tool for device fabrication. Micro-focus SIMS performed with an FIB system will probably provide very valuable additional information.

Among the best developed FIB applications to date are those that involve micro-machining. In mask repair, as shown in Fig.8, the focused ion beam can readily remove opaque defects by localized ion milling. In addition, clear defects can be removed by milling in their place a grating which scatters incident light and renders the local region opaque by preventing the transmission of light. Clear defects can also be removed by ion-induced local deposition of carbon from a hydrocarbon gas. In general, mask repair using an FIB is a clean, fast turn-around process with sub-micron resolution (~ 0.2-0.5um) . Most importantly, it is basically a non-thermal process which results in a minimum damage to the surrounding mask area.

The same micro-machining of focused ion beams can be applied to interconnection materials for circuit "restructuring", in other words disconnecting various sections of an integrated circuit. This is accomplished by ion milling an interconnection until it erodes the conducting material into a reliable open circuit. This technique, shown schematically in Fig. 9(a) has been used successfully to ion mill thin as well as thick metal lines. With thin metal and dielectric layers it is important to terminate the milling properly at the interface between the metal and the dielectric. Sometimes this is not obvious due to hillock growth on the surface of the metal lines. For thick films this is not a serious issue. The cut shown in Fig. 9(b) is through a 5 micron thick layer of Al made with a 75 KV Ga beam with a large beam aperture resulting in a large (2.6 nA) beam current into a large (1 micron) beam spot. Progress during the milling operation (which took nearly 10 minutes!) was readily monitored by observing the secondary electron signal during the repeated milling scans. Fig. 9(c) shows the contrast and shape of the secondary electron signal at the beginning of the ion milling operation. The metal line is embedded in polyimide giving a good contrast, and the sharp peaks delimiting the edges of the metal line are a result of the high efficiency of secondary electron generation at the steep vertical walls of the metal. Fig. 9(d) shows this signal towards the end of the milling operation. The contrast has nearly disappeared and the sharp peaks at the ends of the interconnection are absent indicating the erosion of the steep walls. Ion milling of the surrounding polyimide is more efficient than the metal, requiring good alignment of the milling sweep to the ends of the metal line. Beam charging at the polyimide surface is a problem. Some form of charge neutralization for the incoming ion beam (or a leakage path) is required.

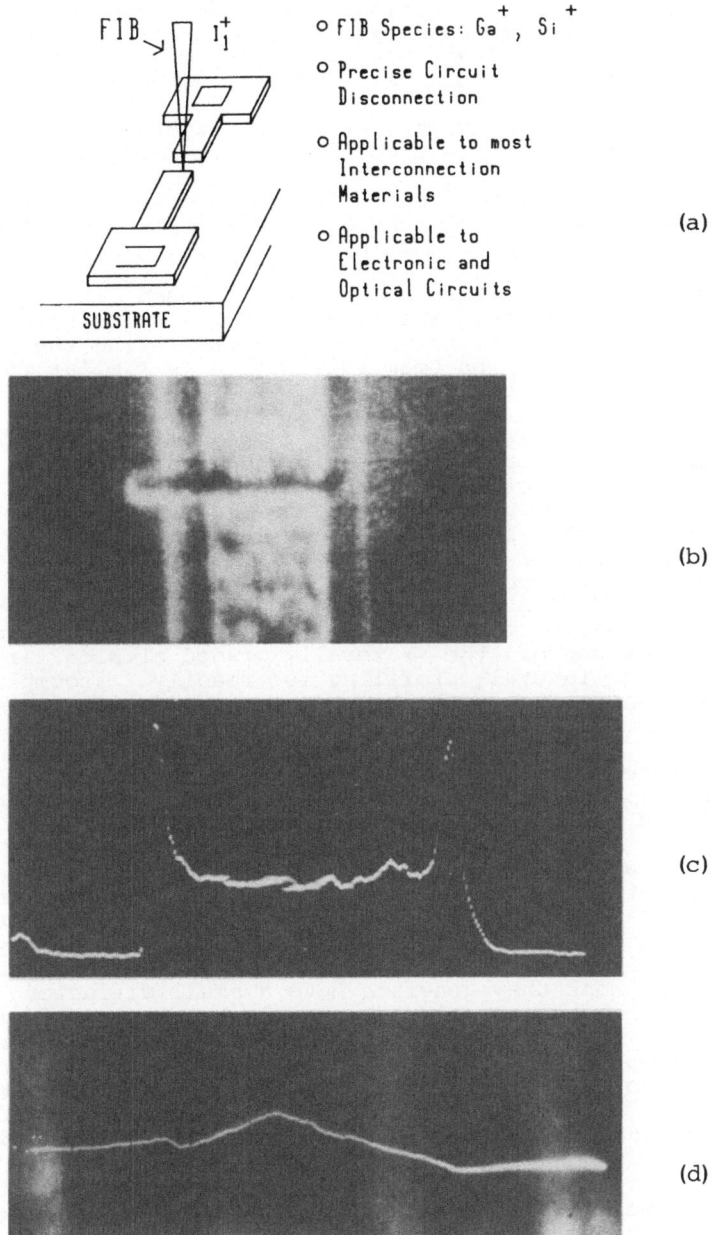

Fig.9. (a) Ion milling of a metal interconnection, (b) SEM photograph of cut metal line embedded in polyimide, (c) secondary electron signal at beginning of cut, (d) secondary electron signal at termination of cut.

In addition to disconnecting faulty sections (link-breaking) by ion milling, the focused ion beam can be used to attach replacement sections. There are at least three direct methods for accomplishing this additive (link-making) action with metals:

a. Ion milling of a "negative" patterning layer such as the stencil or shield layers used in the lift-off process [6].

b. Sputter redeposition of metal from buried reservoirs onto nearby vertical walls thereby forming a short circuit on such surfaces between two conducting layers [7].

c. Raising the stage voltage so that the focused ion beam arrives at the wafer surface at a sufficiently low velocity to stick to the surface rather than implanting or milling the surface. With a decellerated ion beam this method is similar to that employed in the Partially Ionized Beam (PIB) [8].

Ion milling may also play a significant role in the fabrication of optical interconnections. Optical interconnections can cross at right angles without interference. Hence the need for vertical vias is reduced. Nevertheless, multilayered optical interconnections may still be required to handle the quantity of interconnections required or to circumvent areas of high wiring congestion or obstruction. Vertical optical vias require 45 degree tilted mirrors with arbitrary horizontal orientation. Batch fabrication of these micromirrors is impossible because of the laterally graded erosion required. However, this lateral profiling is readily accomplished by programmable variation of the beam dwell time during ion milling. Fig. 10 illustrates the fabrication of a 45 degree tilted mirror fabricated by ion milling in an optical waveguide etched in a polyimide layer prior to filling the guide with an internal dielectric [9]. The beam charging has solved by connecting a thin metal layer on top of the waveguide embedding material to ground.

Ions are known to provide a very effective means for exposing resists due to the combined effect of energy and mass. Focused ion beam lithography, as shown in Fig. 11, is the ion beam equivalent of direct-write electron beam lithography. Since ions have a limited range in the resist material and the secondary electrons they generate have a small diffusion length (compared to those generated by an electron beam of the same energy), ion beam lithography does not experience significant proximity effects. The high resist sensitivity to ions can only be utilized to a certain point, since for uniform exposure a minimum number of particles are required per pixel. An advantage of ion beam lithography is the fact that inorganic as well as organic materials can be used as resists.

The next major area of application of FIB technology involves ion-induced chemical reactions. In ion-induced etching, shown schematically in Fig. 12 , energetic ions are incident on a thin film to be etched simultaneously with the gaseous etching species. The ions greatly increase the etch rate locally . A number of mechanisms can contribute to this effect : (a) enhancement of the reaction rate due to local heating of the surface and to increased surface disorder caused by the ion bombardment; (b) combined chemical and mechanical etching effects; (c) ion milling removal of adsorbed layers which act as

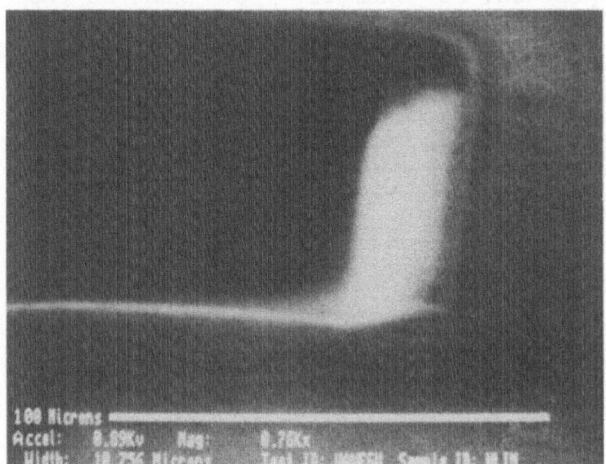

Fig.10. Ion milling of a vertical mirror at the end of an optical interconnection. The embedding material for the optical waveguide is polyimide.

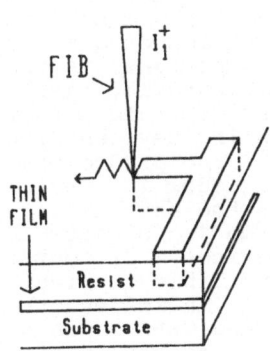

o FIB Species: Ga^+, Si^+

o Resist Layer:
  - Inorganic (cf PMMA)
  - Organic (cf Al/O cermets)

o Substrate: Si or GaAs

o FIB Microlithography
  - Low Proximity Effect
  - Enhanced Resist Sensitivity
  - Very Thin Resists

Fig.11. Microlithography-Exposure

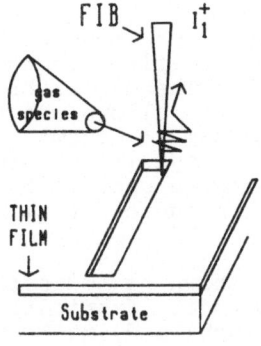

o Energetic Ions Increase Etch Rate Locally by Removing Adsorbed Barrier Layers

o FIB Species Ga^+

o Gases - Etched Lines:
  $Cl_2 \longrightarrow$ GaAs or Si

Fig.12. Ion - Induced Etching

barriers to the etching process. For example, Ga ions in the presence of chlorine gas have been shown to result in the direct etching of patterns in Si as well as GaAs.

A focused ion beam, in the presence of the appropriate gas species, can also be used to induce a chemical reaction at a surface which will result in the local deposition of a thin film. As shown schematically in Fig. 13 , through this localized CVD process the direct-write of thin film patterns can be achieved. Examples of thin film lines deposited by FIB-induced CVD include aluminum from Al(CH3)3 , tungsten from WF6, and as mentioned above carbon from hydrocarbons. Various FIB species have been used for this purpose, including Ga, Si, Be, Au. This technique is viewed as being particularly useful with ion micromachining (or ion-induced chemical etching) to provide a complete circuit restructuring capability through subtractive and additive steps. One problem which has been observed with this type of ion induced CVD of metals from organometallic gasses is the presence of small amounts of C in the deposited metals resulting in high resistivity. However, for short links of metal this may not be serious.

Last, and probably most far-reaching, application of FIB technology to be discussed involves direct implantation. Using a focused ion beam system, direct doping of semiconductor devices can be accomplished, as shown in Fig. 14 , via a maskless and resistless process. This will enable the fabrication of devices with extremely small dimensions. Perhaps most important, is the new category of devices which will now be implementable, namely devices having laterally-varying doping profiles. FIB technology is the ideal, and perhaps only, tool for fabricating devices with laterally-varying doping profiles, such as the graded-base bipolar transistor [10]. This new degree of freedom in device design has hardly begun to be exploited since in the past uniform doping in the horizontal dimension was a " given ". Direct FIB doping can be applied to Si as well as GaAs devices. Of course the appropriate doping species have to be available in liquid metal ion source implementation. Table 3 indicates the availability of LMI sources for various Si and GaAs dopants. While many problems still remain with source lifetime and species current, as discussed in the next section, feasability of most of the desired sources has now been proven.

IV. Liquid Metal Ion Sources

Liquid metal ion sources (LMIS) have been found to be extremely useful in ion beam systems which produce fine focused beams of <1um spot size on the target substrate. The use of high brightness LMIS in focused ion beam (FIB) systems allows one to use standard electrostatic lenses to focus the beam to submicron size with current densities in the 1 to 2 A/cm2 range [11]. The focussing of the ion beam from LMIS to a submicron spot is made relatively easy since the ions at the source emanate from an intense "bright point" (< 500 Å in size).

The combination of LMIS and FIB has been shown to be useful in a variety of applications for microelectronic fabrication processes including micromachining, lithography and maskless direct dopant for device construction . The LMIS usually consists of a ~ 1mm dia needle rod of tungsten or rhenium which has been

TABLE 3 Alloys for Liquid Metal Ion Source I

Dopant	Alloy Composition	Ion Species Intensity			
Antimony	$Sb_{50} Pb_{42} Au_8$	$\dfrac{Sb^+}{6.8}$	$\dfrac{Sb^+}{7.3}$	$\dfrac{Sb^+}{5.5}$	
Arsenic	$As_{10} B_{10} Ni_{40} Pd_{40}$	$\dfrac{As^+}{8.8}$ $\dfrac{As^{2+}}{15.0}$	$\dfrac{^{10}B^+}{-}$	$\dfrac{^{11}B^+}{-}$	
	$As_8 Sn_{68} Pb_{24}$	0.4 0.1	-	-	
	$B_7 As_{20} Pd_{73}$	10.6 5.7	1.6	5.7	
Beryllium	$Be_{14.8} Au_{58.8} Si_{26.4}$	$\dfrac{Be^+}{10.3}$ $\dfrac{Be^{2+}}{12.2}$	$\dfrac{Si^{2+}}{10.1}$	$\dfrac{Al^+}{34.8}$	
Germanium	$Ge_{13.5} B_{30} Pt_{20} Au_{36.5}$	$\dfrac{Ge^+}{4.1}$	$\dfrac{Ge^{2+}}{6.4}$		

Dopant	Alloy Composition	Ion Species Intensity			
Boron	$B_{30} Pt_{20} Au_{36.5} Ge_{13.5}$	$\dfrac{^{10}B^+}{7.0}$ $\dfrac{^{11}B^+}{32.1}$	$\dfrac{^{11}B^{2+}}{0.3}$		
	$B_{60} Ni_{13} Pt_{27}$	8.1 33	0.5		
	$B_{27} Pd_{73}$	5.8 12.1	-		
	$B_{20} Ni_{40} Pd_{40}$	2.3 8.3	-		
Phosphorus	$Cu_{75} P_{25}$	$\dfrac{P^+}{6.6}$ $\dfrac{P^{2+}}{-}$	$\dfrac{^{63}Cu^+}{60}$	$\dfrac{^{65}Cu^+}{30.7}$	
Silicon	$Si_{51} Au_{69}$	$\dfrac{Si^+}{12.2}$ $\dfrac{Si^{2+}}{12.0}$	$\dfrac{Au^+}{60.9}$	$\dfrac{Au^{2+}}{2.9}$	

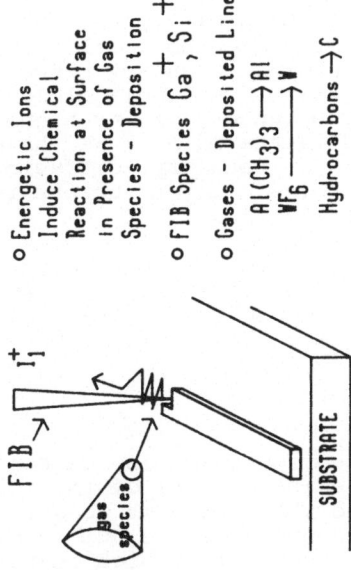

FIB → I_1^+

gas species

SUBSTRATE

o Energetic Ions Induce Chemical Reaction at Surface in Presence of Gas Species - Deposition

o FIB Species Ga^+, Si^+

o Gases - Deposited Lines:

$Al(CH_3)_3 \rightarrow Al$
$WF_6 \rightarrow W$
Hydrocarbons $\rightarrow C$

Fig.13. Ion-Induced Deposition

FIB → I_1^+

GATE

SOURCE   DRAIN

SUBSTRATE

o Direct Implantation
 – Maskless
 – Resistless

o FIB Species: B,As,P,Ge,Si

o Substrates: Si, GaAs

o Fabrication of Very Small Devices

o Fabrication of Novel Devices
 – Laterally Varying Doping Profiles
 – Self-Aligned Structures

o Quick Turn-around

Fig.14. FIB Implantation

sharpened to a fine tip of 2 to 10um radius by mechanical and electrochemical etching. Provision is made for supplying 15 to 70 watts of power to heat the resevoir of the alloy metal (~1-3mm3 volume) to produce a liquid flow to the needle tip. Ions are extracted by means of an intense electric field imposed between the tip and extractor electrode which contain a circular hole (1mm dia) concentric with the tip and ~1-2mm distant from the tip. In Fig. 15 we show a schematic of the source configuration we have developed [12] and which uses a BPd alloy to produce beams of B or Pd in a FIB system. In order to be useful a LMIS must have the following source characteristics: low melting temperature alloy (<1100 C^0); no deleterious chemical reaction of the alloy and needle tip such as dissolution of needle by alloy or fast diffusion of alloy into grain boundaries of needle and tip; no alloying of source material with needle; low vapor pressure of liquid alloy; long lifetime; and good stability over times sufficiently long (~hrs) to complete specific processing steps.

In Fig. 16 are shown typical current-voltage characteristics for a Pd73B27 alloy source which utilized a needle with a 2.5um emitter radius. Note that both the total ion emission current and 11B+ ion current are presented as a function of the extractor voltage. The Pd73B27 alloy was found to have better stability and longer life than another boron-containing alloy of platinum namely Pt58B42. The short term stability exhibited by the Pd73B27 is shown in the test runs presented in Fig. 17. Long term stability tests on the Pd73B27 alloy indicated [12] that current drifts of ~4% (2.5 tip radius) were measured after 20 hrs of operation. The boron current extracted became larger as the source tip radius was decreased from 10um to 2.5um . We were able to achieve a 120 hr lifetime with the Pd73B27 alloy source with no reload of the source resevoir [13].

We have analyzed grain boundary diffusion of liquid metal Pd73B27 into the polycrystalline regions of a rhenium emitter tip of a field emission LMIS. We consider the grain boundary between two adjacent grains to have a planar interface. A single element of Pd was assumed as the main diffusant because of its higher stoichiometry ratio in the PdB alloy used in the LMIS. We have used analytical treatments [14,15] of grain boundary diffusion and we found that the experimentally determined microstructure is convincingly explained by the theoretical analysis.

Thus far we have found the PdB alloy to be the best source of boron ions relative to stability, ease of operation and lifetime considerations. Swanson, Bell and Schwind [16] also have recently reported that PdB alloy is the best alloy to use for boron emission LMIS.

V. Focused Ion Beam Computer Control

The ultimate flexibility of the focused ion beam is achieved though computer control. This permits precise control of the ion beam from a design data base comparable to that which is used for optical or electron beam lithography. Hence a large number of general operations can be performed on a wafer with a high degree of automation. The largest challenge facing the ion beam is the throughput and cost effectiveness of the tool. Already a number

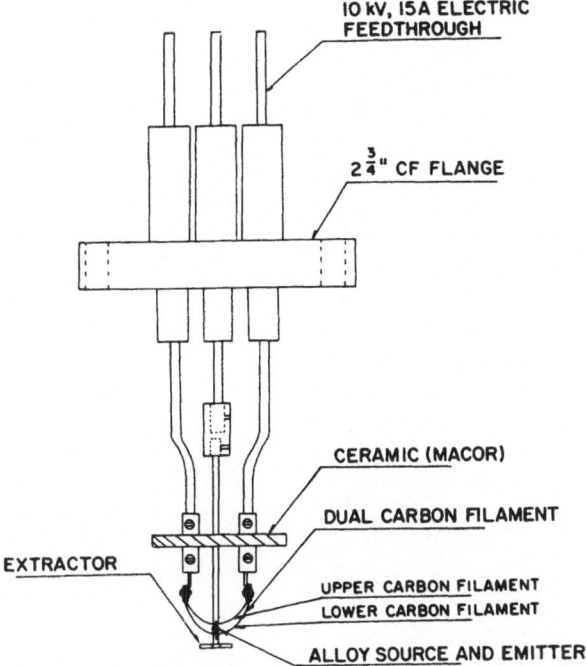

Fig.15. Schematic of the source in a FIB system

10 kV, 15A ELECTRIC FEEDTHROUGH

$2\frac{3}{4}$" CF FLANGE

CERAMIC (MACOR)

DUAL CARBON FILAMENT

EXTRACTOR

UPPER CARBON FILAMENT
LOWER CARBON FILAMENT

ALLOY SOURCE AND EMITTER

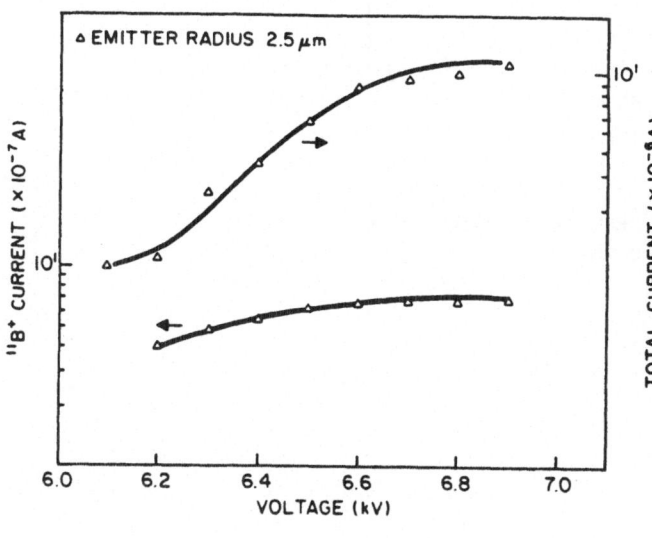

△ EMITTER RADIUS 2.5 μm

Fig.16. Current-voltage characteristics for Pd$_{73}$B$_{27}$ alloy source

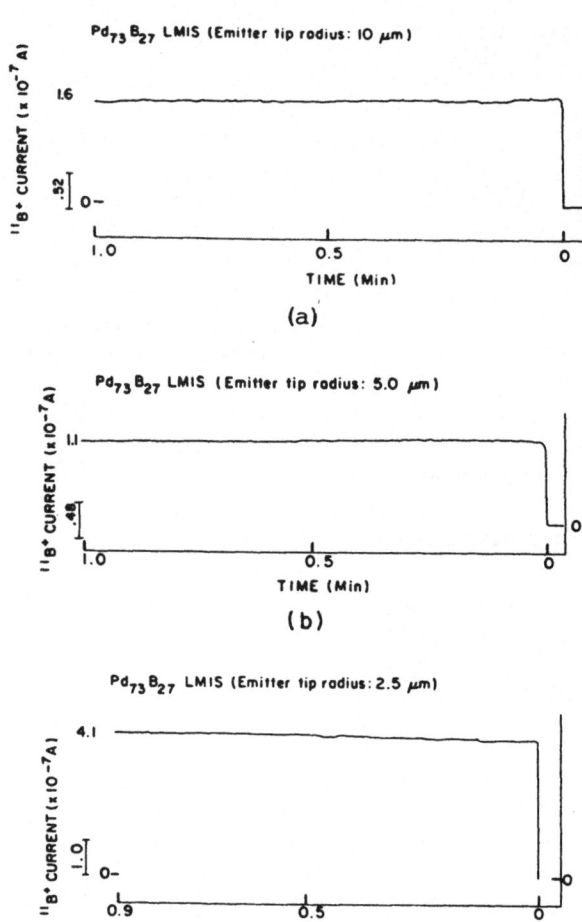

Fig.17. (a–c) Short term stability exhibited by the $Pd_{73}B_{27}$ source in test runs.

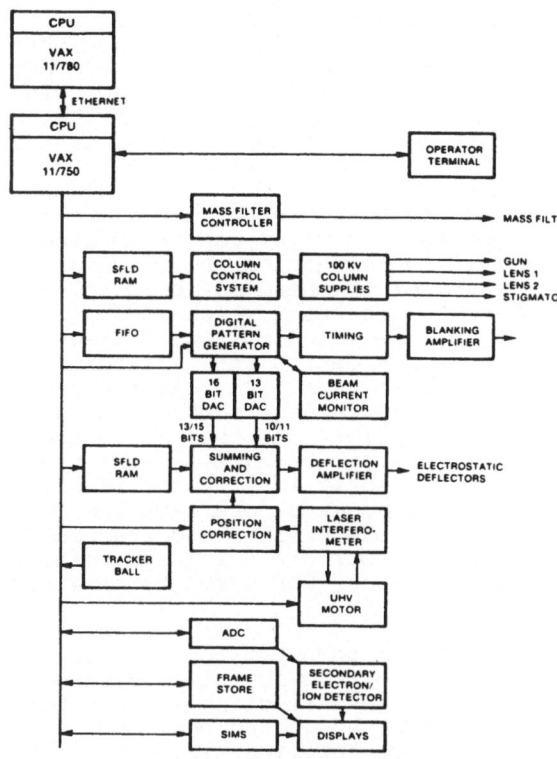

Fig.18. Schematic drawing of computer control of the VG IBL-100S focused ion beam system showing the principal electronics elements over which control can be exercised.

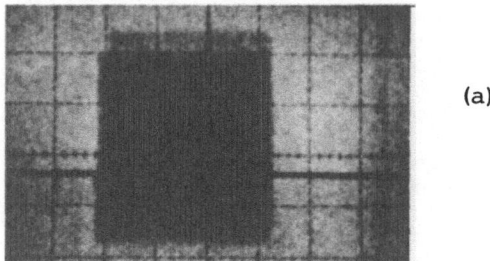

(a)

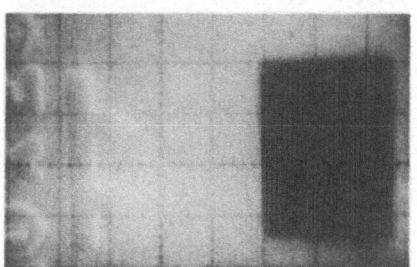

(b)

Fig.19. Secondary electron image of registration mark (a) misaligned prior to registration, and (b) after registration is complete.

of efforts are underway to produce low cost ion beam systems and it is clear that improvements in the magnitude of the focusable ion beam current will favorably affect throughput. The fact that the beam system is under computer control means that the beam may be rapidly reprogrammed to change its function in a manufacturing sequence. Figure 18 illustrates the various features of the VG IBL-100S which can be controlled by the computer (in this case a DEC VAX11/750).

One important function of the computer is the speed and the precision with which registration to the wafer can be obtained. Due to the inaccuracies inherent in the mechanical positioning of the wafer, the limited writing field of the beam may be displaced or rotated from the exact location on the wafer which is desired. In addition, due to distortions in the high voltage deflection amplifiers and distortions in the ion optics the beam position within the writing field must be accurately calibrated and corrected. Figure 19 shows a registration mark viewed by secondary electron imaging on a wafer prior to this correction. The ion beam has "stained" the area near the mark showing where the mark is expected by the computer and it is clearly in error vertically. After correction the stained area overlays the mark properly. Small displacements due to table drift are corrected by an on-line laser feedback system which corrects the high voltage deflection amplifiers so the computer does not have to monitor these signals continuously.

One of the most powerful results of computer control is the possiblity of enhancing the resolution of the ion beam system in certain kinds of operations. The minimum deflection of the beam within a subfield is 1/32 micron. Even though the beam spot size might be only 1/8 micron there are some applications where the mere fact that the beam can be moved in such small increments can be used to advantage. In effect, the Nyquist spatial sampling rate of the system exceeds its native resolution. This may be used to improve the system effective resolution in image enhancement applications [17] or in certain dose precompensation schemes for lithography [18]. The details of this process require that the exact shape of the beam point spread function (including the tails of this distribution) be known, not just the rough spot size. Also necessary in some cases is a nonlinearity in the process such as a high contrast resist. Computerized schemes such as these might extend the usefulness of the ion beam into the sub-tenth micron regime.

VI. Summary and Conclusions

The applications of FIB technology to semiconductor and optoelectronic microfabrication have been reviewed. FIB technology has opened a new window of novel processes and devices which are only now beginning to be explored.

The use of FIB technology as the basis for in-situ processing has been previously explored [1]. The prospects of in-situ fabricated devices and circuits are clearly a reality, the final test of commercial relevance will not be performed until the 1990's.

Acknowledgement

The authors would like to acknowledge the support of the Semiconductor Research Corporation in pursuing this research.

References

[1]   A.J.Steckl, Proc.IEEE,Vol.74,1753 Dec 1986.

[2]   J.Melngailis, JVST B, Vol.5, 469 March/April 1987.

[3]   I.Brodie and J.J.Muray," The Physics of Microfabrication ", New York, NY: Plenum 1982.

[4]   Sample generously provided by T.Morita, OJRL,Tokyo, Japan.

[5]   A.Wagner, Solid State Technology, Vol.26, 97 (1983).

[6]   J. F. McDonald, et al, Proc. 1987 SPIE Conference – Sec. IV pp.206-215

[7]   J. Melngailis, and C. Musil, JVST B4, 1, Jan./Feb. 1986.

[8]   T. M. Liu, et al, Proc. 1987 VLSI Multilevel Interconnection Conf. June 15-17, pp 440-448.

[9]   J. F. McDonald, et al, ibid., pp306-313.

[10] A.Steckl, C.M.Lin, S.D.Chu, J.C.Corelli, Microelectronic Engineering, Vol.5,179 Dec.1986.

[11] R.L. Seliger, J.W. Ward, V. Wang and P.L. Kubena. Appl. Phys. Lett. 34, 310 (1979).

[12] R.H. Higuchi-Rusli, K.C. Cadien, J. Corelli and A. J. Steckl J. Vac. Sci. Technol. B5(1), 190 (Jan/Feb 1987).

[13] R.H.   Higuchi-Rusli, J. Corelli, A. J. Steckl, and K.C. Cadien, to be published in J. Vac. Sci. Technol. (1987).

[14] J.C. Fisher, J. Appl. Phys. 22, 74 (1951).

[15] R.T. Whipple, Phil. Mag. 45, 1225 (1954).

[16] L.W. Swanson, A.E. Bell, 31st Int'l. Symp. on Electron, Ion and Photon Beams, May 26-29, 1987,   Woodland Hills, Calif to appear in J. Vac. Sci. Technol. (1988).

[17] W. O. Saxon, "Computer Techniques for Image Processing in Electron Microscopy," Academic Press, 1978.

[18] M.Haslam and J. F. Mcdonald, Proc. 1986 Cust. Int. Circ. Conf., May 12-14, pp. 632-635.

LASER DIRECT WRITING FOR DEVICE APPLICATIONS.

G. Auvert, Y. Pauleau and D. Tonneau
Centre National d'Etudes des Télécommunications
BP 98, 38243 Meylan, France.

Abstract.

In silicon semiconductor technology, a one micron size probe is required for the modification of interconnection networks. The laser direct writing technique using a focused laser beam is a possible solution in the building of new connections in the circuit. The complete process is composed of three elementary laser – assisted steps: etching of the insulator, deposition of an insulator on conducting materials and deposition of a conductor on an insulator.

Introduction.

The interaction of a laser beam with a solid surface in the presence of a reactive gas has been extensively studied using various high – power lasers (1,2,3). The chemical reaction induced by the laser beam is either photolytic or pyrolytic in origin and the corresponding reaction rate can be measured. In order to apply these available results to modify or build a new connection network of a VLSI circuit, a selective reading of the literature has to be made. The aim of this paper is to clarify the most practical choice for the laser direct writing of interconnections in an integrated circuit.

The numerous fabrication steps are shown in Figure 1 for a new integrated circuit. To process the first prototype, it takes several weeks. Any way which can indicate if the prototype will work, will save fabrication time. For MSI, the fabrication of a circuit board can simulate the circuit. For VLSI, this solution is too complex and some computer simulation programs are under development . An alternative is to use a gate array and to build all the interconnection lines using a laser direct write system. Then, one prototype is available and the circuit can be checked on a real scale. In the following, we will describe the process and the equipment which seem to be the most powerful to build such a prototype.

1) Basic choice.

a) Pyrolytic regime.

The reaction rate induced by a laser at a solid surface may be

201

D. J. Ehrlich and V. T. Nguyen (eds.), Emerging Technologies for In Situ Processing, 201–211.
© 1988 by Martinus Nijhoff Publishers.

measured for several mechanisms. It has been found that the reaction rate observed during some pyrolytic style processes exeeds the complementary photolytic mechanism. The pyrolytic reaction may typically allow deposition or etching of the chosen substrate at rates of 0.1 – 100 μm/s. These speeds are suitable for making the numerous circuit connections required. Thus, this mechanism has been favoured in this work.

Market Evaluation	**Market evaluation**
Circuit Conception	**Circuit Conception**
Mask Fabrication	
Circuit Processing	**Circuit mounting**
Circuit Mounting	**Laser Fabrication**
Test	**Test**
Conventional Process	Laser – assisted Process

Fig:1. Fundamental steps for integrated circuit fabrication using a gate array. The conventionnal process is useful for a series of more than 200 pieces. Below 10 circuits, the time saved in the laser assisted process can be very attractive.

b) Gaseous phase.

To maximise the supply of reactants to the solid surface of a substrate, to enable modification to occur, rapid species diffusion to and from this region is necessary. It is for this reason, that reactions occurring at the gas/solid interface may be of more use than solid/liquid junctions.

Another argument against the use of liquids with silicon microelectronics is the necessity for low temperature. The liquid must not boil in order to have a controllable process. Hence, the temperature must be below 100 – 200°C. For silicon, temperatures as high as 1400°C may be used during processing and only a gaseous phase will allow us to use the corresponding high reaction rates involved in such pyrolytic reactions.

c) Continuous wave lasers.

Due to the availability of very high power lasers, the limiting process in a laser – induced pyrolytic reaction is not the photon flux or the number of photons but the number of reacting molecules (4). Therefore, as the reaction takes place during the presence of photons, it seems reasonable to suppose a higher average reaction rate with a cw laser than with a pulsed laser even in the case of a high repetition pulsed laser.

Furthermore, compared to photolytic deposition processes, pyrolytic processes are less sensitive to process parameters such as partial pressure, adlayer formation and laser wavelength. However, they are strongly dependent on surface parameters such as thermal conductivity, absorption coefficient and reflectivity coefficient of the substrate and deposited material. Consequently, a special well – controlled laser power is required for pyrolytic reactions. All these constraints can be satisfied using cw lasers.

In order to reach the high temperature necessary to induce pyrolytic reactions, an absorbing substrate and high power lasers are required. For a silicon substrate, visible and UV photons are necessary and the corresponding high – power lasers (above 0.1W) are available. The corresponding wavelengths are 0.6 μm for a Kr^+ laser, 0.5 μm for Ar^+, 0.33 μm for Ar^{++} and 0.25 μm for doubled argon ion laser. Using these wavelengths in a high magnification microscope, a good spatial resolution is obtained. For example, deposited lines as small as 0.2 μm have been drawn on a silicon wafer (5).

2) Laser direct writing.

a) Instrumentation.

The key elements of the laser direct writing apparatus are illustrated schematically in Figure 2.

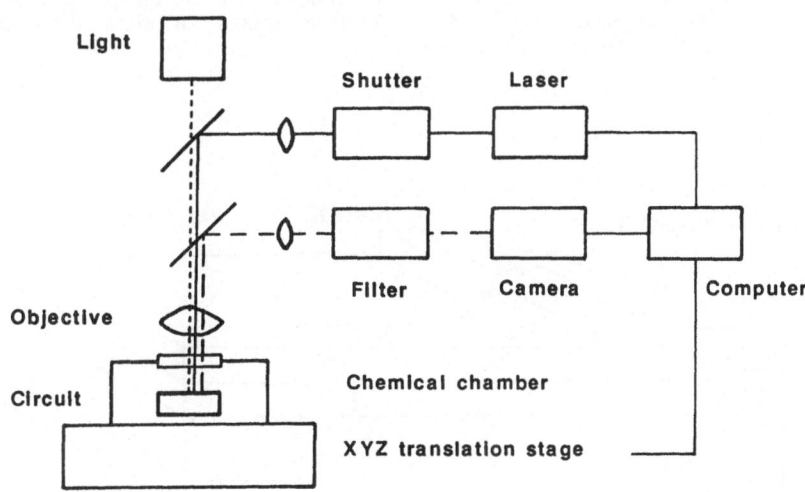

Fig:2. Apparatus for laser direct writing. The visible or near UV laser beam is mixed with the optical path of a high magnification microscope. A computer drives the laser, the circuit position and the gas fill system.

They consist of a visible or UV laser, a high – magnification optical microscope, an XY translation stage for substrate positioning, a vacuum and gas – fill system and a dedicated computer controller. For the laser source, the principal considerations are: average power, transverse mode quality, wavelengths.

The microscope is equipped with coated beam splitters to collimate the laser beam along the axis of the final objective lens. The microscope is used to focus the laser beam onto the workpiece as well as to view the substrate for accurate positioning and process monitoring. For high – resolution work, below one micron, high numerical aperture objectives are used. Typical average laser – beam intensities onto the substrate are around 1 MW/cm^2 for high temperature pyrolytic reactions on a silicon surface.

The precision translation stage for high resolution direct writing is usually a stepper motor driven in steps between 0.1 and 1µm. In order to scan at high speeds, around one centimeter per second, while maintaining such a high resolution, a dc motor has to be associated with optical encoding for feedback control. The stages can be driven in patterns by a microcomputer for simple drawing or by a minicomputer that can accept data from standard mask generators for complete drawing of a VLSI circuit.

The subtrates themselves are either mounted – integrated circuits or 4 inch wafers. They are mounted in a several – millimeter pathlength windowed vacuum cell filled with an appropriate vapor. Most pyrolytic reactions can be induceded in a static pressure. But for very low pressure and very reactive gases, a slow flow has to be generated in the reaction chamber.

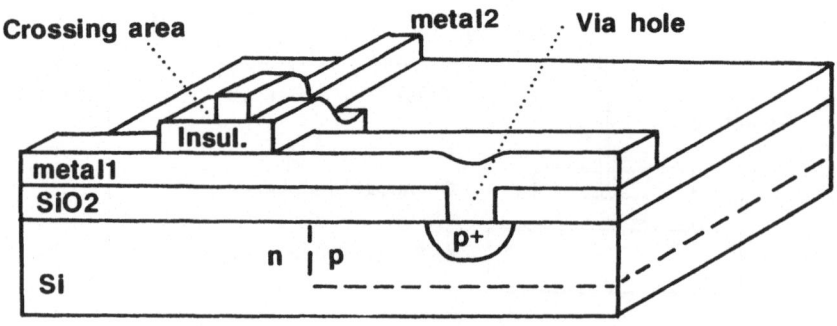

Fig:3. 3D view of a two level crossing line and a via hole obtained by using a high temperature interconnecting process associated with a local deposition of an insulator.

b) Performances.

During the laser induced reaction, a material is deposited or the surface is etched. In both cases, a growth or etching rate can be measured either in situ or after the experiment. This growth rate represents the time evolution of the surface level. If we can measure the growth rate V_g, the scanning speed V_s necessary to achieve a given thickness h can be approximated by the equation: $V_s = V_g * 2 * R/h$. R is the spot radius. This formula is based on the reasonable hypothesis which supposes a laser induced temperature independent of the scanning speed. For the laser irradiation of silicon using a spot diameter of one micron, the preceding formula can still be valid for a scanning speed of up to 1cm/s.

3) Process.

a) High temperature.

When high temperatures are required in a laser assisted process, numerous reactions can be used. Figure 3 shows the different levels which are necessary to make all the connections in an integrated circuit. The starting circuit contains all the transistors. The via holes for connections are previously opened. The first metallic layer is deposited on the insulator and into the via holes. A good contact resistance with no diffusion of the metal into the silicon must be achieved. As a laser pyrolytic reaction gives a good step coverage no particular treatment has to be carried out for the via. During this step, nearly 90% of the connections are drawn. For the others, an insulator must be deposited on the first metallic layer

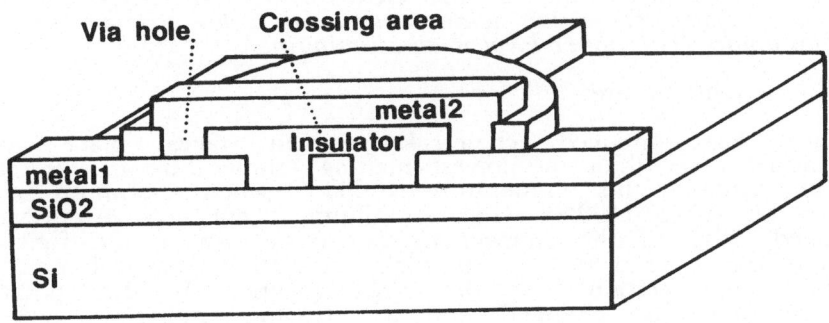

Fig:4. 3D view of two crossing interconnection lines in the case of a wide area deposited insulator. Two vias are necessary to ensure the continuity of the upper level connection.

where the second metallic level will pass. This insulator deposition must be induced within a localized area. Then the upper level of interconnect lines can be drawn using the same reaction as for the first level. This process uses two deposition reactions and no etching reaction.

Using a non – localized reaction for the deposition of insulator, opening of the via in the insulator has to be performed (see Figure 4). This etching step is a supplementary step in the high temperature process.

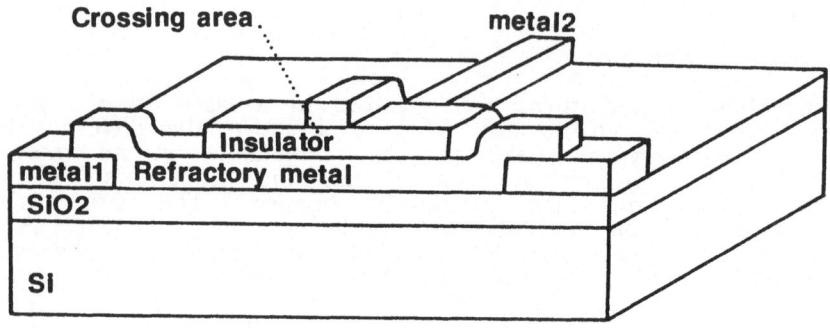

Fig:5. In a low temperature process, the aluminum lines must be changed into a nickel line at the cross – over. The refractory metal must have a low deposition temperature and a low contact resistance with aluminum. The upper level is all aluminum.

b) Low temperature.

Each laser – induced chemical reaction takes place at a temperature lower than the lowest melting temperature of the circuit material. Actually, this temperature is 600°C when aluminum is used and can be up to 1400°C when aluminum is not used. As previously mentioned, 90% of the connections can be drawn in the first level. If aluminum is used due to its high conductivity, special care must be taken into account for the crossing line. At this point, the aluminum must be changed into a refractory metal in order to withstand the deposition temperature of the insulator. This is shown in Figure 4. In this case, the second metallic level can also be in aluminum. If the circuit is completely processed using a refractory metal the high temperature process can be used.

4) Chemical reactions.

a) Nickel deposition.

The deposition of nickel using a laser may be achieved via the decomposition of nickel tetracarbonyl (6,7,8). This reaction may be

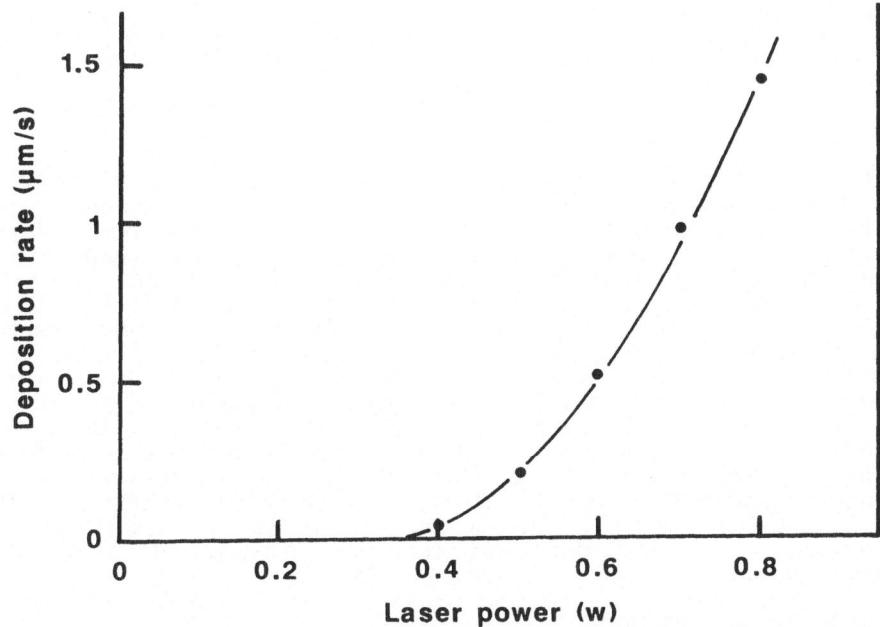

Fig:6. Deposition rate of nickel versus the argon ion laser power. The carrying gas is nickel tetracarbonyl and the laser spot size used in the experiments is 200 μm but can be reduced to one micron.

driven at high rates thus leading to the rapid growth of the nickel layer. Figure 6 shows the dependence of the nickel deposition rate versus the laser power of an argon ion laser emitting in the visible range. The exponential dependence confirms that the deposition process is thermally activated. The activation energy, calculated from Figure 6, is around 15 kcal/mole. The deposition process can be observed from 300°C and can be considered as a low temperature process. At temperatures around 500°C, the deposition rate is above 1 μm/s and this reaction seems to be useful for laser interconnections. It has been proved that the rate limiting reaction takes place in the adsorbed layer (8), which requires no further increase in the deposition rate to be made by increasing the carbonyl pressure. As nickel is a refractory metal, the laser – assisted decomposition of $Ni(CO)_4$ can be included in a high temperature interconnection process.

b) Aluminum deposition.

Integrated circuits often rely upon an aluminum interconnection network. This metal has the advantage of a good conductivity and a low contact resistance with the silicon substrate. Therefore, the deposition of this metal using various lasers has been extensively studied (9). Unfortunately, the pyrolytic decomposition of trimethyl aluminum gives a deposition of aluminum carbide with no conducting property (10,11,12). It seems that in recent works, by using a photolytic process, the deposition of pure aluminum with TMA or tri – isobuthyl aluminum is induced (13). As no deposition rate studies are yet available, no comparison with the deposition of nickel can be made. As the melting temperature of aluminum is around 600°C, the deposition process of aluminum is one elementary step for a low temperature process.

c) Etching of aluminum.

The Etching of metals has not been studied as extensively as that of semiconductor materials (14). A contributing factor to this is the presence of a thin but relatively inert oxide layer. Therefore, laser assisted etching processes must be capable of etching controllably two different materials. Most of the time, the thin oxide is chemically inert and only after the diffusion of the etching agent through this region, can the metal etching reaction occur. This reaction may give rise to a gaseous compound. The gas is accumulated below the oxide up to a pressure where it mechanically breaks. At this point, the etching barrier disappears. The etching rate increases sharply and the etch area extends drastically. Therefore, good spatial control of the metal etch area is difficult. In the case of a high temperature interconnection process, as no etching is necessary, this difficulty is avoided. However, this etching problem has to be solved in order to transform the interconnection process into the repairing process of a finished integrated circuit.

d) Etching of insulator.

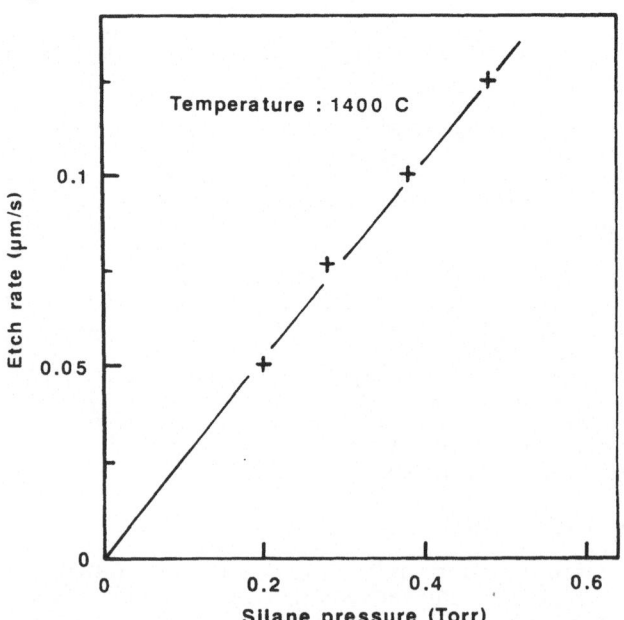

Fig:7. Etching rate of silicon dioxide versus silane pressure. The laser induced temperature is 1400°C. For a quartz substrate, a CO2 laser has been used to heat the area to be etched. For a silicon substrate covered by an oxide layer, the etch reaction is induced by an argon ion laser which heats up the silicon oxide by conduction.

Insulators tend to be the most inert materials and therefore the most difficult to etch. In most cases, the mechanisms, even for standard planar processes, are not well understood. In the literature, SiO_2 and SiN have been etched at a high rate, about 200 um/s using a pulsed CO_2 laser and freon ambient (15). Unfortunately, this high speed is obtained during the pulse and a multiphoton absorption is supposed for the etching reaction. The corresponding average etching rate is very low and no equivalent mechanism can be achieved using a cw laser (Argon ion or CO_2) (15). As a result, no high etching rate reactions have been described in the literature (16).

As an alternative, we have studied the etching of SiO_2 using a silane ambient and CO_2 laser irradiation. A reasonably high etch rate may be obtained as shown in Figure 7. The process follows the chemical reaction: $Si + SiO_2 - > 2 SiO$. As SiO is a gaseous molecule above 1000°C, this etching step can be integrated into a high temperature interconnect process.

e) Insulator deposition.

Numerous experiments describing non – patterned deposition of insulators such as SiO_2 and Si_3N_4 have been reported (2). Excellent material properties are reported and the deposition rate is not very

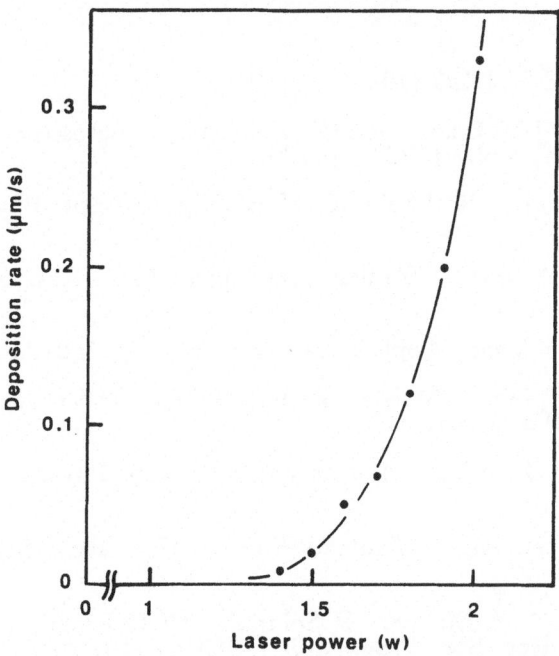

Fig:8. *Deposition rate of silicon nitride versus the argon ion laser power. The ambient is a silane – ammonia mixture with 1 Torr of silane and 5 Torr of amonia. The deposition area is localized to the laser heated zone. The process is controlled by laser – induced silane decomposition.*

important because a longer time will be taken for the opening of via holes by laser induced etching. However, localized reactions allowing patterning are required for laser direct writing. Unfortunately very few examples have been reported to display high deposition rates (17).

The localized deposition of silicon – nitride has been carefully studied in our laboratory. The ambient was a silane – ammonia mixture and the irradiating laser was an argon laser tuned into the visible wavelengths. Figure 8 shows the dependence of the deposition rate on the laser beam power. The deposition rate is found to be reasonable and seems to be controlled by the silane decomposition. Therefore, a further increase may be obtained by increasing the silane decomposition at the substrate level.

5) Conclusion.

The laser interconnection of a gate array or an integrated circuit may use either a high temperature or low temperature process. The presence of aluminum in the circuit necessitates the low temperature process. By using a refractory metal such as nickel, and the localized deposition of Si_3N_4 as an insulator, the high temperature process can be performed by the direct writing of one micron size lines all over the circuit and at several interconnect levels.

References.

1) F.A.Houle SPIE Vol.459 (1984), p.110.

2) D.J.Ehrlich, J.Y.Tsao, VLSI Electronics:microstructure Science, Ed. N.G.Einspruch, Vol.7 (1983), p.129.

3) Y.Ritz – Froidevaux, R.P.Salathé, H.H.Gilgen, Appl. Phys. A, 37,121 (1985), p.121.

4) D.Tonneau, G.Auvert, Y.Pauleau, European Mat. Res. Soc. Proc. V4 (1985), p.125.

5) D.J.Ehrlich, J.Y.Tsao, Appl. Phys. Lett. 44, (2), 1984, p.267.

6) S.D.Allen, R.Y.Jan, S.M.Mazuk, K.J.Shin, S.D.Vernon, Mat. Rec. Soc. Proc. V29, (1984), p.1.

7) W.Krauter, D.Bäuerle, F.Fimberger, Applied Physics A31 (1983), p.13.

8) G.Auvert, D.Tonneau, Y.Pauleau Europ. Mat. Rec. Soc. Proc. V4, (1986), p.109.

9) P.K.Boyer, C.A.Moore, R.Solanki, W.K.Ritchie, G.A.Roche, G.J.Collins, Mat. Res. Soc. Proc. V17, (1983) p.119.

10) Y.Rytz – Froidevaux, R.P.Salathe, H.H.Gilgen, Physics Letters, Vol.84A, N1, (july 1981), p. 216.

11) G.S.Higashi, G.E.Blonder, C.G.Fleming, Mat. Res. Soc. Proc. (1986).

12) J – E.Bourée, J.Flickstein, Y.I.Nissim, Mat. Res. Soc. Proc. (1986)

13) G.E.Blonder, G.S.Higashi, C.G.Fleming, Appl. Phys. Lett. 50, (12),1987, p.766.

14) M.Rothschild, J.H.C.Sedlacek, D.J.Ehrlich, Appl. Phys. Lett. 49(22)1986, p.1554.

15) J.I.Steinfeld, T.G.Anderson, C.Reiser,D.R.Denison, L.D.Hartsaug, J.R.Hollahan, Jour. Electrochem. Soc. 127, (1980), p.514.

16) B.T.Dai, B.S.Agrawalla, S.D.Allen, Mat. Rech. Soc. Proc. Extended Abstract, "Beam induced chemical processes", (1985) p.143.

17) S.Szikora, W.Krauter, D.Bäuerle, Materials lett.V2, 4A, (1984) p.263.

LASER-INDUCED PHOTOETCHING OF SEMICONDUCTORS WITH CHLORINE

J.L. PEYRE, D. RIVIERE, Ch. VANNIER and G. VILLELA

LABORATOIRES DE MARCOUSSIS, Centre de Recherches de la C.G.E., Route de Nozay, 91460 MARCOUSSIS - FRANCE

1. INTRODUCTION

Pulsed-laser processing of semiconductors for a number of device applications is now well established. Recent studies have demonstrated a potentially widening role for dry etching technologies which make use of photochemical reactions induced by laser irradiation. In this context, excimer lasers provide interesting specifications such as the possibility of working at different wavelengths in the ultraviolet range : 193 nm (ArF), 248 nm (KrF), 308 nm (XeCl) and 351 nm (XeF) for the most important ones. Moreover, high-energy pulses, variable repetition rates and incoherent light increase the efficiency of excimer-induced processing.

The purpose of this paper is to report several results on laser induced photoetching of silicon and aluminum with chlorine and to demonstrate the possibilities of excimer laser in semiconductor processing.

The first experimental result concerning the photoetching of a silicon single crystal induced by a high power argon laser in bromine gas has been reported in 1977 [1]. High resolution etching of single-crystal and polycrystalline silicon has been demonstrated in 1981 by D.J. EHRLICH & al. [2] : an argon-ion laser has been used to etch silicon in gas phase chlorine and hydrochloric acid. Photothermal and photochemical reactions are probably responsible for very high etch rates (greater than 5 μm/s). UV light excited poly-Si etching employing chlorine has also been investigated by M. SEKINE & al. [3] in 1983. The influence of doping on the photoetching process and the role of the electron-hole pair generation have been shown. Another experiment using a XeCl laser has demonstrated the dependences on conduction type, sheet resistance and orientation in single-crystal silicon etching characteristics [4].

To our knowledge, the last experimental results obtained with excimer lasers have been reported by T. BALLER & al. [5] concerning the measurements of mass spectra and time-of-flight (TOF) distributions of the particles desorbed from a chlorinated silicon target during laser irradiation with 308 nm and 248 nm photons. Finally, the influence of laser power, scanning velocity and beam polarization on deep trench formation has been demonstrated by G.V. TREIZ & al. in 1986 [6].

In the case of metals, the enhancement of the reaction rate in the wet etching of aluminum thin films by a tightly focused argon laser beam has been reported [7]. On the other hand, J.E. ANDREW & al. have studied the possibility of patterning thin metal films based on photoablative mechanisms [8]. The first investigations on excimer laser etching of aluminum in chlorine environment have been reported by G. KOREN in 1985 [9] : etch rates of up to ~ 1 μm/pulse were obtained in several Torr of Cl_2 and with fluences over

213

D. J. Ehrlich and V. T. Nguyen (eds.), Emerging Technologies for In Situ Processing, 213–220.

0.25 J.cm^{-2}, using a XeCl laser. Different mechanisms of the etching process have been suggested to explain the observed results [10]. Etch products, reaction products and etching mechanisms for the reactions of chlorine with Al<100> have also been reported by H.F. WINTERS [11].

2. EXPERIMENT

The experimental set up consists of a modified ion etching reactor GIR 160 provided by ALCATEL and an excimer laser built at the Laboratoires de Marcoussis. A schematic diagram of this apparatus is shown in Fig. 1.

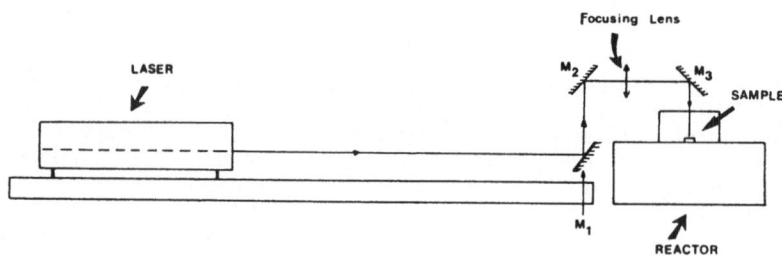

FIGURE 1. Experimental set-up

The laser produces up to 50 mJ pulses of 20 ns duration at low repetition rates (several Hz). The beam cross-section is approximately 1 cm x 1 cm with fluence uniformity typically 60 % over 90 % of the beam area. One of the most important characteristics of the laser is its compactness (the dimensions are 65 cm x 12 cm x 13 cm) particularly useful in experimental studies. On the other hand the change of wavelength is easy, by an appropriate change of gas mixture and cavity mirrors. The laser beam is focused on the sample with a Suprasil lens and through the reactor windows. The influence on the sample can be varied by the introduction of attenuators on the beam or modifying the position of the lens. The GIR 160 reactor provided from ALCATEL has been modified in order to accomodate the beam from the excimer laser. Different windows allow to introduce the UV light and to observe the sample during the irradiation. The system is pumped by a turbo-pump ALCATEL 5041 CP and a primary pump ALCATEL 2033 CP1, providing a residual vacuum of 3.10^{-5} mbar. The chlorine gas from Matheson used is 99.99 % pure and the pressure in the cell is monitored with two capacitance manometers (MKS Baratron), in order to obtain pressure ranges from 10^{-2} to 1 mbar and from 1 to 100 mbar. The maximum gas flow is 200 sccm for the chlorine gas.

A quadrupole mass spectrometer (BALZERS) has been mounted on the system, in order to analyse the residual gas in the cell, the quality of the chlorine used in the experiment and the products of desorption induced by the laser photoetching.

Measurements of depth profiles are performed, whenever needed, with the aid of a mechanical stylus (Tencor). Scanning electron microscopy has also been used for the determination of the etching quality and surface morphology.

3. RESULTS AND DISCUSSION
3.1. Silicon etching

Si was irradiated principally with photons of 308 nm (sometimes 248 nm) in chlorine atmosphere. Experiments were performed on Phosphorus or Boron

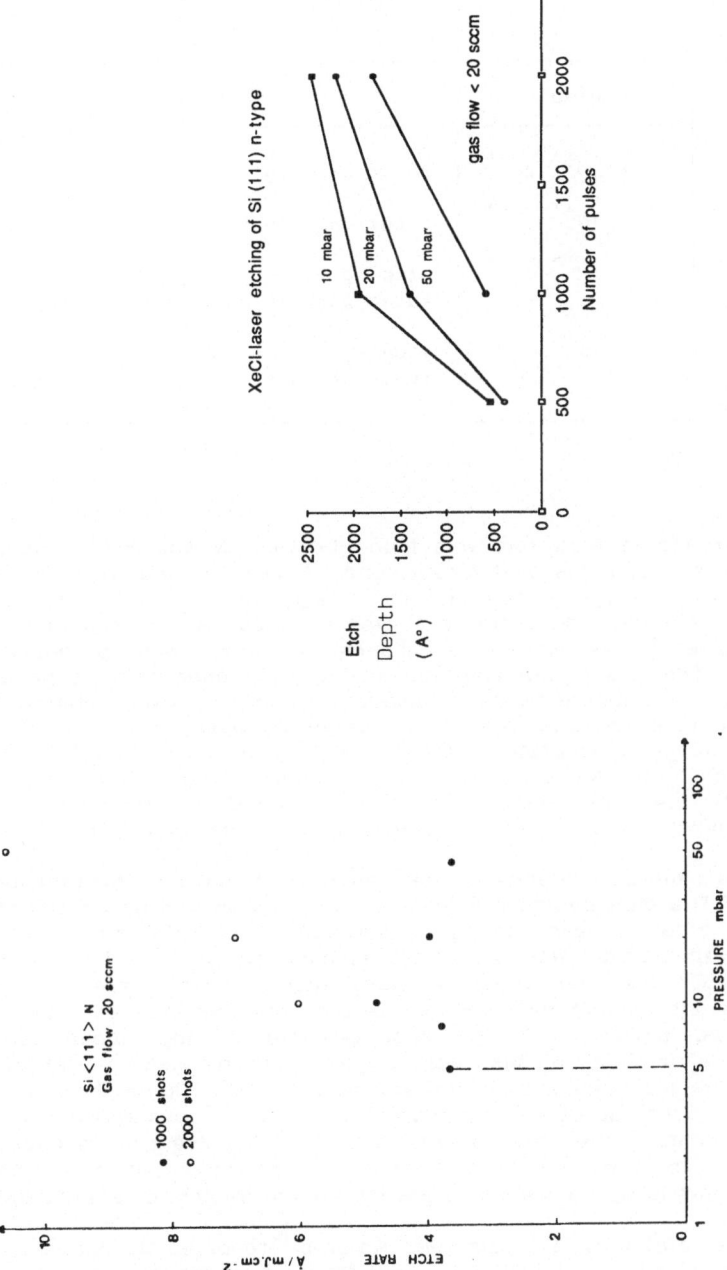

FIGURE 3. Etch depth as a function of the number of laser pulses for different chlorine pressures

FIGURE 2. Influence of the chlorine pressure on the etch rate

doped single-crystal silicon. The single-crystal wafers used are described in table 1. Native oxide was eliminated by dipping wafers in HF solution before introduction in the reactor.

Crystal	Doping	Results
Si < 100 >	P 6.10^{14} cm^{-3}	ablation only
Si < 111 >	P 10^{15} cm^{-3}	ablation only
Si < 111 >	N 10^{15} cm^{-3}	etching laser fluence threshold : 500 mJ/cm^2
Si < 111 >	N 3.10^{16} cm^{-3}	etching laser fluence threshold : 150 mJ/cm^2

Table 1

A clear difference in etch rate was found between N and P-type polished samples and between samples of different conductivities. Greater laser fluences are required to etch P-type silicon and the energy threshold decreases when donor atom concentration increases in N-type samples. An explanation to this phenomenon, already well-known in plasma etching, was proposed by M. SEKINE [4]. The role of the electrons in the conduction band is generally admitted. Cl radicals generated by irradiation would react with electrons to produce negative chlorine ions, which can penetrate easily into the Si and form volatile products of desorption ($SiCl_x$). With regards to the effect of the crystal orientation, our statistics are not yet sufficient to support our first observations. On the other hand, effects were found for variations in the experimental conditions, such as gas pressure, gas flow and laser intensity.

Figure 2 shows the dependence of the etch rate on the gas pressure for N-type Si < 111 >. The etch depths are given for normalized energy irradiation on the sample, in order to take the Cl_2 absorption of the 308 nm beam into consideration. Reproducible and detectable etching occurs only for pressures above 5 mbar. On the high pressure side, etching rates were limited by absorption along the optical path. When experiments are stopped after 1000 pulses one could believe that the most effective etching occurs around 10 mbars. But after 2000 pulses etching seems to be favoured at higher pressures, near values mentioned in the bibliography [3] Figure 3 shows etch rates obtained for 3 pressures with comparable laser pulse average energy (30 mJ) at different times of the process. Etch rate of 2 Å/pulse can be deduced, which corresponds to previous published values. Increased repetition rate (more than 100 Hz) seems to be necessary to get interesting depths for applications in microelectronics.

A laser impact profile on a 3" diameter wafer is presented in Figure 4. The disturbed surface of the bottom with two main "basins" reflects the inhomogeneity of the laser beam. Homogeneity may be greatly improved by using a UV optical fiber, but its damage threshold turned out to be too low for reliable use. Improvements to the excimer laser itself are under study.

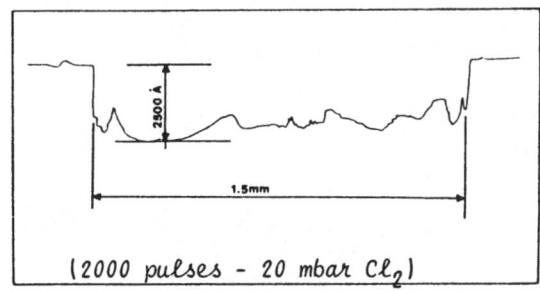

(2000 pulses - 20 mbar Cl_2)

FIGURE 4. Impact profile of a laser etched area on silicon

Moreover, the existence of a deposited film around etched regions seems to be caused by a too long time of presence of the gas in the chamber. Gas flow was increased to improve the results. For N-type Si < 111 > no improvement was observed at pressures lower than 20 mbars, but etch depths are clearly increased at higher pressures (Table 2), probably because desorption products are more easily evacuated.

Signals detected on the quadrupole mass-spectrometer for the masses 63 (SiCl) and 98 ($SiCl_2$) seem to confirm such an hypothesis.

Pressure (mbar)	3	5	10	30	50
low gas flow (20 sccm)	-	1000	1950	850	600
high gas flow (200 sccm)	-	650	800	1000	850

Table 2 : Etch depths (Å) for 1000 pulses (30 mJ) on N-type Si<111>

The fastest reaction occurs at 100 mbars with the reduced optical path of 1 cm : 2,5 Å/pulse. For P-type <100> Si, that had not previously been etched except by thermal ablation, the same result was finally obtained (2,5 Å/pulse) but with higher fluence (650 mJ/cm²).

The dependence of the etch rate on the fluence for N-type Si<111> samples is shown in Fig. 5. The rate first rises quickly with increasing fluence from the threshold mentioned above, reaching a level where the surface reaction is not greatly accelerated by further increase in laser power. Above melting point (about 0,75 J/cm²) thermal effects become preponderant and etch rates rise again.

Other experiments were also performed on polycrystalline silicon wafers with CVD SiO_2 mask (about 6500 Å thickness), that is not damaged during silicon etching. Comparable results were obtained with XeCl and KrF lasers whereas Cl_2 absorption and therefore photogeneration of Cl radicals are less important at 248 nm. Photothermal effect and lattice photo-excitation by 248 nm photons [5] seem to prevail in the etching mechanisms.

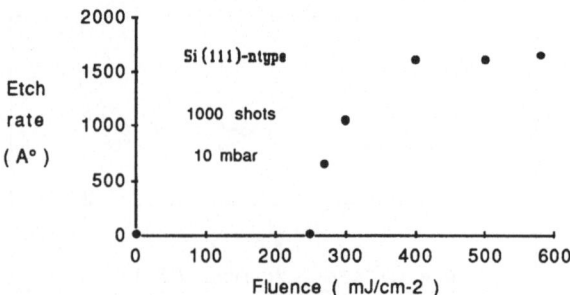

FIGURE 5. Etch rate as a function of the laser fluence

A profile of a pattern drawn on the Alphastep shows Si etching of 3000 Å in 23 μm large trenches after 4000 KrF shots of 400 mJ/cm² (fig. 6). With 6000 additional pulses no deeper etching was reached and only the homogeneity of the bottom was improved. However we could observe with scanning electron microscope that the edge definition of the etching remains insufficient, probably because of the inhomogeneous laser beam.

Further work will include anisotropy and resolution improvements.

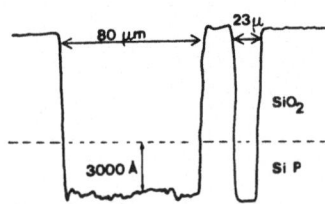

FIGURE 6. Etching through SiO_2 mask

3.2. Aluminum etching

The second part of our study deals with photo-assisted etching of aluminum in chlorine environment. Samples to be investigated were evaporated 2600 Å Al films on glass substrates. Etch rates were determined by counting the number of pulses required to etch through these films. Two wavelengths (248 and 308 nm) were used in the experiments. Some etch rates are given as a function of the chlorine pressure in Fig. 7. One can notice at first that no detectable etching was obtained in the absence of gas under similar experimental conditions. Total removal without gas occurs only with pulses higher than 3 J/cm² and the substrate is also damaged, which is not tolerable for microelectronics applications.

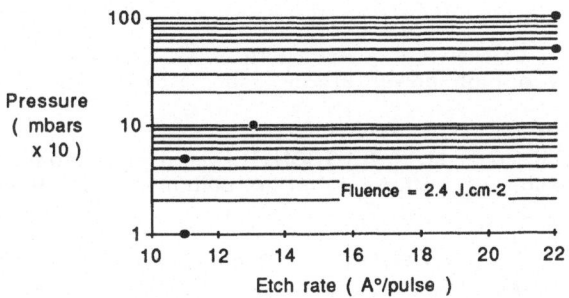

FIGURE 7. XeCl laser etching of aluminum

In Cl$_2$ atmosphere XeCl assisted etching rises from 10 Å/pulse at low pressure to 20 Å/pulse beyond 5 mbar/ Moreover etching seems to be easier with KrF laser : required fluences are lower (250 mJ/cm^2) and yield higher etch rates (40 Å/pulse).

Our observations are in good agreement with previously suggested hypothesis [9] : the laser role seems to be essentially the ablation of a very thin layer created on the surface by Cl$_2$ gas (AlCl$_3$ or Al$_2$Cl$_6$). The fact that etching is faster with KrF laser whereas Cl$_2$ absorption is very weak at 248 nm demonstrates that photogenerated Cl radicals are not involved in the primary reaction. Etch rates are rather determined by the kinetics of surface reactions with Cl$_2$. The role of Al$_2$O$_3$ films in surface was not investigated. Other experiments at higher pressures (200 mbars) give no better results but deposited films appear near the impact in the direction of the pumping.

4. CONCLUSIONS

In summary UV light enhanced etching of Si and Al exposed to chlorine with a flexible and reliable excimer laser built at the Laboratoires de Marcoussis has been investigated.

Although the mechanisms are complex, promising results have been obtained that show the influence of doping, gas pressure, gas flow and laser fluence on Si etching. The importance of the chlorine gas for thin Al film etching has been demonstrated.

Improvements in the technique and a better understanding of the etching mechanisms are under study.

ACKNOWLEDGMENT

This work was supported by the Direction Générale des Télécommunications (Direction des Affaires Industrielles et Internationales).

The authors would also like to thank B. LACOUR for his helpful assistance in the laser experimentation and for valuable discussions on the subject.

REFERENCES

1. SVESHNIKOVA L. L. , DONIN V. I. , REPINSKII S. M. : Initiation of bromine-silicone reaction by a high-power argon laser, Sov. Tech. Phys. Lett. 3 (6), 223 (1977)

2. EHRLICH D. J. , OSGOOD R. M. , DEUTSCH T. F. : Laser chemical technique for rapid direct writing of surface relief in silicon, Appl. Phys. Lett. 38 (12), 1018 (1981)

3. SEKINE M. , OKANO H. , HORIIKE : Photo-excited poly-Si etching – conduction type and resistivity dependencies, 1983 Dry Process Symposium Proceedings, 97 (Tokyo, 1983)

4. ARIKADO T. , SEKINE M. , OKANO H. , HORIIKE Y. : Single-crystal silicon etching characteristics using excimer laser Cl_2 gas, Mat. Res. Soc. Symp. Proc. 29, 167 (1984)

5. BALLER T. , OOSTRA D. J. , de VRIES A. E. , Van VEEN G. N. A. : Laser-induced etching of Si with chlorine, J. Appl. Phys. 60 (7), 2321 (1986)

6. TREYZ G. V. , BEACH R. , OSGOOD R. M. : Rapid direct writing of high-aspect-ratio trenches in silicon, Appl. Phys. Lett. 50 (8), 475 (1987)

7. TSAO J. Y. , EHRLICH D. J. : Laser-controlled chemical etching of aluminu, Appl. Phys. Lett. 43 (2), 146 (1983)

8. ANDREW J. E. , DYER P. E. , GREENOUGH R. D. , KEY P. H. : Metal film removal and patterning using a XeCl laser, Appl. Phys. Lett. 43 (11), 1076 (1983)

9. KOREN G. , HO F. , RITSKO J. J. : XeCl laser controlled chemical etching of aluminu in chlorine gas, Appl. Phys. A40 (13), 23 (1986)

10. KOREN G. , HO F. , RITSKO J. J. : Excimer laser etching of Al metal films in chlorine environments, Appl. Phys. Lett. 46 (10), 1006 (1985).

11. WINTERS H. F. : Etch products from the reaction on Cl_2 with Al (100) and Cu (100) and XeF_2 with W (111) and Nb, J. Vac. Sci. Technol. B3 (1), 9 (1985).

RECENT ADVANCES IN PHOTO-EPITAXY FOR INFRARED DETECTOR FABRICATION

S.J.C. IRVINE, M.C. WARD and J.B. MULLIN
Royal Signals and Radar Establishment
St Andrews Road, Malvern, Worcestershire WR14 3PS, UK

1. INTRODUCTION

The fabrication of infrared photon detectors from narrow band gap compositions of the $Cd_xHg_{1-x}Te$ (CMT) alloy has been a relatively small scale, high skill and high cost process. Compared with silicon the yield stress of $Cd_xHg_{1-x}Te$ is very low, 10-15 MN m^{-2} (1) and solid state diffusion rates are high; dopant diffusion rates at 200-300°C in $Cd_xHg_{1-x}Te$ being comparable with diffusion rates in Si at 900°C. For device processing these properties make CMT an extremely difficult material to handle, requiring exacting polishing techniques to avoid damage and very low temperatures (< 100°C) throughout processing to avoid Hg loss.

The attraction of CMT is in its fundamental properties which are near ideal for infrared detection especially at the long wavelength band around 10 μm. The band gap is tunable with composition from 0 to 1.6 eV and has a direct valence to conduction band transition at the Brillouin zone centre over the entire range. The direct gap transition enables a high quantum efficiency for photon absorption. The high electron/hole mobility ratio (~ 100) is ideal for photoconductor response and the high electron mobility (~ 250,000 cm^2/V-s for x = 0.2) is suitable for very fast transistor operation (2).

The applications of CMT detectors has been largely confined to either specialist scientific uses, high cost military thermal imagers and a relatively small number of medical thermal imagers. These first generation thermal imagers use a small linear array of photoconductive detectors where the scene is mechanically scanned, using germanium optics, over the detectors. Size and cost of thermal imagers could be considerably reduced by using a 2-D array (staring array) of infrared detectors thus removing the mechanical scanning. An example of a 2-D array of CMT detectors is shown in Fig. 1. TV compatible imagers, however, require on the order of 10^5 photovoltaic detectors with enormous focal plane complexity in order to read out and correct signals. The complexity of staring arrays is compounded in the manufacturing technology by the inherent properties of CMT, outlined above, which places severe constraints on the processing.

It is partly for reasons of high product cost and extreme processing conditions that an in-situ processing technology for CMT detector arrays is so attractive. At present the manufacture of a detector array includes the steps of damage free polishing to 10 μm thickness, passivation of the surface with sputtered ZnS, glueing onto a substrate, ion beam processing steps, and metallisation. Each step can expose surfaces to contamination, heat and damage, and is largely uncharacterised before proceeding to the next step. The whole process is very wasteful in high cost material; most of it being polished or etched away to form the final structure. Something in the region of 40 definable steps are involved, each with the potential for decreasing the ultimate detector performance by introducing electrically active point defects through damage, contamination or Hg loss.

221

D. J. Ehrlich and V. T. Nguyen (eds.), Emerging Technologies for In Situ Processing, 221–232.

222

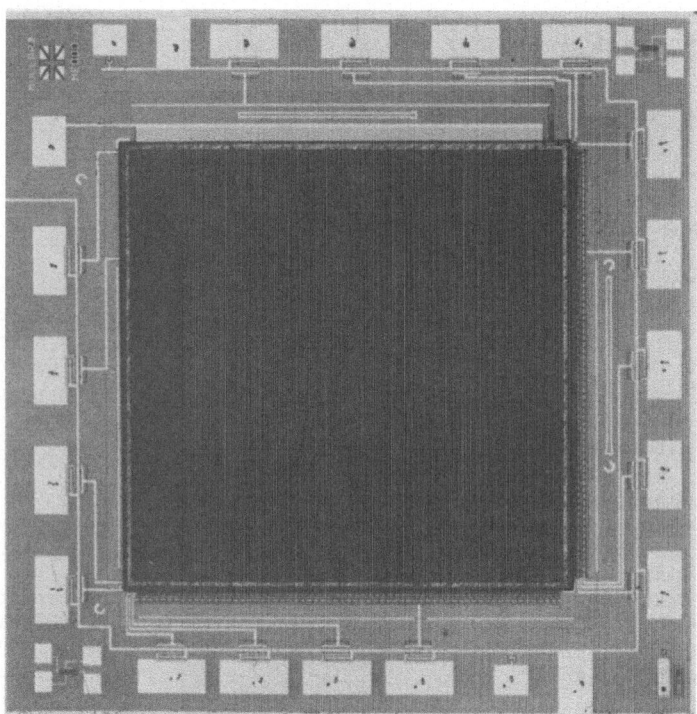

FIGURE 1. A 64x64 element array of CMT detectors on a silicon readout chip (courtesy of Mullard Ltd).

An in-situ processing technology for CMT must have the following features in order to avoid the limitations of current technology.

(a) Low energy processing, ie avoid ion damage which introduces electrically active defects.

(b) Low temperature processing – around 250°C or below for epitaxial growth and 100°C for subsequent processing.

(c) Selective area processing – using non-contact masks.

(d) Maintain clean surfaces and control alloy surface composition.

(a) and (b) exclude the possibility of using high energy ion beams and concern about point defect formation would probably exclude even low energy ion beam techniques. However, reactive ion etching has been used successfully for thinning sections of CMT for TEM studies without introducing damage such as dislocation loops (3) UV photo-processing involves energies of ~ 5 eV, far less than the ion energies and can actually enhance crystalline quality when used in epitaxial growth (4).

Photochemical processes could be used for epitaxial deposition of a CMT detector structure onto a suitable substrate such as GaAs or Si which contains the signal processing circuits. They can also be used for deposition of dielectrics (5), for etching (6) and metallisation (7). This paper will concentrate on epitaxial growth using photo-Metal Organic Vapour Phase Epitaxy (Photo-MOVPE) as part of an in-situ processing technology for advanced CMT detector arrays.

The photolytic reactions in epitaxial growth can influence the growth
process in a variety of ways ranging from photo-modified epitaxy ie doping
where there is no change in growth rate with illumination (see Kukimoto (8))
to photo-initiated epitaxy where no deposition occurs in the absence of UV.
The specification in in-situ processing for low temperatures and selective
area deposition requires the latter category of photo-epitaxy.

2. CURRENT CAPABILITIES OF PHOTO-MOVPE FOR INFRARED DETECTOR ARRAY PROCESSING

This setion will deal with the current state of progress in photo-MOVPE of
CMT and related binaries towards the ideal in-situ processing. The first
section will deal with reaction mechanisms and the prospects for preparing
the desired detector structures. The second section will consider the properties of current photo-epitaxial layers and highlight areas where improvements need to be made.

2.1 Reaction mechanisms

The photo-initiated decomposition process of diethyl-telluride (Et_2Te), for
the growth of HgTe and CdTe appears to be quite different for the two compounds. For HgTe it is primarily a surface decomposition and for CdTe it is
a vapour decomposition. In the case of HgTe, no deposition occurs onto the
substrate at low Hg pressure and under high intensity illumination ($\sim$ 1 W/
cm^2) from a Hg arc lamp (9). This means that the direct photodissociation
of Et_2Te does not play an important part in HgTe epitaxy. However, a strong
surface reaction occurs at high Hg pressures, close to the Hg rich solidus
($> 10^{-2}$ atm). This has been described as a surface photosensitisation reaction (9) and is shown schematically in Fig.2.

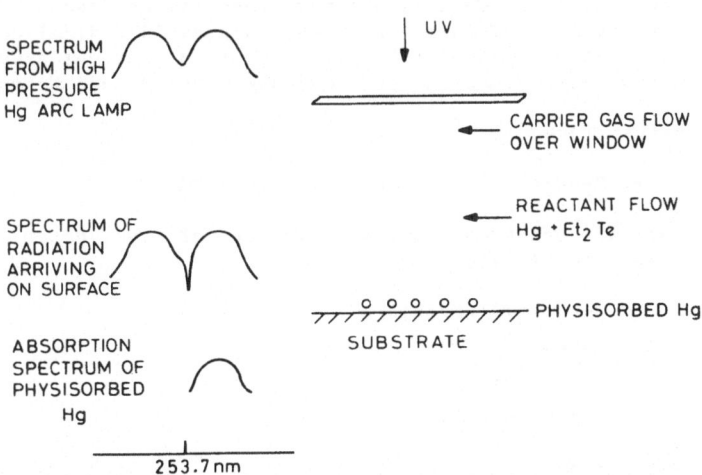

FIGURE 2. Schematic representation of the photo-sensitisation process for
HgTe epitaxy, showing selective UV absorption in an adsorbed layer of Hg.
The actual shift in the absorption spectrum is not known for this sytem, in
the diagram it is shown as a long wavelength shift.

Growth rates in the region of 1 μm/h can be maintained on the (100) and (110)
substrate orientations over a wide range of temperatures from 200°C to 300°C.
Thermodynamic model calculations by Mullin et al (10) of the electrically

active Hg vacancy concentration would indicate that at 250°C, under these growth conditions, the undoped carrier concentration should be below 1×10^{15} cm^{-3}. This, therefore, satisfies one of the requirements of the photo-epitaxial process that control of electrical properties should be by means of dopants rather than native defects. The growth rate of HgTe on an InSb or CdTe substrate does not vary significantly on changing the carrier gas from hydrogen to helium (11) but the amount of deposit on the silica reactor wall does decrease, allowing the flushed window region to be kept clear for up to 2 hours for helium compared with only 1 hour for hydrogen. These observations can be rationalised with the following possible reactions. Firstly using H$_2$ as the carrier gas (11)

$$h\nu + Hg + (C_2H_5)_2Te + H_2 \rightarrow Hg^*(\text{surface}) + (C_2H_5)_2Te + H_2 \tag{1}$$

$$Hg^*(\text{surface}) + (C_2H_5)_2Te + H_2 \rightarrow 2C_2H_5^\circ + H_2 + HgTe \tag{2}$$

$$Hg^*(\text{surface}) + H_2 \rightarrow Hg + 2H^\circ \tag{3}$$

$$C_2H_5^\circ + H^\circ \rightarrow C_2H_6 \tag{4}$$

Reaction (4) relies on the dissociation of diatomic hydrogen by photo-sensitisation (reaction 3). Product mass spectrometer analysis using the liquid N$_2$ sampling and distillation technique described elsewhere (13) shows that from this experimental arrangement the predominant product is ethane, as shown in the spectra in Fig. 3. The hydrogen quenching of the free alkyl radicals must be rapid otherwise large amounts of ethylene would be seen from a β elimination reaction. This is observed when low Hg pressures are used. When He carrier gas is used in place of H$_2$, this rapid radical quenching reaction cannot take place and it has been shown in previous studies that the predominant volatile reaction products are ethylene and butane (13).

$$h\nu + Hg \rightarrow Hg^*(\text{surface}) \tag{5}$$

$$Hg^*(\text{surface}) + (C_2H_5)_2Te + He \rightarrow 2C_2H_5^\circ + HgTe + He \tag{6}$$

The ethyl radicals can then react in a variety of ways, depending on the nature of the surface (11,13).

$$C_2H_5^\circ + HgTe \rightarrow Hg + C_2H_5Te^\circ(\text{etching}) \tag{7}$$

$$C_2H_5^\circ + C_2H_5^\circ \xrightarrow{\text{(substrate)}} C_4H_{10} \tag{8}$$

$$C_2H_5^\circ \rightarrow C_2H_4 + H^\circ \ (\beta \text{ elimination}) \tag{9}$$

In the absence of hydrogen, the longer free radical lifetimes will permit the etching type of reaction shown in Eqn. (7). Reactions such as (8) are likely to shorten the free radical lifetime by product surface catalysis which probably explains why, under these circumstances deposition of HgTe onto silica is more difficult. It is assumed that the product catalysed reactions would occur predominantly on the substrate.

These free radical mechanisms are probably very important in photo–MOVPE where control of the surface chemistry is needed for high quality epitaxial growth, and maintaining a clear window for illumination is an important technological consideration.

The photo–MOVPE of CdTe does not follow the same behaviour as HgTe. One critical implication is that rapid decomposition of the precursor alkyls

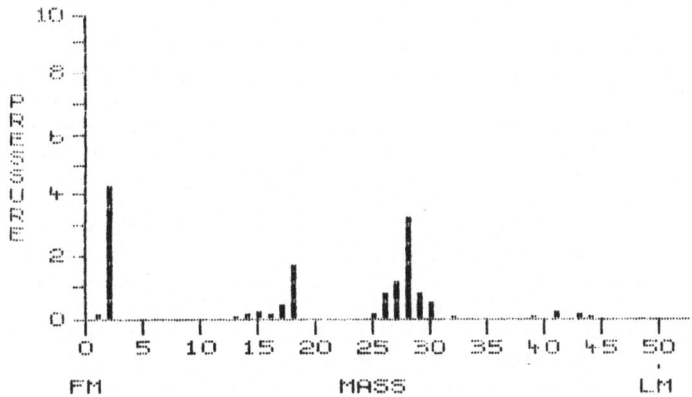

FIGURE 3. Product mass spectrum after HgTe photo-epitaxy in H_2 carrier gas at 25C°C. The mass peaks between 26 and 30 show the characteristic cracking pattern for ethane (C_2H_6).

dimethyl-cadmium (Me_2Cd) and Et_2Te can occur together in the vapour and the reaction is not so sensitive to the nature of a surface. The reasons for this are (a) Me_2Cd will decompose rapidly in the vapour to yield Cd as a consequence of just one photon absorption (11,18), and (b) the free radicals released can cleave off radicals from other alkyls such as Et_2Te or dissociate H_2 if this is being used as a carrier gas (11). It has been shown by Kisker et al (4) using a low pressure Hg arc lamp that the quantum efficiency for CdTe photo-MOVPE can be as high as 3; this has been attributed to the hydrogen radical effect. Using the high intensity UV source, that was used for HgTe photo-MOVPE, yields very high concentrations of Cd and Te in the vapour. This can be sufficient to homogeneously nucleate CdTe, precipitating the familiar "dust" on the substrate surface.

The problem occurs with the growth of the CMT ternary in that HgTe requires high alkyl concentrations and high UV source illumination intensity, and CdTe requires low alkyl concentrations and low intensity UV in order to avoid homogeneous nucleation. Early attempts at growing CMT under high intensity UV conditions resulted in severe homogeneous nucleation, even at Me_2Cd partial pressures as low as 0.04 torr in hydrogen (12). The change of carrier gas from hydrogen to helium reduced homogeneous nucleation and permitted a surface reaction to occur. The free radical mechanism in photo-MOVPE has been described in detail elsewhere (12) and will only be considered here insofar as it is relevant to in-situ processing.

The very low Me_2Cd partial pressures allowed, even when helium carrier gas is used, makes reproducible control of the alloy composition, x, more difficult than in pyrolytic MOVPE. The maximum permissible partial pressure can place a ceiling on the maximum x that can be grown without causing homogeneous nucleation, although x between 0.2 and 0.3 is possible (13). An implication of this observation, which has been subsequently confirmed by CdTe photo-MOVPE at 200°C, is that the presence of the Hg vapour not only enhances the decomposition of Et_2Te but also of Me_2Cd. The precise mechanism is not clear but the implication is that the alloy CMT growth is actually more difficult than either of the binaries.

An observation which has been used to support the free radical mechanism is that using a high intensity strip of illumination 0.5 cm wide on a low intensity background can give high quality epitaxy within the high intensity region but a "dust" deposit on either side (13). An increase in the free radical concentration, achieved by adding dimethyl-mercury (Me$_2$Hg) which will readily photo-dissociate to yield CH$_3^*$ radicals, can further suppress homogeneous nucleation (13). The width of the epitaxial region will expand and the effective ceiling on Me$_2$Cd concentration will increase but the penalty is a reduced CMT growth rate. A series of three sets of secondary ion mass spectrometry (SIMS) profiles in Fig. 4 illustrates the effect of adding 10 torr of Me$_2$Hg to the reactants. Fig. 4a shows for comparison major element profiles for a photo-MOVPE HgTe layer. The CMT layer grown at x ($<$ 0.1) in a helium carrier gas is shown in Fig. 4b and a comparison with Fig. 4c shows that adding Me$_2$Hg clearly has the effect of increasing the Cd content (to x = 0.61) and correspondingly reducing the Hg content (15). The presence of Hg in Me$_2$Hg is just incidental in these experiments as the yield of Hg is far smaller than the ambient Hg partial pressure of approximately 30 torr.

The ability to grow epitaxial layers of CMT by photo-MOVPE at temperatures below 250°C has been demonstrated but the precision in the control of alloy composition is not yet sufficient for in-situ deposition of CMT detector elements. The need for a purely photo-initiated deposition reaction has been satisfied as under "dark" conditions at 250°C and below, no deposition occurs. Precision in local deposition of epitaxial CMT may be achieved using a UV laser source. High intensity CW UV radiation ($\sim$ 100 mW) is now available from intracavity frequency doubling the main argon ion (514 nm) line. A laser source also offers the advantage over a lamp in being more quantitative (in calculating the photochemical quantum efficiencies) as a result of monochromaticity.

Preliminary laser deposition experiments using an unfocussed beam of 35 mW at 257 nm, depositing CdTe on to InSb gave the thickness contour plot shown in Fig. 5. The oval shape is due to the angle at which the beam was striking the substrate and shows that the deposition rate follows the radiant intensity.

A CMT detector structure will require the growth of a number of layers in-situ with both n and p type doping. Multilayer structure growth using photo-MOVPE has been demonstrated (14), but no doping has yet been reported. Alkyl sources of the group IIIs will photo-dissociate under similar conditions. The group V hydrides absorb UV at $<$ 200 nm and would not be suitable for p-type doping. Other group V-organic precursors, such as alkyl substituted PH$_3$ and AsH$_3$, would be needed with high UV absorption cross sections at longer wavelengths ($>$ 200 nm).

2.2 Characterisation of photo-MOVPE layers

The particular properties of the photo-MOVPE layers relevant to this paper are crystal quality, electrical characteristics, interface composition control and purity.

A single crystal X-ray diffractometer has been used to establish epitaxial growth (12) but little detailed information can be deduced from these results. Either double crystal X-ray diffraction or cross sectional TEM can provide this more detailed information. The best epitaxial growth has been onto good quality CdTe substrates and photo-MOVPE layers with double crystal rocking curve widths of $\sim$ 40 arc seconds have been obtained (15). Cross sectional diffraction contrast TEM shows that the substrate layer interfaces are not ideal giving rise to a larger interfacial dislocation density than that due to lattice mismatch (14). The dislocation density does decrease rapidly from

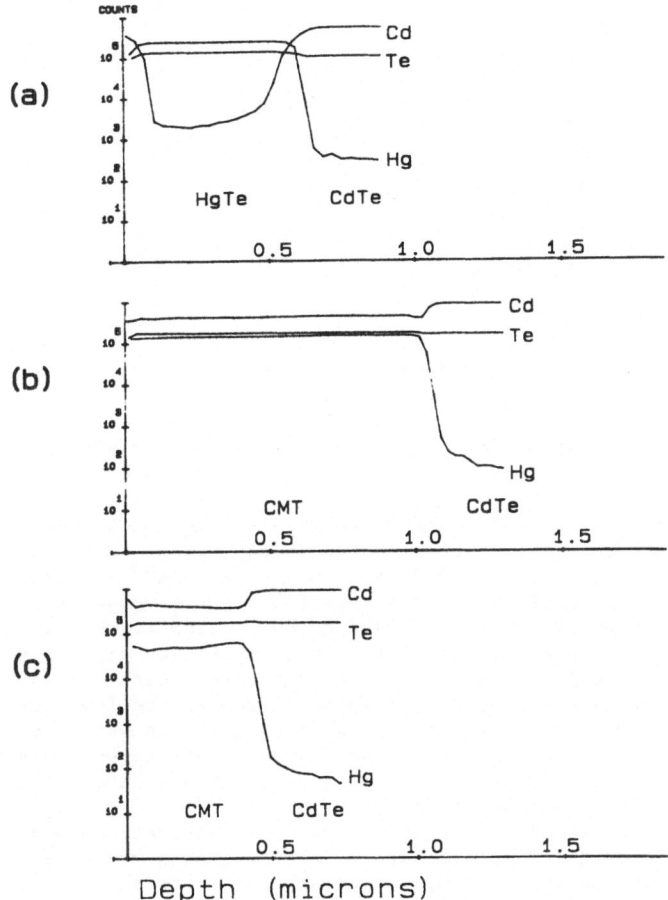

FIGURE 4. Major element SIMS depth profiles from a Cameca 3F lMS for (a) a HgTe layer grown at 250°C on to a CdTe substrace and with a CdTe capping layer, (b) a $Cd_x Hg_{1-x}Te$ layer (x < 0.1) and (c) a $Cd_x Hg_{1-x}Te$ (x = 0.61) layer grown with the addition of Me_2Hg to the vapour stream.

the interface on (100) oriented substrates which indicates that buffer layers would have to be grown for best detector performance. On III-V substrates additional problems of interfacial phases can occur which also would require the growth of a suitable buffer layer. The cross sectional TEM micrograph in Fig. 6 shows a (100) HgTe photo-MOVPE layer grown onto an InSb substrate. The interfacial layer is probably due to an indium rich surface which has formed indium telluride phases. This is a near lattice matched system with mismatch of only 0.2% but as with the CdTe substrate interface, the interface disruption is considerable with defect concentrations decreasing rapidly from this interface.

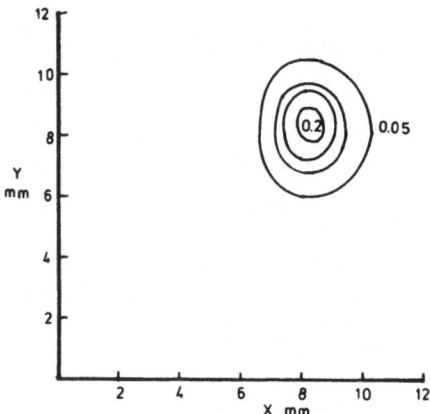

FIGURE 5. Contour plot of CdTe thickness (in μm) on an InSb substrate at 200°C using 35 mW at 257 nm. The contours are in steps of 0.05 μm.

Electrical measurements on photo-MOVPE layers have mostly been confined to Hall measurements on HgTe. A detailed study of variable temperature concentrations in the region of 2×10^{16} cm^{-3} to 1×10^{17} cm^{-3} and maximum Hall mobility in the region of 55,000 cm^2/V-s (14). These values are typical for epitaxial HgTe grown by pyrolytic MOVPE or MBE. SIMS trace element depth profiling showed a number of possible dopant impurities diffusing from the substrate/layer interface (14). This was consistent with the layers 1 μm thick or more giving the best Hall results. Even at the low growth temperature regime of 230-250°C, fast diffusing impurities can still contaminate the epitaxial layer, so suitable buffer layers would need to be grown as part of a detector structure. These layers would probably not have to be much thicker than 1 μm at a growth temperature of 250°C.

Interface abruptness of photo-MOVPE grown heterostructures has been reported elsewhere in detail (14) and for the chosen growth range of 200-300°C typical values of 300-400 Å given. The sharpest interface was for a HgTe/CdTe heterostructure grown at 230°C which gave an interface width of 105 Å as measured by SIMS. It is not clear whether the limitation on interface width is currently interdiffusion or compositional grading due to gas flow switching. The predicted interdiffusion coefficient from these results is 100 times larger than that measured by Arch et al (17) for a superlattice. One major difference between pyrolytic and photolytic MOVPE is that in the former any changes in gas composition will automatically have an effect on the composition of the growing layer whereas in the latter, no growth occurs when the UV is off. Therefore, changes in gas composition can be made independently of the growing layer. This should remove all reliance of interface abruptness on the gas flow switching. However, because of the method by which Hg vapour is introduced, by means of a heated Hg source inside the reactor cell, the Hg flux has to be continuous and changes of composition made in the presence of ~ 30 torr of Hg. The effect of Hg partial pressure on growth rates has already been discussed in section 2.1 and the usual method of saturating a Te vapour with Cd to achieve CdTe composition may just result in the formation of dust. A buffer layer has been grown with the Hg zone cool and then the UV lamp turned off while the Hg zone was warming up. Subsequent CdTe layers have been grown under low alkyl concentration conditions with the Hg zone on and without any dust formation. The penalty is

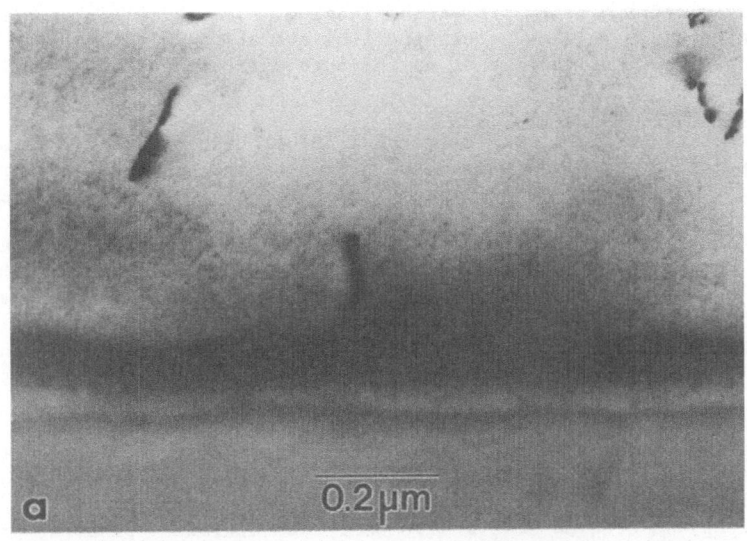

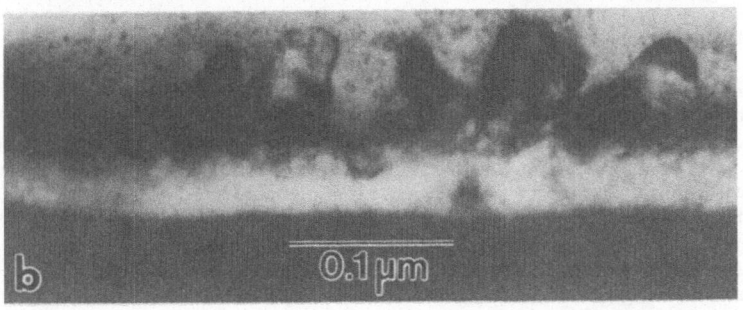

FIGURE 6. Cross sectional TEM micrograph of a HgTe(100) layer on InSb(100) showing (a) low dislocation density in layer but with poor interfacial quality and (b) of the interface region showing second phase precipitates.

clearly in the growth rate achievable for the CdTe deposited under these conditions. Despite this, the added flexibility of photo-initiated deposition, makes the prospects for a high degree of control at interfaces very good. Modifications to interface composition profiles to tailor them for specific device requirements may be important in heterojunction devices which will not tolerate subsequent heating to interdiffuse the interface.

3. FUTURE PROSPECTS FOR IN-SITU PROCESSING OF INFRARED DETECTORS

In this section the use of photo-epitaxy in combination with other photolytic processing techniques will be considerd in a single process fabrication of an infrared detector array. The objectives are in removing some of the uncertainties of interface control, introduced by current processing techniques, and to have a high throughput industrial processing capability. Photo-MOVPE would play an essential part in this processing facility and would have to be compatible with other processing steps either carried out in the same chamber or in a multiple interconnecting chamber system. The first issue to be considered is whether the selective area epitaxy should

be by laser writing or by projection imaging, i.e. serial or parallel. The
steps involved in in-situ processing from the epitaxial deposition through
to the contact metallisation of an infrared detector structure will then be
described.

3.1 Serial or parallel processing?

The relative merits of serial or parallel processing are (a) parallel
requires masks, serial does not, (b) ultimate line width resolution may be
better for serial than parallel, and (c) processing rates are far more rapid
for parallel processing.

Using non-contact masks for parallel processing will require mask alignment in multi-step processing but this should be no more complex than in
current process technology. Serial processing will also require careful
alignment and its advantages in this respect would only be in "one off"
structures where mask fabrication could be costly and time consuming.

The ultimate linewidth resolution for CMT detectors is not an important
issue. Detector pixels are typically on a 50 μm pitch hence it is difficult
to envisage that submicron resolution would be necessary.

Processing rates are an important issue, especially considering the physical dimensions of even quite modest arrays. A 32 x 32 element array would
occupy 1.6 x 1.6 mm^2 and the size of arrays required in the future 256 x 256
would occupy an area of 12.8 x 12.8 mm^2. The ultimate epitaxial processing
rates of course depend on the growth rates for the CMT structure. As already
mentioned, growth rates of ~ 1 μm/h are possible at present, limited by
either the surface reaction rate, as in the case of HgTe, or vapour nucleation in the case of CdTe. With the latter, it is conceivable that higher
growth rates could be achieved if the beam volume was small, hence reducing
the probability for vapour nucleation. Taking an optimistic view, the growth
rate would ultimately be limited by the transport rate of the alkyls or dissociation products to the growth surface. For alkyl partial pressures in the
region of 1 torr and for beam saturation conditions, i.e. all the alkyl
vapour is dissociated by the beam, the maximum growth rate is in the region
of 50 μm/h. These growth rates have been achieved for CdTe epitaxy by
pyrolytic-MOVPE under transport limited conditions. Taking this growth rate
for laser writing of a 32 x 32 element array, total thickness of 10 μm, will
give a deposition time of 200 hours. As the figures taken here have been
optimistic, the laser writing route can be ruled out on the grounds of processing rates being too slow. For a step and repeat imaging technique where
a complete array is deposited in parallel a slower maximum growth rate of
5 μm/h, will be used because of the increased probability for vapour nucleation. One array could be deposited in 2 hours and has the added advantage
of not increasing processing times going to larger array sizes. A number of
arrays could be deposited by a step and repeat process. Step and repeat
could be a good compromise between processing rates and compositional uniformity over a wafer. The most rapid process would be depositing arrays over a
whole wafer, which may be 3-6 cm diameter, in parallel. However, the range
of alloy composition which can be tolerated by the specification of detector
cut-off wavelength would have to be met over the whole wafer irrespective of
size. This presents a problem for scaling up the fully parallel process and
is essentially the problem encountered in current ex-situ CMT detector fabrication. To maintain the absolute variation in alloy composition over a wafer
within specification, the gradient in composition must decrease as wafer
size increases. This is not the case for step and repeat processing where
no improvement in uniformity gradient would be required for increased wafer
size. The prospects for scaling this technology are, therefore, very good.

A final comparison should be made with current CMT epitaxy by photo-MOVPE
and a step and repeat in-situ process. At present, in order to grow 10 μm

of CMT, the growth time is approximately 10 hours, to which further processing time such as mesa etching would have to be added. The current scale is
approximately 1 cm^2 which would make ten 32x32 arrays or just one 256x256
array. This would place the step and repeat processing time at between 2
and 20 hours.

3.2 In-situ fabrication of a heterojunction detector

The 40 definable steps in current infrared detector processing could be
reduced to something like 7, all performed in-situ. A proposed heterojunction detector on a silicon circuit for contact and readout is shown schematically in Fig. 7. The substrate would be passivated with SiO$_x$ for loading
into the processing cell and the following in-situ steps would then be
performed.

(1) Laser etching of windows through SiO$_x$ to expose clean Si surface.

(2) Photo-MOVPE growth of CdTe buffer and narrow band gap p-type CMT.

(3) Photo-MOVPE growth of wide band gap n-type CMT.

(4) Photo-MOVPE of CdTe passivation layer providing conformal coverage of
structure.

(5) Etch holes through CdTe to n- and p-CMT layers.

(6) Etch holes through SiO$_x$ to contact tracks.

(7) Deposit metallisation to contact structure.

These steps are indicated with the numbers in Fig. 7. The removal of the
SiO$_x$ for exposing Si substrate (step 1) or to expose the contact tracks
(step 6) could be achieved using photo-thermal processes, provided that the
total heat dissipated is small. The other steps have to be photolytic in

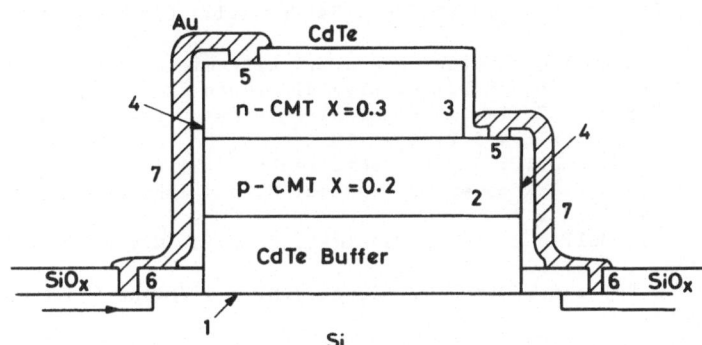

FIGURE 7. A proposed CMT infrared detector that could be completely fabricated by a seven step in-situ laser process. The processing steps are indicated by the numbers shown (see text).

order to avoid excessive heating of the epitaxial structure. An experimental in-situ processor could be connected to a UHV surface diagnostic system,
via a load lock to examine surface cleanliness and surface structure after
certain steps. Alternatively optical surface diagnostic techniques such as
reflection absorption infrared spectroscopy (RAIRS), ellipsometry or surface
Raman spectroscopy could be used. As the whole process is in-situ, interfacial contamination would be avoided and no damage should be introduced as
the energy source (UV) is ~ 5 eV; also low temperatures will minimise stress.

4. CONCLUSIONS

The current state of CMT infrared detector fabrication and requirements for future large focal plane arrays have been outlined. Photo-MOVPE for selective area deposition of CMT epitaxial structures is ideally suited to these requirements and could be integrated with other laser processing techniques for a complete in-situ fabrication technology. The high cost of infrared detector arrays and the difficulties in processing CMT makes this a prime example of an area where laser in-situ procesing could make a rapid impact.

The present position, towards this eventual goal, is that high quality CMT and CMT heterostructures can be grown but improvements are needed in control of alloy composition, growth rates and background carrier concentration. Early results on laser deposition of CdTe look encouraging and are consistent with the need for selective area epitaxy. By in-situ fabrication of CMT focal plane arrays, very large, complex arrays would not only become a reality but could also be cheap.

ACKNOWLEDGEMENTS

The authors gratefully acknowledge the following in the preparation of this manuscript, Dr A.G. Cullis and Mr N. Chew for the TEM micrographs; Mr G.W. Blackmore for the SIMS profiles; Mrs O.D. Dosser for EDX thickness mapping and Mrs J. Clements for technical assistance.

REFERENCES

1. Cole S.: J. Materials Science 15 (1980) 2591.
2. Ashley T., Crimes G., Elliott C.T., Harker A.T.: Electronics Letters, 22 (1986) 611.
3. Cullis A.G., Chew N.G., Hutchinson J.L.: Ultramicroscopy 17 (1985) 203.
4. Kisker D.W. and Feldman R.D.: J. Crystal Growth 72 (1985) 102.
5. Boyd I.W.: Springer Series in Chemical Physics 39 (1984) 274.
6. Houle F.A.: Proc. SPIE 459 (1984) 110.
7. Ehrlich D.J., Tsao J.T.: J. Vac. Sci. Technol. B1 (1983) 969.
8. Kukimoto H., Ban Y., Komatsu H., Takechi M. and Ishinati M.: J. Crystal Growth 77 (1986) 223.
9. Irvine S.J.C., Mullin J.B. and Tunnicliffe J.: J. Crystal Growth 68 (1984) 188.
10. Mullin J.B., Royle A., Giess J., Gough J.S. and Irvine S.J.C.: J. Crystal Growth 77 (1986) 460.
11. Mullin J.B. and Irvine S.J.C.: J. Vac. Sci. Technol. A4 (1986) 700.
12. Irvine S.J.C., Giess J., Mullin J.B., Blackmore G.W. and Dosser O.D.: J. Vac. Sci. Technol. B3 (1985) 1450.
13. Irvine S.J.C. and Mullin J.B.: J. Crystal Growth 79 (1986) 371.
14. Irvine S.J.C., Giess J., Gough J.S., Blackmore G.W., Royle A., Mullin J.B., Chew N.G. and Cullis A.G.: J. Crystal Growth 77 (1986) 437.
15. Irvine S.J.C., Mullin J.B., Blackmore G.W., Dosser O.D. and Hill H.: Mater. Res. Soc. Proc. Symp. R (1986) (in press).
16. Irvine S.J.C., Mullin J.B., Hill H., Brown G.T. and Barnett S.J.: (to be published).
17. Arch D.K., Faurie J-P, Chow P., Staudenmann J-L and Hibbs-Brenner H.: 1985 US Workshop on Physics and Chemistry of Mercury Cadmium Telluride.
18. Chen C.J. and Osgood R.M., J. Chem. Phys. 81 (1984) 327.

SYNTHESIS AND CHARACTERIZATION OF LASER DRIVEN POWDERS

E. Borsella, L. Caneve*, R. Fantoni
ENEA, Dip. TIB, Unità Settoriale Fisica Applicata, CRE Frascati, C.P. 65 -
00044 Frascati, Rome, Italy

1. INTRODUCTION

At the present time, there is a considerable interest in the potential use of ceramics as high temperature engineering materials in applications where metals have traditionally been employed, e.g. gas turbines and diesel engines[1]. This is because the resistance to thermal shock and oxidation is combined to higher operating temperatures (up to 1500 °C) compared with metals.

Ideal sinterable powders for fabricating high strength ceramics must be extremely fine ($\leq$ 1 µm), pure and uniform in size[2]. In principle, high quality powders could be produced in laser induced reactions between suitable gaseous precursors[3]. In fact, laser driven processes take place in a clean environment, with cold non reactive chamber walls and the final powder characteristics can be controlled by a careful choice of process parameters (such as reactant pressures, laser fluence and wavelength).

Recently there has been a considerable amount of activity in the field of laser driven powder production ; Si, Si_3N_4 and SiC have been synthesized in several laboratories[3-5]. However, successful applications rely on a detailed understanding of the nature of laser induced reactions and their effect on the properties of powders.

To this aim, in our laboratory the process of laser driven powder production has been carefully investigated by means of optical in-situ diagnostics. Si, Si_3N_4 and SiC have been produced by TEA CO_2 laser induced chemical reactions starting from SiH_4 alone or mixed with NH_3 and hydrocarbons. The formation of silicon oxynitride powders from SiH_4 and nitrogen oxides has been studied as well. In this later case, the possibility of driving the chain reaction initiated by laser excitation by varying the reactant partial pressures has been demonstrated. Several complementary off-line diagnostics (X-ray diffractometry, mass spectrometry, I.R. spectrophotometry) have also been employed.

The main results will be presented and discussed in the following.

2. EXPERIMENTAL APPARATUS

In our experiment, excitation of reactant gases was carried out by use of a Lumonics 102 TEA CO_2 laser operating in the 10 m range. The laser beam enters the reaction cell through a ZnSe window after having been mildly or tightly focussed by a 2 m focal length NaCl lens or a 2.5" focal length ZnSe lens. In the first case ("parallel optics") laser intensity was varied in the range 10-30 MW/cm^2, while in the second configuration ("focussing optics") the laser intensity in the centre of the cell was of about 500 MW/cm^2. SiH_4 was chosen as silicon precursor since spectroscopic data indicate strong absorption in the 10 µm range.

Si_3N_4 powders have been synthesized from gas phase reactions between silane and ammonia ; SiC powders have been formed after reaction of silane

D. J. Ehrlich and V. T. Nguyen (eds.), Emerging Technologies for In Situ Processing, 233–240.

with C_2H_4, C_3H_6 or C_2H_2. Attempts have also been made to obtain SiC powders from direct photolysis of tetramethylsilane.

Silicon oxynitrides have been produced starting from SiH_4 and N_2O, NO, NO_2 mixtures at different partial pressures.

It is well known[6] that light emission from electronically excited fragments is observed when silane is irradiated by a focused or unfocused CO_2 laser beam. In order to get an insight into the silane decomposition process and reaction with additives, the entire luminescence spectrum was detected with an optical multichannel analyzer (EG&G OMA III Mod. 1420).

As sketched in Fig. 1, the luminescence was collected perpendicularly to the laser beam by a 12 cm focal length quartz lens and focused on the entrance slit of a 0.32 m focal ISA spectrograph (Mod. HR-320) supplied with a 150 grooves/mm grating.

An intensified silicon photodiode array detector (512 elements) was mounted at the exit of the spectrograph and the luminescence spectrum was acquired with a resolution of approximately 5 Å per channel.

In our experiment, gas excitation takes place either by direct radiation absorption or through V-V collisional transfer. The last mechanism is responsible for the activation of non absorbing species.

Gaseous compounds react as soon as the threshold temperature is reached. Nucleation starts when the monomer concentration [M] exceeds the critical supersaturation level. The subsequent growth process is condensation:

$$[M]_n + M \rightarrow [M]_{n+1}$$

where $[M]_n$ is a nucleus with n momomers.

Particles in the laser beam act as sinks of monomers thus reducing the level of supersaturation below the critical value and only existing particles continue to grow. The narrow distribution of final particle sizes is due to this effect. Finally a dynamical equilibrium is reached between the

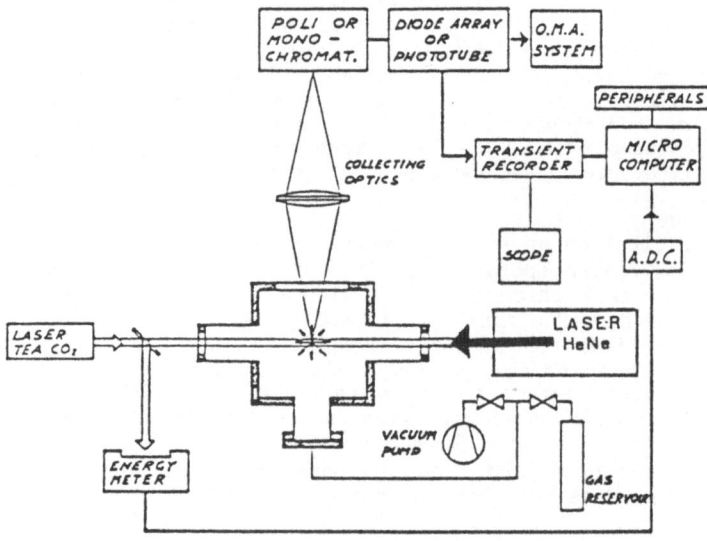

FIGURE 1. Experimental apparatus.

process of particle growth and removal by convective diffusion and gravitational settling. Detection of HeNe laser light scattered by particles can be used to monitor the process of nucleation and growth.

3. RESULTS AND DISCUSSION

Si. Polycrystalline, dark brown silicon powders are formed when silane is irradiated with the P(20) (944 cm^{-1}) line of a focussed or unfocussed TEA CO_2 laser.

At low laser fluence we have observed an orange luminescence (in the range 400-800 nm) which was attributed to an emission from singlet and triplet electronic excited states of SiH_2[5]. This result confirms the occurrence of the initial gas phase decomposition reaction:

(1) $SiH_4 \rightarrow SiH_2 + H_2$

Final mass spectrometric product analysis performed on photolized samples shows that under the present experimental conditions molecular hydrogen is the only gaseous reaction product. Powder was formed mostly by polycrystalline silicon (as proven by X-ray diffraction spectra); only small peaks due to the presence of Si-H bonds have been found in the IR spectra of powders pressed in KBr pellets. Thus, reaction (1) should be followed by predissociation of electronically excited silylene radicals (mostly in the triplet state)[7]:

$SiH_2 \rightarrow Si + H_2$

The global reaction occurring with an increase in total pressure turns out to be:

$SiH_4 \rightarrow Si(s) \downarrow + 2H_2 \uparrow$

The nucleation and growth of polycrystalline silicon powders is evident in the HeNe scattering measurements. The increase in powder production vs silane pressure is shown in Fig. 2(a).

When the laser radiation is focussed in the centre of the cell, a bright pink luminescence is emitted at p(SiH_4) > 3 Torr. Typical spectra consist of sharp peaks (in the range 200-700 nm) which have been assigned to Si, SiH, H, H_2 emissions[6]. A different fragmentation process is involved in this high fluence regime. The presence of electrons, detected between two electrodes, led to the conclusion that CO_2 laser induced silane breakdown takes place in the focal region. In spite of the different initial decomposition mechanism, final powders exhibit the same characteristics as in the low fluence regime, since silicon nucleation and growth take place as soon as supersaturation level is reached.

Si_3N_4. Brown powders are formed after irradiation of mixtures of SiH_4 and NH_3 at high laser fluence. IR absorption spectra of these powders exhibit features in the range 830-880 cm^{-1} (Fig. 3a) and around 490 cm^{-1} which are due to asymmetric and symmetric stretching of Si-N bond, respectively. X-ray diffraction spectra show the formation of Si_3N_4-β and the presence of free silicon. The TEM picture of laser synthesized Si_3N_4 is shown in Fig. 4. The content of free silicon decreases as the NH_3 partial pressure is increased, NH_3/SiH_4 ratios larger than 5 tend to produce the stoichiometric Si_3N_4-β compound in the ground state. In order to investigate the reaction mechanism we have added NO to the mixture looking for the luminescence from HNO* (at 734.1-771.8 nm) due to the following reactions:

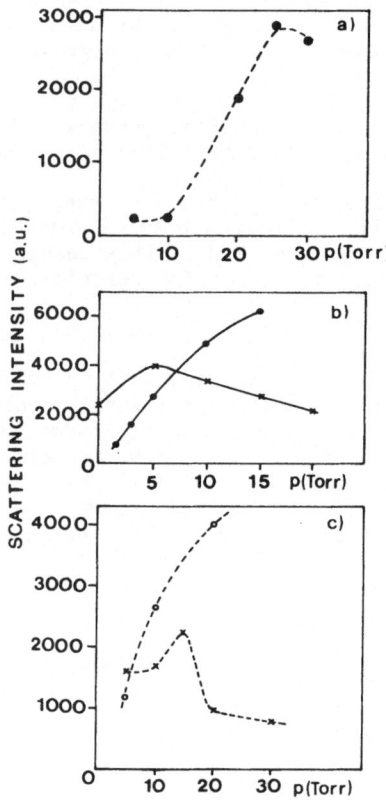

FIGURE 2. HeNe laser scattering intensity vs reactant pressure: a) • SiH_4; b) • p_{SiH_4} variable and p_{NH_3} = 15 Torr, × p_{NH_3} variable and p_{SiH_4} = 5 Torr; c) ○ p_{SiH_4} variable and $p_{C_2H_4}$ = 10 Torr, × p_{SiH_4} = 15 Torr and $p_{C_2H_4}$ variable.

$$NH_3 + nh\nu \rightarrow NH_2 + H$$

$$H + NO \rightarrow HNO^*$$

Luminescence from HNO* has not been detected at $p(NH_3) \leq 10$ Torr. Furthermore, no relevant difference has been found in powder production when the reactant gases have been irradiated at 944 cm^{-1}. On the basis of these results,we hypothesize that primary products of silane decomposition attack vibrationally excited NH_3 molecules. This hypothesis is further confirmed by measurements of HeNe light scattering at different partial pressures of SiH_4 and NH_3 (Fig. 2b). The scattered intensity at a fixed silane pressure (p_{SiH_4} = 5 Torr) is almost constant vs NH_3 pressure in the range 1-20 Torr, while it increases linearly vs silane pressure in the range 1-15 Torr at a fixed ammonia pressure (P_{NH_3} = 15 Torr).

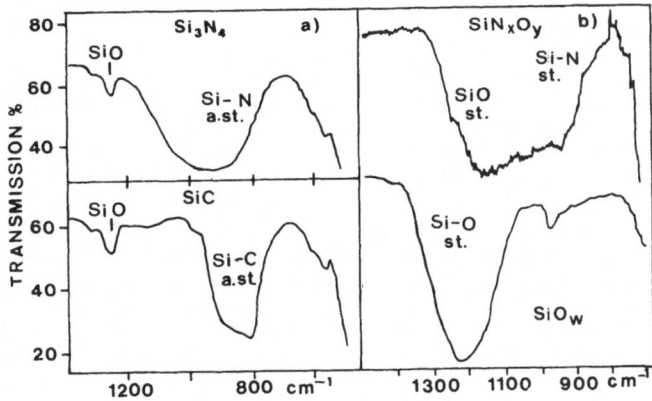

FIGURE 3. IR spectra of silicon containing powders; Si_3N_4 from SiH_4 (15 Torr) + NH_3 (45 Torr), SiC from SiH_4 (20 Torr) + C_2H_4 (20 Torr), SiN_xO_y from SiH_4 (17.5 Torr) + NO (2 Torr), SiO_w from SiH_4 (17.5 Torr) + NO (35 Torr).

FIGURE 4. TEM picture of laser synthesized Si_3N_4 (x 330 K).

Attempts to produce silicon nitride starting from mixtures of silane and nitrogen were unsuccessful.

At low laser fluence, the reaction between SiH_2 and vibrationally excited NH_3 occurs with a low rate, yielding a negligible production of powders.

SiC. Grayish-black silicon carbide powders have been produced starting from 1:1 mixtures of silane with several hydrocarbons i.e. C H , C H , C H. IR absorption spectra exhibit a band in the range 760-860 cm ascribable to asymmetric stretching of SiC bond (Fig. 3a). At high laser fluence, emission from C_2 (Swann bands) is observed at $p(C_3H_6) > 5$ Torr, $p(C_2H_4) \geq 7.5$ Torr and $p(C_2H_2) \geq 40$ Torr. As in the case of silicon nitride, SiC powders are formed even at pressures for which luminescence from excited hydrocarbon fragments is not observed, thus suggesting that reactions take place between silane fragments and vibrationally excited hydrocarbons. HeNe scattering measurements performed on irradiated mixtures of SiH_4 and C H give results similar to those obtained from mixture of SiH_4 and NH_3 (Fig. 2c). Powder production efficiency is dependent on the chemical additives according to the following trend : $C_2H_2 \gg C_3H_4$. As far as the stochiometry of powders is concerned, X-ray diffraction spectra show an increase in free silicon content going from C_2H_2 to C_2H_4.

Attempts to produce silicon carbide powders by adding CH_4 to SiH_4 have been completely unsuccessful. This is certainly due to the large mismatch between vibrational frequencies of SiH_4 and methane. On the other hand, C_2H_4 and C_3H_6 show absorption bands in the 10 μm range (940-950 cm^{-1}), hence these molecules can be excited either by laser radiation or by efficient resonant V-V transfer. A very special case is represented by C_2H_2 which does not absorb 10 μm radiation, but proves to be the most effective additive in SiC powder production both at high and low laser fluence. Efficient V-V collisional energy exchange with silane molecules could be favoured by quasi-resonance between frequencies of $2V_4$ levels of silane and V_2 or $3V_4$ vibrorotational levels of acetylene. Since at low laser fluence, as already mentioned, electronically excited SiH_2 is the primary product of SiH_4 photolysis, we can hypothesize that this species interacts strongly with C_2H_2. The possibility of a reaction mechanism involving collisional electronic excitation of acetylene is at present under investigation in our laboratory. Some remarks on this point will be reported in the following in connection with analogous processes occurring in oxynitride formation. Finally, it is worth mentioning that attempts have been made to obtain SiC powders by direct photolysis of $Si(CH_3)_4$. The final yield was too low to consider a process based on this compound as starting material; in fact the $Si(CH_3)_4$ laser induced decomposition proved to be quite ineffective at 10 μm, vibrational modes being too off-resonant with respect to CO_2 laser emission.

Silicon-oxynitrides. A very efficient process for the production of silicon oxynitride and silicon oxide powders takes place when mixtures of silane and NO (at different partial pressures) are irradiated by focused or unfocused CO_2 laser. Silicon oxynitride powders with different oxygen/nitrogen contents are formed when NO/SiH_4 ratio is varied as shown by the change in IR spectra features and color of the powders. Going from low NO/SiH_4 ratio (1:10) to high NO/SiH_4 ratio (4:1), the color of the powder changes from brown to white. Light brown, yellowish and grayish powders are formed at intermediate ratios. The corresponding spectra show either the domain of the SiN peak (835-880 cm^{-1}) over the SiO feature (1020-1090 cm^{-1}) or the reverse situation (Fig. 3b).

An analogous trend in silicon oxynitride formation is found when mixtures of SiH_4 with NO_2 or N_2O are irradited, although the reaction yields are lower.

As in the case of hydrocarbons, the possibility of effective V-V exchange ($2V_4$ [1890 cm^{-1}] of $SiH_4 \sim \gamma$ [1900 cm^{-1}] of NO) is responsible for the higher reaction yield in the case of NO with respect to NO_2 and N_2O.

For all these additives, a strong luminescent flash (Fig. 5a) is observed in the reaction volume either at low or high laser fluences (as in the case of C_2H_2). In the same spectral region the luminescence from pure SiH_4 decomposition is observed (Fig. 5b). From the comparison between the spectra reported in Fig.5, the luminescence observed in the presence of NO has been attribued to emission from triplet excited states of SiH_2. Paramagnetic molecules (such as NO and NO_2) are known to affect singlet-triplet mixing. The silylene radical is expected to be formed in the $^1B_1$ electronically excited state, while efficient predissociation of SiH_2 should take place from $^3B_1$ triplet state[7]. The role of NO could be to favour the mixing between $^1B_1$ electronic excited state of SiH_2 and $^3B^1$ state in the activated complex ($SiH_2...NO*$). Non paramagnetic species such as N_2O, anc C_2H_2 might play the same role provided that these species have been collisionally excited to low lying electronic triplet state. Work is in progress to prove this hypothesis.

In the case of laser assisted reaction of SiH_4 with NO, an increase in total pressure has been observed which mainly corresponds to formation of gaseous hydrogen, as proven by mass spectrometric final product analysis. At low NO/SiH_4 ratios, the global reaction is :

$$SiH_4 + NO \rightarrow SiNO + 2H_2$$

while at high NO/SiH_4 ratios, the global reaction seems to be :

$$SiH_4 + 4NO \rightarrow SiO_2 + 2H_2 + 2N_2 + O_2$$

followed by water formation and subsequent condensation (as shown by slow pressure decrease). Analogous results have been found for laser initiated

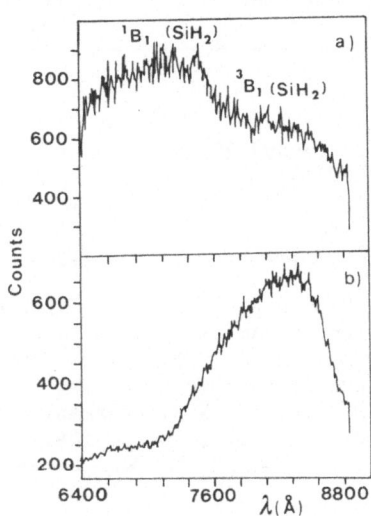

FIGURE 5. Chemiluminescence spectra measured after low intensity laser excitation ($\Phi = 1$ J/cm^2) : a) 20 Torr SiH_4, b) 17.5 Torr SiH_4 + 2 Torr NO

reactions of SiH_4 with N_2O and NO_2, the threshold being higher for the first than for the second ($p_{thr}(N_2O)$ = 6 Torr and $p_{thr}(NO_2)$ = 4 Torr added to 18 Torr of SiH_4).

Before ending the discussion on the formation of silicon oxynitrides, it has to be mentioned that ultrafine silicon nitride powders undergo spontaneous oxidation which make them very similar to the silicon oxynitrides produced at low NO/SiH_4 ratio.

The spontaneous oxidation of silicon nitride powders occurs on the grain surface as it appears on the pictures taken by electron microscopy (see Fig. 4).

4. CONCLUSIONS

Si, Si_3N_4, SiC and silicon oxynitride powder formation by laser induced gas phase homogeneous reaction has been demonstrated. The process has been investigated by several diagnostics and attempts to clarify the reaction mechanism have been made. Laser synthesized powders have specific surface areas about 100 m^2/g which is roughly ten times greater than the surface of commercially available powders, as shown by BET results (Table 1).

X-ray diffraction and electron microscope analyses indicate that uniform grains are produced with sizes of the order of 100 nm.

TABLE 1. Specific surfaces of laser synthesized powders.

Sample	Reactants	Specific Surface (m^2/g)
Si	15 Torr SiH_4	105
Si	30 Torr SiH_4	114
Si_3N_4	15 Torr SiH_4 + 45 Torr NH_3	160
SiC	20 Torr SiH_4 + 20 Torr C_2H_4	93
SiC	20 Torr SiH_4 + 20 Torr C_3H_6	51
SiC	30 Torr $Si(CH_3)_4$	132

FOOTNOTE AND REFERENCES
* ENEA guest
1. Ferretti M.: Sciences & Techiques, 22 (1986) 40.
2. Kent Bower H.: Mat. Sci. Eng. 44 (1980) 1.
3. Suyanna Y., Marra R.M., Haggerty J.S., Kent Bower H.: Am. Ceram. Soc. Bull., 64 (1985) and references therein.
4. Kizaki Y., Kandori T., Fujitami Y.: Japan. Journ. of Appl. Phys., 24 (1985) 800.
5. Cauchetier M., Croix O., Luce M., Michon M., Paris J., Tistchenko S.: Proc. 6th World Congress on High Techn. Ceramics, Milano, June 1986.
6. Borsella E., Caneve L., Giardini-Guidoni A., Mele A.: "Mechanism of Laser Induced Decomposition of Gaseous Polyatomic Molecules" in "Photons and Continuum States of Atoms and Molecules", N.K. Rahaman, C. Guidotti and M. Allegrini (Springer-Verlag, Berlin 1987), 16, p. 134.
7. Thoman J.W., Steinfeld J.I., Mc Ray R., Knight A.E.W.: to be published on J. Chem. Phys.

PHYSICAL PROPERTIES OF LASER WRITTEN CHROMIUM OXIDE THIN FILMS*

C. ARNONE AND C. ZIZZO

Dipartimento di Ingegneria Elettrica, Palermo (Italy)
and
Centro per la Ricerca Elettronica in Sicilia, Monreale (Italy).

1. INTRODUCTION

Deposition of chromium oxide films by Ar^+ laser photolysis of chromyl chloride vapors has been recently demonstrated[1]. Photolytic reactions in the adsorbed layer allowed the deposition of CrO_2 films on dielectric and semiconducting substrates using 488 nm radiation, with resolution down to 1 μm. Also, by photoinitiated pyrolysis of the vapor phase, single Cr_2O_3 filamentary crystals were grown. In this contribution a novel technique for film deposition is presented, together with optical and electrical characterization.

Chromyl chloride is a compound whose characteristics can be usefully exploited for setting up a general purpose laser deposition system. First of all, its unique property of allowing visible light photodeposition greatly simplifies alignment and optimization of the writing apparatus and the optical monitoring system. Moreover the evident pollution and corrosion it creates inside the vapor phase handling apparatus (source, reaction cell, exhaust lines and vacuum pumps) leads to designing a safety system suitable for most chemicals used in laser photodeposition. The experimental apparatus used for photolysis is summarized in Fig. 1. Several auxiliary components have been added to the conventional direct writing setup:
- an autofocus system or a video processor, connected to a vertical (z) platform, allows the automatic search of best focusing condition according to actual position (for optically finished surfaces) or best microcontrast (for rough and diffusing surfaces);
- a probe beam and a ratio detector are used for optical measurements on deposited films;
- a beam conditioner spatially filters the writing beam and makes the optical matching with the reduction optics (a microscope head is used).
Two fast galvanometer scanners have also been added, as explained below.

2. FAST SCANNING DEPOSITION

Certain materials (chromyl chloride is among them) need special precautions when being photodissociated, in order to obtain clean, evenly deposited films and repetitive quality results. The cases of radiation-adsorbate interaction and photo-pyrolytic deposition are particularly sensitive to the deposition procedure[2]. The experiment with chromyl chloride is discussed below, but it can be considered of general application.

D. J. Ehrlich and V. T. Nguyen (eds.), Emerging Technologies for In Situ Processing, 241–247.

242

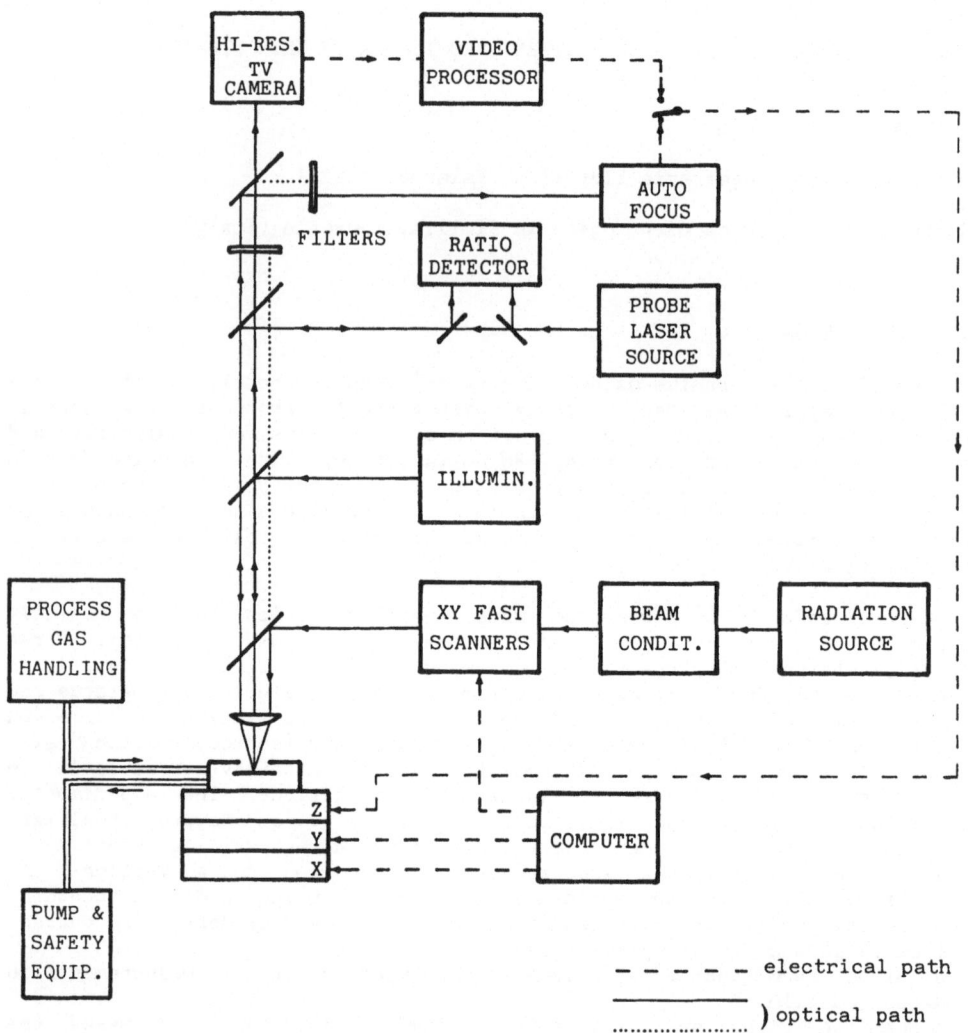

FIGURE 1. Experimental setup for laser direct writing.

An ideal photodeposition process leads to a linear relation between deposition rate and light power density. However several non-linearities can occur because of different limiting or enhancing mechanisms. Three major factors, influencing each other, must be taken into account when depositing from adsorbates: surface diffusion speed, surface temperature and radiation absorption of the deposit. An efficient and fast deposition process requires a diffusion speed proportional to laser fluence, that is an adsorbate flow toward the reaction area driven only by a concentration gradient. For the photolysis of chromyl chloride on silicon substrate a maximum laser fluence of 0.15 mJ/μm^2 at 100 mTorr has been measured. Exceeding the saturation fluence results in a loss of resolution, together with consistent substrate contamination outside the writing area. Fig. 2a shows the transversal profile of a 10 μm wide line, deposited in fluence limited conditions. It follows the gaussian shape of the beam, indicating a linear regime. Distortion appears in Fig. 2b, where saturation occurs together with spurious deposits.

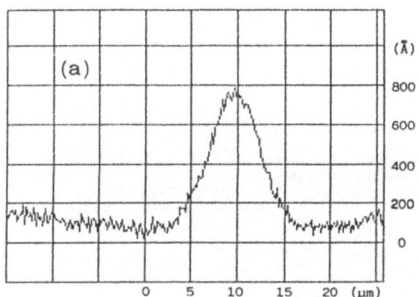 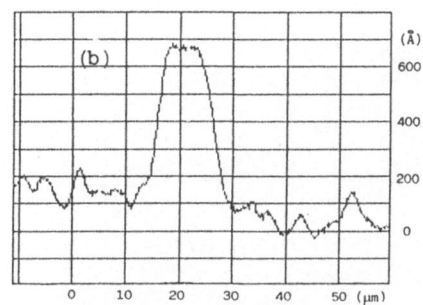

FIGURE 2. Profilometer plots of chromium oxide laser written lines: linear (a) and saturation (b) regime.

An insufficient surface diffusion speed creates uneven deposits, usually with a smaller thickness in the center of the writing spot. The peripheral part of the writing beam, in this condition, photodissociates all the adsorbed material diffusing toward the reaction area, before it reaches the center of the spot. Moreover surface temperature affects diffusion and, more relevantly, fixes the amount of coverage, that is the number of molecules per unit time and unit surface that can be dissociated and deposited. For this reason a deposition in the linear regime, while allowing a well controlled deposition speed, necessarily creates a film whose thickness follows the beam profile (usually quasi-gaussian). Difficulties then arise when making large area deposits by writing more lines side by side.

Another source of non-linearity is the beam induced temperature increase of the deposited film, related to the possibility of a photolytic to pyrolytic transition in the writing process. To minimize the non-linearities and get very uniform deposits a slow and radiation fluence limited process should be adopted. In order to limit the number of photons per second interacting with the adsorbate a low intensity beam is necessary, if slow mechanical stages are used for the movements of the substrate. This is not practical for non-laboratory applications.

Fast scanning technique, coupled with conventional stage translation, represents a very good compromise between quality and deposition speed, especially when paths larger than laser spot size must be written.

A conventional x,y (translation) and z (focus) stage is used for substrate slow scanning or step and repeat operations, while a couple of position servo-controlled mirrors allows fast beam scanning on limited areas, with a short dwell time but several repetitive passes (Fig. 1). Beam induced temperature rise of the substrate or deposited layer under these conditions becomes negligible, together with deposit edge irregularities. Fig. 3 shows the results obtained by a step (substrate translation) and repeat (beam scanning) deposition of chromium oxide on a metallic substrate, with 2 µm resolution. A 3 sec interval between steps and 5 scans/sec were adopted, at 50 mTorr vapor pressure. Variable thickness or width lines can be also created with this technique with a single "slow" scan, combined with a transverse "fast" scan, as shown in Fig. 4.

FIGURE 3. CrO_2 step and repeat writing on steel.

FIGURE 4. Variable thickness and width CrO_2 deposit on Si.

To some extent beam scanning for direct writing can be compared with projection lithography[3], where a mask is uniformly illuminated by a large beam and reduced by the microscope optics. With beam scanning a virtual mask is created by the writing beam in the image plane conjugated with the substrate plane. Also, a variable density mask can be simulated by properly modulating the laser beam.

3. OPTICAL CHARACTERIZATION.

From optical measurements on laser written dielectric film information relative to possible applications of the deposits and to the deposition process itself can be obtained.

The same writing equipment can be usefully adopted for refractive index and absorption coefficient measurements at different wavelengths. As shown in Fig. 1, a probe beam is sent inside the area where a film is being deposited by fast scanning. The reflected beam reaches a dual detector, where its ratio with the probe beam is calculated. In such a way the fluctuations of the probe source are minimized. Alternatively an optical stabilizer can be used[4]. With this method real-time thickness monitoring or delayed spatially resolved measurements is possible. This technique has been applied to chromium oxide films deposited on quartz substrates by 488 nm Ar^+ laser light. In situ probing at 633 nm gives the thickness versus reflectivity plot of Fig. 5.

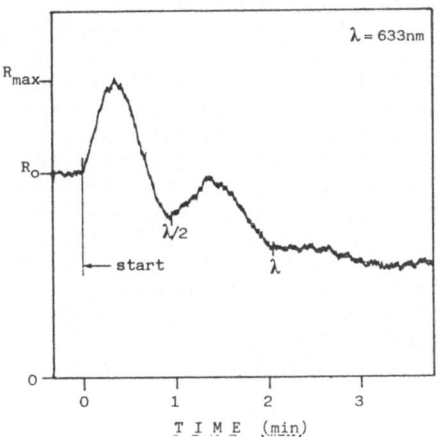

FIGURE 5. Optical reflectivity vs. deposition time.

R_o being the substrate reflectivity, one can immediately observe that a reflectivity increase occurs as soon as the film starts growing. This indicates a refractive index higher than that of the substrate. It is possible to evaluate the index from the ratio R_{max}/R_o. For a non absorbing film[5]:

$$n = \left[n_o \frac{1 + \sqrt{C}}{1 - \sqrt{C}} \right]^{\frac{1}{2}}$$

with

$$C = \frac{R_{max}}{R_o} \left[\frac{n_o - 1}{n_o + 1} \right]^2$$

Actually the reflectivity plot asymptotically tends to a constant reflectivity after a thickness of a few wavelengths, indicating a strong absorption at 633 nm. Exact relations including the complex index should be used in this case. However the above equations can still be adopted as a first order approximation, giving a refractive index of 1.52, close to that of the quartz substrate used for the measurement.

The optical absorbtion length can be also easily evaluated, from the reflectivity data, around one micron thick (at 633 nm). From the same plot one can observe a proportionality between optical thickness (multiples of half wavelengths and time, indicating a linear link of laser fluence and deposition speed.

A straight deposition rate measurement as a function of dwell time can be easily obtained by combining fast scanning and microscope visual inspection. A properly shaped deposit has to be prepared for the purpose. By a computerized control of the beam scanning width in one direction (y), hyperbolic profile can be determined for the film width. This is obtained by applying a variable sawtooth or triangular signal to the fast scanning system, while the substrate is being slowly translated in the orthogonal (x) direction. Since for a fluence limited process the product between dwell time and scanning length is constant, a hyperbolic envelope is necessary in order to get an x scale proportional to thickness-i.e. deposition rate. Calibration of this x axis is readily obtained by observing the film through a low power microscope and illuminating the sample with monochromatic light, like a sodium vapor source (Fig. 6). From the refractive index, the spacing of interference fringes and their number, film thickness can be readily calculated as a function of laser fluence.

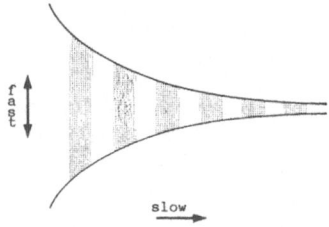

FIGURE 6. Film pattern for deposition diagnostics.

4. GAS INDUCED ELECTRICAL CONDUCTIVITY

Chromium oxide films are known to exhibit rather good dielectric properties[6]. For a half micron thick film a resistance in excess of 500 MΩ/square has been determined. However when exposed to air a resistance lower than 1 MΩ/square can be measured, while the original dielectric properties can be readily re-estabilished by vacuum pumping the sample or exposing it to a dry gas. With this simple experiment ambient water vapor is seen to be responsible for the conductivity. A suggested explanation of this phenomenon is the trapping of non photolyzed molecules in the chromium oxide film. Chromyl chloride is a strong hydrolyzing agent and tends to capture water vapor for originating chromium oxide and hydrochloric acid. Because of this reaction an increase in surface conductivity must be expected. To evaluate this interesting behavior some samples were prepared, by fast scan writing an oxide layer in the gap between two aluminum vacuum deposited fingers. Electrical conductivity was measured whenever the samples were exposed to a carrier gas containing a percentage of water vapor. At the same time a change in surface morphology could be observed, with the formation of a green and porous oxide layer. Both increases in conductivity and surface changes appeared only in presence of fresh or flowing gas, supporting the hypothesis of surface reaction.

The very fast ($\approx$ 0.1 sec) and reversible conductivity change that can be obtained from these laser written sensors suggests their possible use for dry gas monitoring applications.

(*) Work supported by Italian CNR and MPI.

REFERENCES

1. C.Arnone, M.Rothschild, J.G.Black and D.J.Ehrlich, Appl. Phys. Lett. 48, 1018 (1986).
2. D.J.Ehrlich and J.Y.Tsao, J. Vac. Sci. Technol. B1,969 (1983).
3. D.J.Ehrlich, Polymer Engineering and Science 26,No.16 (1986).
4. S.Sciortino, Thesis, Dpt. of El. Eng., Univ. of Palermo (1986).
5. C.Calì, V.Daneu, S.Riva Sanseverino, Optica Acta 27,1267 (1980).
6. G.V.Pankina, D.I.Leikis, E.S.Sevastyanov, Elektrokhimiya 16,213 (1980).

UV LASER INDUCED OXIDATION OF SILICON IN SOLID AND LIQUID PHASE REGIME

E. FOGARASSY, A. SLAOUI, C. FUCHS

CENTRE DE RECHERCHES NUCLEAIRES (IN2P3), Laboratoire PHASE (UA du CNRS 292)
23, rue du LOESS, 67037 STRASBOURG CEDEX (FRANCE)

ABSTRACT

We present and discuss in this paper new results on laser induced oxidation of silicon irradiated with an intense UV light provided by a pulsed excimer laser working at 193 nm. The two different regimes of oxidation which take place either in solid or in liquid phase following the conditions of irradiation will be detailed.

INTRODUCTION

One of the most important steps in the fabrication of electronic devices is the formation of good quality SiO_2 films. From the point of view of device performances, it is very desirable to form these oxide films at a low substrate temperature. By contrast to conventional thermal treatment, lasers offer the possibility to confine high temperature processing to only the surface (<0.5 µm), within well-defined regions of good spatial resolution during very short periods of time. Different ways have been explored in this field using both CW /1-4/ and pulsed / 5 - 13 / lasers emitting infrared, visible or ultraviolet light.

We present and discuss in this paper new results on laser induced oxidation of silicon irradiated with an intense UV light provided by a pulsed excimer laser working at 193 nm. Two different regimes of oxidation are observed which depend on the conditions of irradiation : at low energy, the oxidation process takes place in solid phase, with a specific contribution of photolytic reactions induced at gas/SiO_2 and Si/SiO_2 interfaces. At high energy, the oxidation of silicon results from the fast diffusion of oxygen from air (or O_2 atmosphere) into liquid silicon. The incorporation of oxygen and the formation of a stoichiometric oxide layer depend on several factors such as the oxygen pressure, the intensity of the incident light, and the existence and nature of impurities which may be present in the silicon. These different parameters will be analysed in detail and the optimization of the oxidation process both in the solid and liquid phase regime will be discussed.

EXPERIMENTAL

UV oxidation of silicon was carried out in air or in dry O_2 atmosphere, at pressures in the range 10^{-1} to 200 torr, using repetitive UV pulses (60 Hz) of 20 nsec FWHM, provided by an ArF excimer laser ($\lambda = 193$ nm). The laser energy density could be varied between 10 and 500 mJ/cm². Irradiations were performed directly on virgin Si (FZ) of <100> orientation, and after implantation of different impurities such as Si (30 keV), As (40 keV), F (35 keV) and N_2 (5 keV) to doses in the range 2×10^{15} to 5×10^{16} cm^{-2}.

The crystal damage, oxygen incorporation and As redistribution following laser irradiation were analysed by Rutherford Backscattering and ion channeling techniques (^{4}He+, 1-2 MeV).

D. J. Ehrlich and V. T. Nguyen (eds.), Emerging Technologies for In Situ Processing, 249–256.
© 1988 by Martinus Nijhoff Publishers.

The bonding character,stoichiometry and density of the oxides were characterized by infrared absorption using a PERKIN ELMER 938G double beam grating spectrometer at room temperature.

The film thickness and refractive index were measured by ellipsometry (632.8 nm wavelength, 70° angle of incidence).

SOLID PHASE REGIME

Several groups have recently observed a specific influence of visible and ultraviolet light on oxidation process of silicon /14-16/. In these experiments, additional non thermal effects have been directly attributed to the use of ultraviolet photons of high energy. Especially with UV photons of 6.4 eV provided by the ArF excimer laser, it is possible to induce internal photoemission of electrons, at the $Si-SiO_2$ interface, directly from the valence band of Si into the conduction band of SiO_2 which needs photons of 4.2 eV /17/. These photo-injected electrons are able to combine with oxygen dissolved in SiO_2 to form negatively charged species which may play an important role in the oxidation process. It has been also suggested that the ultraviolet photons may interact in a non thermal fashion on chemical bonds in   the solid at $Si-SiO_{x<2}$ interface/12/ inducing not only bond rearrangements between oxygen and silicon atoms ($h\nu \geqslant 4.5$ eV) but also stable structural defects such as oxygen vacancies in SiO_2. These processes could be helped by the creation of a high density of electron-hole pairs ($> 10^{21}$ cm^{-3}) in the Si lattice by intense pulsed light provided by the excimer laser, which may weaken Si-Si bonds. We can notice finally that the diffusion of oxygen through the SiO_2 layer toward the $Si-SiO_2$ interface may be promoted by the UV photo-dissociation in the gas phase of O_2 molecules into atomic oxygen ($h\nu \geqslant 5.1$ eV). In this field, Fiori /12/ showed the possibility of using a pulsed KrF excimer laser ($\lambda = 248$ nm) emitting photons of 5 eV to induce oxidation of Si at very low temperature. However, the major difficulty in the interpretation of photonic oxidation processes is to evaluate accurately the thermal contribution resulting from the interaction between the incoming light and the solid /18/. This interaction could be very strong owing to the high absorption coefficient of Si at UV wavelength ($\geqslant 10^6$ cm^{-1}). We have evaluated numerically the surface temperature and its time dependence for Si irradiated with a pulsed ArF excimer laser of 23 nsec (FWHM) pulse duration having roughly trapezoidal shape. The numerical data, which are reported in Fig. 1. show that for a laser energy density below 200 mJ/cm^2, the surface temperature does not exceed 650K. For an energy density per pulse of 130 mJ/cm^2, a 30 Å thick oxide layer was grown at the Si surface after 10 min exposure time (3.6×10^4 pulses at 60 Hz repetition rate) in air, as deduced from the differential infrared absorption spectrum of Fig.2. The peak position corresponding to the stretching vibration of the Si-O bond which is located at 1050 cm^{-1} close to the value found for very thin thermally grown SiO_2 (1060-1065 cm^{-1}) could be partly related, as suggested by Boyd et al /19/ to the compressive stress in the oxide arising from the structural transition from crystalline Si to the amorphous SiO_2 for very thin films (less than 100 Å). The width of the IR band (~ 100 cm^{-1} at FWHM) is however significantly larger than for thermally grown thin oxides (~ 70 cm^{-1}). This behaviour could indicate a higher degree of disorder in the laser oxide layer grown at low temperature.

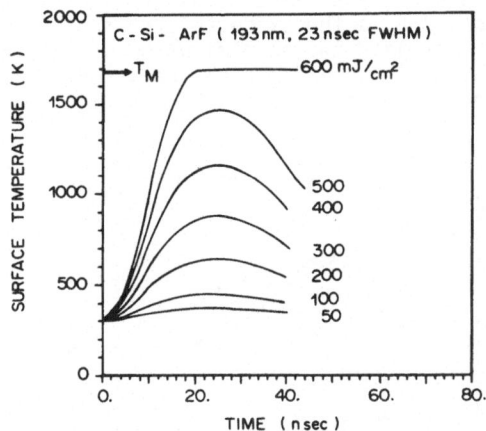

FIGURE 1. Calculated temperature at surface of virgin Si irradiated with a pulsed ArF laser (193 nm, 23 nsec FWHM) : thermal evolution as a function of the pulse energy density.

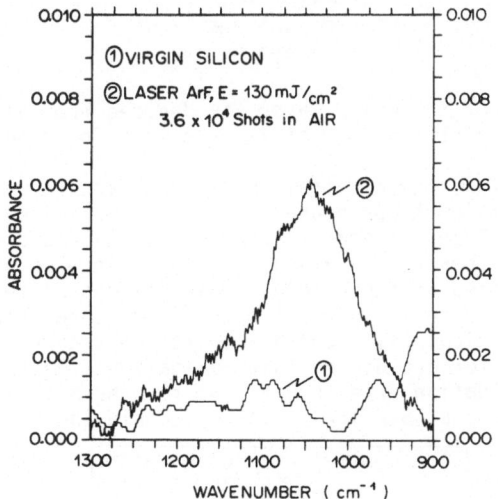

FIGURE 2. IR absorption spectrum (Si-O stretching band) of a 30 Å thick oxide layer grown on virgin Si by ArF laser induced oxidation in solid phase regime. (the influence of the native oxide grown on both the front and back of the sample has been considered in the evalutation of the oxide thickness).

The thickness of the oxide grown on the surface of Si as a function of the number of laser pulses, reported in Fig. 3 for $E \simeq 150$ mJ/cm², has been deduced from ellipsometry experiments assuming a refractive index of 1.46 for SiO_2. Oxides grown in dry O_2 environment are thicker than those obtained in air. We observe also a significant influence of the residual O_2 pressure in the oxidation chamber since a maximum thickness of about 50 Å is measured

after 10^5 laser shots at 10^{-1} torr. This behaviour is not yet totally understood, however the reduction in the oxide thickness grown at 200 torr (40 Å
after 10^5 shots) may be related to the diminution of the intensity of photon beam interacting with Si surface due to the strong absorption of the UV
light by ozone created in the chamber at 193 nm.

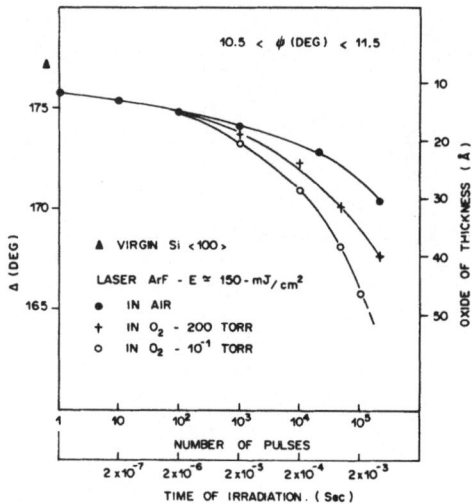

FIGURE 3. Thickness of the oxide layer (as deduced from ellipsometry) grown
on Si as a function of the number of laser pulses in solid phase regime.

Under the present experimental conditions, the oxide thickness grown at the surface of virgin Si at low temperature appears to be limited to around 50 Å for
reasonable times of laser treatment (30 min for 10^5 laser shots at 60 Hz).
As for conventional thermal oxidation, the laser induced oxidation rate is
significantly increased in the case of doped silicon since oxide thicknesses
exceeding 100 Å have been measured at high doping level (5×10^{21} As/cm², Fig.
4.). It would be particularly important to follow the redistribution of the
dopant at Si-SiO₂ interface during laser oxidation to determine if the enhancement in the oxidation rate results, as for thermal oxidation, from the
segregation of arsenic at Si/SiO₂ interface within the silicon /20/.

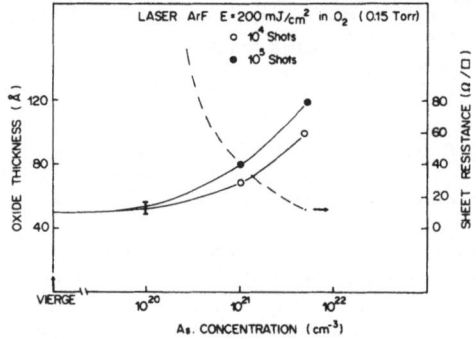

FIGURE 4. Influence of the doping level on the thickness of the oxide layer
grown on Si by laser induced oxidation in solid phase regime.

LIQUID PHASE REGIME
Under suitable conditions,   laser irradiation leads to melting of the
Si surface to a depth which may be several thousand angstroms, and solidi-
fication by liquid phase epitaxy from the underlying single crystal substrate with very high solid-liquid interface velocity of several meters/sec
/21/. The melting threshold of virgin crystalline silicon which is close on
600 mJ/cm² as deduced from the calculations of Fig.1. for a pulsed ArF laser, is considerably lowered for amorphous silicon (100-200 mJ/cm²) mainly
as a result of the reduced thermal conductivity of a-Si against c-Si. By
working close to the melting threshold of Si ($E \simeq 150$ mJ/cm²), we observed
a strong influence of the presence and type of impurity in the silicon, on
the oxidation rate (Fig. 5).

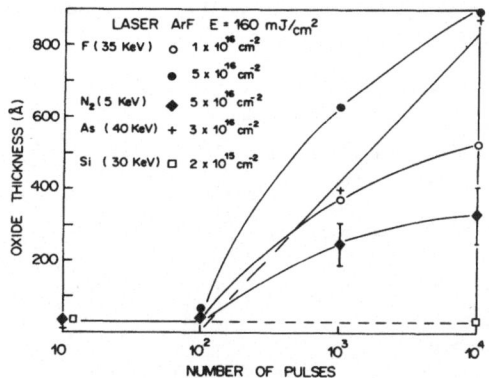

FIGURE 5. Influence of arsenic, nitrogen and fluorine on the thickness of
the oxide grown on Si irradiated with a pulsed ArF laser working in liquid
phase regime.

For As, F and N_2, thick oxide layers (up to 900 Å) are grown after 10^4 repetitive laser pulses. By contrast, we do not detect any significant oxygen
incorporation in   Si implanted samples.

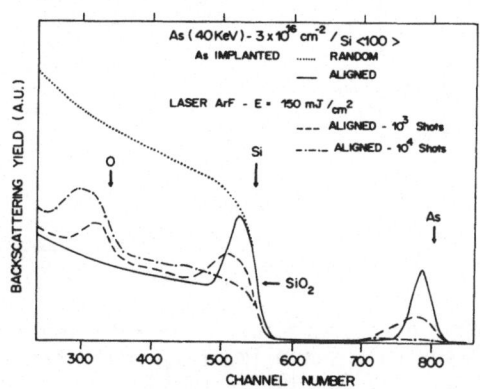

FIGURE 6. RBS spectra in random and channeling conditions for As (40 keV)
implanted Si before and after ArF laser (150 mJ/cm², 10^3 and 10^4 shots).

254

RBS spectra of Fig. 6., recorded in random and channeling conditions for As implanted Si, show clearly the influence of the number of pulses on oxygen incorporation and the formation of a quasi stoichiometric SiO_2 layer after 10^4 shots. In the same figure, we observe a strong redistribution of As, which extends both into the grown oxide layer and into silicon at a depth which corresponds to the existence of a melting phase regime. Evidence for the formation of a stoichiometric and low disordered SiO_2 on the surface of As implanted Si is also provided by the position ($\simeq$ 1075 cm^{-1}) and the width ($\simeq$ 70 cm^{-1}) of the IR band ($\simeq$ 1080 cm^{-1} and $\simeq$ 70 cm^{-1} for thermally grown oxide) following 10^4 repetitive laser irradiations(Fig. 7.). Very similar behaviour is found with N_2 and F into silicon. Infrared absorption spectra recorded in Figs 7. and 8. respectively for arsenic and fluorine, show that the oxide thickness, which may be easily deduced from the height of the IR peak, increases as a function of the number of laser shots and the concentration of impurities.

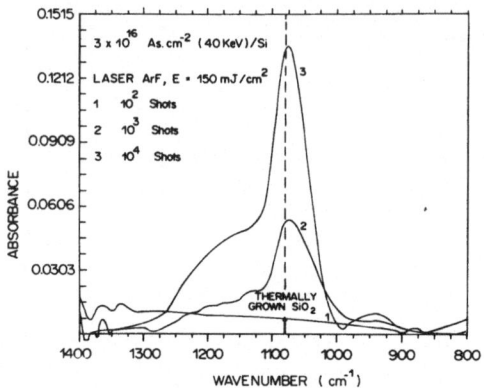

FIGURE 7. IR absorption spectra for As (40 keV) implanted Si as a function of the number of laser pulses (ArF, 150 mJ/cm², 10^2, 10^3 and 104 shots).

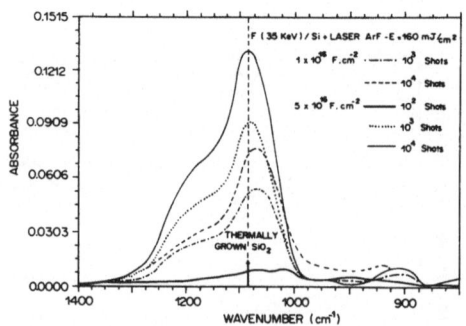

FIGURE 8. IR absorption spectra for F (35 keV) implanted Si as a function of the number of laser pulses and impurity concentration (ArF, 160 mJ/cm², 10^2, 10^3, 10^4 shots, 1 and 5 x 10^{16} F.cm^{-2}).

We have also examined the influence of laser energy density on the oxidation process.

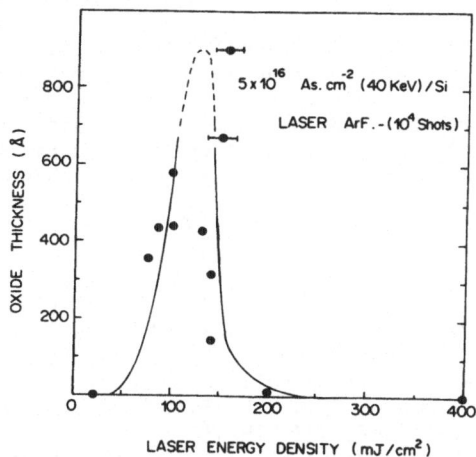

FIGURE 9. Influence of the laser energy density on the oxide thickness grown on surface of As (40 keV, 5 x 10^{16} cm^{-2}) implanted Si.

The results of Fig. 9. giving the thickness of the oxide grown on the surface of As implanted Si after 10^4 shots as a function of the laser energy density show that the oxygen incorporation presents a maximum value around the melting threshold (E $\simeq$ 150 mJ/cm^2). Below this value, the solid phase diffusion of oxygen into Si is limited by the duration of the irradiation which is very short (2x10^{-3} sec after 10^5 shots as reported on Fig. 3.). In the liquid phase regime, above 300 mJ/cm^2, we have a complete annealing of the As implanted surface layer after the first pulses. At the same time, the As profile is redistributed in depth into the molten zone by liquid phase diffusion, reducing a possible surface influence of the high doping level on the oxidation process. Among the different possible explanations we may suggest that the range of energies where strong incorporation of oxygen is observed (E $\simeq$ 150 to 300 mJ/cm^2) could be correlated to the surface formation of a fine and large polycrystalline region which was clearly identified by Narayan et al /22/ in the amorphous Si layer for energy densities comprised between the melting threshold and that needed for complete annealing. However, more experiments are needed in order to better understand the specific role of these different impurities in the oxidation process. In particular, it is important to examine the redistribution of As, F and N at Si-SiO$_2$ interface during the process. On RBS spectra of Fig. 6., arsenic appears to be redistributed both into the oxide and silicon substrate. On the contrary, preliminary experiments performed by SIMS and AUGER spectroscopy show that fluorine and nitrogen for one part escape from the solid to gaseous ambient through the oxide and for another part appear to be accumulated at Si-SiO$_2$ interface during the growing of oxide.

We may suggest that the large electronegativity of N and above all F when compared to that of Si could play a role in the oxidation process. Finally, due to the large differences in the thermal and optical properties of SiO$_2$ against Si, the behaviour of the grown oxide surface layer when submitted to the laser pulse above the melting threshold for silicon is questionable.

CONCLUSION

These experiments have clearly demonstrated that pulsed UV excimer lasers are able to induce oxidation of silicon both in solid and liquid phase regime.

In solid phase, thin oxide layers not exceeding 50 Å in virgin Si and 100-120 Å in doped Si have been grown at low temperature by repetitive laser irradiation (10^5 shots).

In liquid phase, we observed the formation of thick oxide layers (up to 900 Å after 10^4 shots) in presence of different impurities such as arsenic, fluorine and nitrogen which appear to play a major role in the oxidation process.

REFERENCES

1. Gibbons JF : Jpn. J. Appl. Phys. Suppl. 19, 121 (1980).
2. Boyd IW, Wilson JIB, WEST JL : Thin Solid Films 83, L173-L176 (1981).
3. Boyd IW, WILSON JIB : Appl. Phys. Lett. 41(2), 162 (1982).
4. Boyd IW : Appl. Phys. Lett. 42(8), 728 (1983).
5. Battaglin G, Della Mea G, Drigo AV, Foti G, Bentini GG, and Servidori M: Phys. Status Solidi(a) 49, 347 (1978).
6. Hoh K, Koyama H, Uda K, Miura Y : Jpn. J. Appl. Phys. 19, n° 7, L375-L378 (1980).
7. Chiang SW, Liu YS, Reihl RF : Appl. Phys. Lett. 39(9), 752 (1981).
8. Liu YS, Chiang SW, Bacon F : Appl. Phys. Lett. 38(12), 1005 (1981).
9. Cros A, Salvan F : Appl. Phys. A28, 241 (1982).
10. Orlowski TE, Richter H : Appl. Phys. Lett. V45, 2, 241 (1984).
11. Fogarassy E, White CW, Lowndes DH, Narayan J : "Beam Solid Interactions and Phase Transformations", edited by Kurz H, Olson GL, Poate JM, V51, 173, MRS 1985.
12. Fiori C : Phys. Rev. Lett. V52, n°23, 2077 (1984).
13. Siejka J, Srinivasan R, Perrière J, Braren R, Lazare S : "Dielectric Layers in Semiconductors : Novel technologies and devices", EMRS, edited by Bentini GG, Fogarassy E, Golanski A, Les Editions de Physique, 213 (1986).
14. Oren R, Ghandhi SK : J. of Appl. Phys. V42, n°2, 752 (1971).
15. Schafer SA, LYON SA : J. Vac. Sci. Technol. 19, 494 (1981) and J. Vac. Sci. Technol. 21, 422 (1982).
16. Young EM, Tiller WA : Appl. Phys. Lett. 42(1) 63 (1983), Appl. Phys. Lett. 50(1) 46 (1987) and Appl. Phys. Lett. 50(2) 80 (1987).
17. Williams R : Phys. Rev. 140, A569 (1965).
18. Fogarassy E, Unamuno S, Regolini JL, Fuchs C : Phil. Mag. B, V55, n°2, 253 (1987).
19. Boyd IW, Wilson JIB : Appl. Phys. Lett. 50(6), 320 (1987).
20. Grove A.S., Leistiko Jr. O, Sah CT : J. of Appl. Phys. V35, n°9, 2695 (1964).
21. Fogarassy E : "Energy Beam-Solid Interactions and Transient Processing" edited by Nguyen VT and Cullis AG, Les Editions de Physique, V4, 153 (1985).
22. Narayan J, White CW, Aziz MJ, Stritzker B, Walthuis A : J. Appl. Phys. 57(2), 564 (1985).

LASER ASSISTED PLASMA ETCHING OF SILICON DIOXIDE

O. Chevrier & A. Boudou
Direction Technique Groupe - Bull
78340 LES CLAYES SOUS BOIS

1.0 INTRODUCTION

The first step in the repairing or the reconfiguration of a circuit by metal straps is the opening of holes in the passivation layer below these straps. Plasma etching is a common process in microfabrication where the reactive gases are generated in the glow discharge, but the localization of the etching in this case requires prior masking of the surface by a patterned resist.

Laser assisted dry etching processes, on the other hand, offer a solution which does not require masking, as the laser beam, which is itself localized, is used for the etching. This type of technique has been extensively investigated in recent years using a wide variety of materials and light sources. Etching by laser radiation enhances the etch rate through several different chemical processes, such as photodissociative production of gas phase radicals, surface heating and light generated electron/hole pairs (1).

The combination of laser assisted chemistry and glow discharge allows the localized etching of materials with an otherwize slow dark etch rate.

In addition, it may be possible to accomplish localized metal deposition just after etching by laser assisted PECVD in the same apparatus (2). Little has been published on this method (3-6) and these papers are devoted mainly to the semiconductor etching.

In this paper, we describe the local etching of SiO_2 on aluminium in a DC plasma of CF_4/O_2 gas with an argon laser beam fixed at 514.5nm. We will show that there is an acceleration of the etching due to the laser light.

2.0 EXPERIMENT

For the experiment, specific Si wafers were processed and then cut in 1cm by 1cm squares. The samples consisted of a silicon substrate (figure 1) on which was grown a thermal SiO_2 layer of thickness 4400Å. Magnetron sputtered aluminium (7000Å) was then deposited, followed by PECVD SiO_2, processed at a temperature of 220°C and under a pressure of 0.9 Torr. For the etching experiment, each sample was attached to an aluminium holder with silver paint which is a good heat and charge conductor. A direct current discharge in a CF_4/O_2 gas, 20% O_2, 230 mTorr pressure, was performed in a planar type etcher (Figure 2). The upper electrode was connected to the positive high voltage, and the bottom electrode was grounded. The sample holder was mechanically clamped to the ground electrode.

A 15mm focal length sapphire lens was inserted into the upper electrode. This was used as other lenses were found to be attacked by the plasma. An argon laser beam (Spectra Physics 171) set at 514.5nm was focussed on the sample through this lens. In the outer optical path, a 300mm focal length lens allowed the optical adjustment of the system

D. J. Ehrlich and V. T. Nguyen (eds.), Emerging Technologies for In Situ Processing, 257–264.
© 1988 by Martinus Nijhoff Publishers.

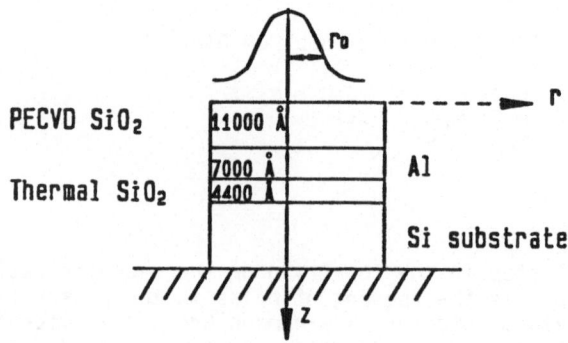

FIG.1 Sample structure

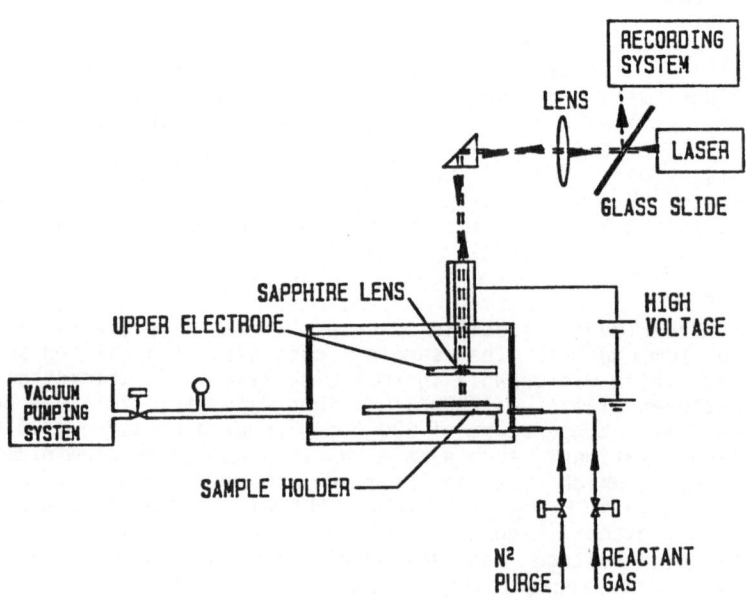

FIG.2 Experimental arrangement

(figure 2). In order to measure the laser spot diameter under these working conditions, a Si sample coated with photoresist was used. By burning the resist with the minimum necessary power, the diameter of the spot was found to be 15 μm. To obtain a smaller diameter, it would be necessary to use a lens with a shorter focal length, and to place it closer to the electrode. However, this would limit the quality of the plasma.

The SiO$_2$ thickness on the aluminium was measured with a Nanospec AF T250, using the bare aluminium as reference. Measuring the thickness before and after etching gave the dark etching rate. During the etching, a part of the beam reflected by the sample surface was chart-recorded with a photometer. As the intensity of this reflected beam is a function of SiO$_2$ thickness, through the equation :

$$R = \frac{R1+R2 + 2\sqrt{R1R2}\ cosa}{1 + R1R2 + 2\sqrt{R1R2}\ cosa}$$

where R = reflectivity of the sample (= SiO$_2$ on Al)
R1= reflectivity of the aluminium alone
R2= reflectivity of SiO$_2$ PECVD alone
a = 2πne/λ   n=reflectivity of the SiO$_2$ PECVD
                  λ=514.5 nm

the etching rate all through the experiment is known because the periodicity of R with the thickness of the oxide is
e = λ/2n = 171.5 nm in our case.

The etching rate was deduced from the time between sucessive maxima and minima.

3.0 RESULTS AND DISCUSSION

To obtain good values for the gas pressure and current in the DC discharge, etching rate versus pressure of CF$_4$/O$_2$ and current were explored. It was found that under 25mA current, the etching rate depends greatly on the pressure, the etching rate becomes constant when the pressure is more than 200 mT, at a current of 25mA. The conditions of etching finally chosen were 25mA current discharge, and 230 mTorr pressure. Figure 3 shows the SiO$_2$ thickness variation when etching in these conditions, giving an etch rate of 70 Å/min. During the experiment, the electrode temperature increased and a thermocouple gave an average temperature of 60°C. No stopping effect due to accumulated electrical charges on the sample surface was detected during the DC discharge.

In order to estimate the temperature attained locally during irradiation, a simple model was used (7), which considers :

- a layer of thermal conductor, the aluminium,
- a laser beam of power P, as the source of heat, of radius r0 at the aluminium layer
- no absorption of the beam by the PECVD SiO$_2$ deposited on the aluminium, and no loss by radiation or convection at the surface
- the only thermal loss is the loss into the silicon substrate, and the upper layer of silicon plays no role at all in the heat repartition, except in modulating the coefficient R.

This loss is mainly characterized by the contact between the silicon substrate and its support. If the substrate-sample holder thermal contact is

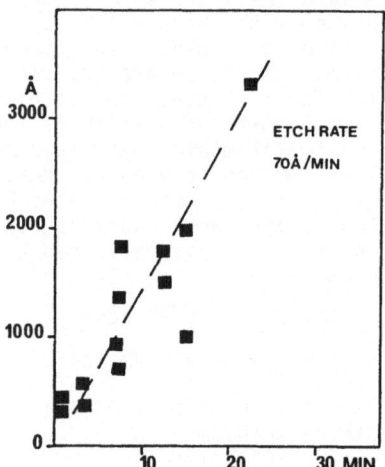

FIG.3 Etched SiO_2 versus time in CF_4/20 % O_2 plasma.
Current = 25 mA. Pressure : 230 mT.

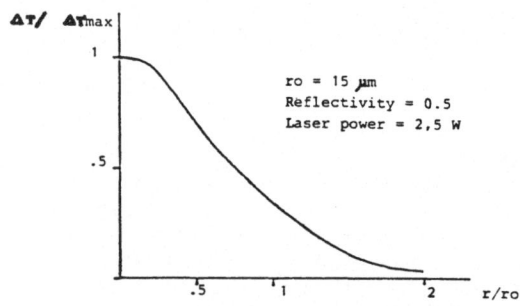

FIG.4 Distribution of temperature
on the surface of aluminium.

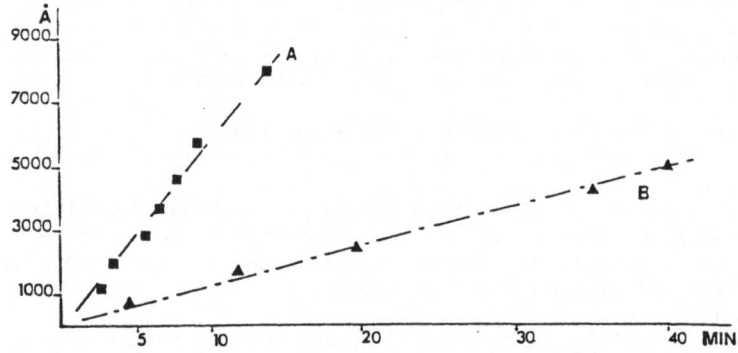

FIG.5 Etched SiO_2 versus time in plasma plus laser
A : power = 2,85 Watts - B : power = 0,25 Watt

considered to be perfect, the sample holder clamped and the ground electrode act as heat sink which governs the temperature of the base of the silicon. The repartition of the temperature at the surface (figure 4) shows that if r=2r0, the increase in temperature is neglegible. It is thus possible to deduce that the power necessary on the aluminium layer to melt the metal is 5W minimum in these ideal conditions, with a thermal oxide layer thickness of 4400 Å, and a reflectivity of 0.5. The conditions chosen for this experiment are below this limit. Using this thermal model, it is assumed that the temperature of the PECVD layer is that of the aluminium.

Furthermore, it has been shown that, without a good thermal contact with the sample holder, the melting point of the aluminium is easily reached, and this creates an opening in the PECVD SiO2 by explosive evaporation of Al. This condition of good thermal contact with the support, was verified before each experiment.

The experimental procedure in each case involved first starting the D.C. discharge, and admitting the laser beam five seconds later. Using chart recorded reflectivity, it was seen that the etching commences immediately after the switching-on of the laser. This is to be expected as the plasma is started first. From these recordings, knowing the initial oxide thickness, and using the time between maximum and minimum, the etching rate can be estimated. The etch rate was found to be constant i.e., there was always the same delay between maxima and minima, irrespective of the etching time This is shown by the straight line between experimental points in Figure 5. Given the method of measurement, and the fact that the etch rate depends on reflectivity, it is not possible to know precisely the etch rate at each point in the interval between each maximum and minimum, and thus it is not possible to say whether the rate is modulated or not by the reflectivity.

As the initial thickness of the SiO_2 was measured with a Nanospec measurement system, it is possible to estimate the thickness of the film etched up to the first maximum (or minimum). The points thereafter correspond to an exact given thickness and give a straight line when drawn in Figure 6. The fact that the extrapolation back to time zero passes through the point zero on the y-axis indicates that the supposition of constant etch rate even from time t=0 is correct.

In Figure 5, the points on the line A show the enhancement due to the power of the laser. 500 Å/min etching is obtained with a power of 2.85 W, which corresponds to more than five times the dark etching rate. In this particular case, the temperature at the middle of the laser spot is estimated to be 480 °C for an average reflectivity of 0.5 (the aluminium melting point is 660°C). During the course of the experiments, it was found that there was a dispersion from set to set due to the difference in reflectivity of the aluminium. It should be noted that no slow-down in the etch rate was seen before the film was totally consumed (which would be expected from film charging considerations), but a discontinuous change in reflectivity was seen when the SiO_2 was etched to a few hundreds of Angstroms from the aluminium surface. At this point, an explosion of this final layer occurs due to electrostatic discharge.

It was found, upon investigation of the the etching rate versus laser power, that the etch rate increases as the power increases, as is shown in Figure 6. It was found that there was no threshold minimum power for accelerated etching. Likewise, it was found that there was no maximum power - except the melting of the aluminium itself. The dispersion in the results was found to be no greater than the dispersion for etching in plasma

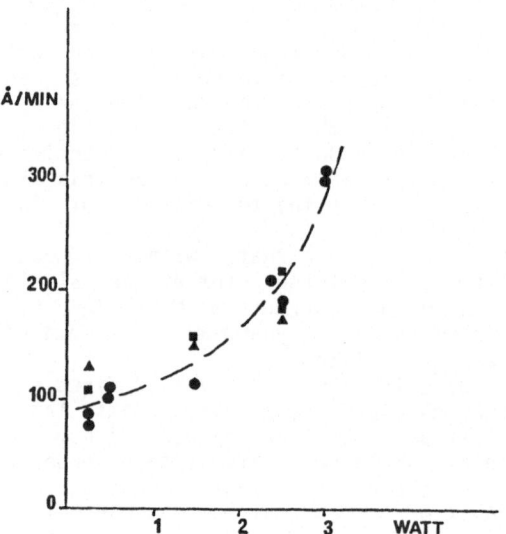

FIG.6 Etch rate versus incident power.

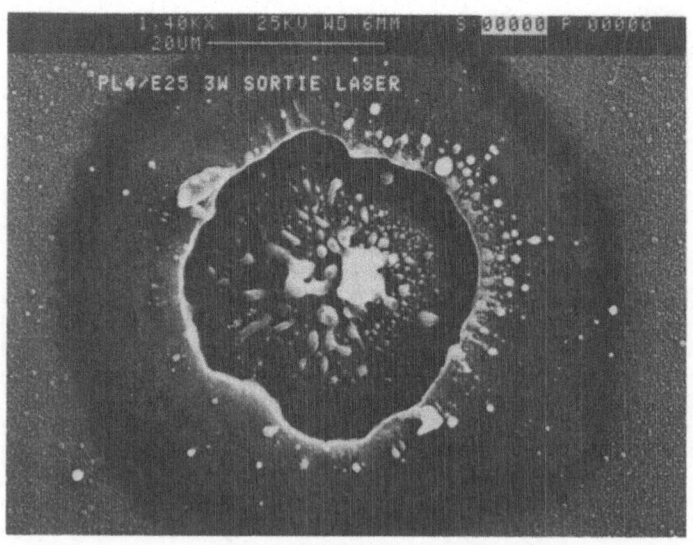

FIG.7 S.E.M. microphotography
incident power : 1.5 Watt

discharge alone.

Plotting the rate of etch against 1/T, the etch rate was found to obey Arrhenius' law with activation energy $0.11eV \leqslant Ea \leqslant 0.36eV$, which is in agreement with published results (8-11). It should be noted that the points depart from this law at higher laser powers ($\leqslant 2.5$ W).

SEM photomicrographs (figure 7) show the presence of a hole whose depth depends on etching rate and etching time. In this particular case, the sample was etched at 1.5 W for 20 minutes. Calculations of the expected depth and α-step measurements correlate well.

With respect to the shape of the holes themselves, it can be seen that they have flat bottoms and sharp edges, with reentrant sidewalls. In the centre there is a deposition of material. Chemical wet etching shows that these are SiO_x redepositions. Thus, it seems that localized deposition takes place on the sample, where the surface is hottest. As the edges of the hole are quite clean, this might indicate the presence of fluorosilicon groups that are dissociated in the central part of the incident beam.

This deposition causes a decrease in the average of the recorded reflectivity and makes a point-to-point interpretation difficult, and prevents a detailed analysis of the evolution in time of the reflectivity. Nevertheless, the reflectivity was found to vary periodically with time.

It can be seen in Figure 7 that the holes are quite circular in shape, and have diameters going from 25μm for 1.5 W laser power, to 55 μm at 3 W. As the temperature varies from Tmax to ambient going from the centre of the beam to 2r0, it is to be expected that the diameter of the hole be less than 4r0 (28μm). This, and the fact that the rate of etch departs from Arrhenius' law for large laser powers, suggests that the etching here is not only a thermal process. It is possible that this is due to diffusion of active species on a larger area than the beam spot (12), or light activation of adsorbed fluorine on the surface, which enhances the etching (12), or the activation of desorption of activating species (1), can induce an etching over a larger area.

A better understanding of the physical and chemical processes involved would require the use of different gases. It is possible that gases with a higher activation energy for etching of SiO_2 would give information of the thermal aspect of the etch, while gases with low thermal activation energy would allow us to separate out the non-thermal process involved in laser assisted plasma etching of silicon dioxide.

4.0 CONCLUSIONS

In this presentation, the localized etching of SiO_2 in a CF_4/O_2 D.C discharge using an argon laser has been studied. It is shown that the etching rate increases fivefold in comparison to the D.C. discharge alone. It is found that the etch rate is constant with time, and shows no effect of saturation at long times. It is further found that the size of the hole etched is much greater than the beam size, suggesting that the process is not solely thermally activated. Given the size of the holes etched (up to 55 μm), the time required to etch 1um SiO_2 (20 minutes) and the localized deposition occurring during etching, some improvements are needed before this method can be adopted for serial use.

Acknowledgements

The authors would like to thank D. Lévy for his help in the set up of the experiment and M. Héritier for temperature calculations. They are also grateful to B. Doyle for his critical readings of the manuscript.

References
1. P.D. BREWER, G.M. REKSTEN, R.M. OSGOOD : Solid State Technology
    April 1985 p. 273.

2. J.M. GEE, P.J. HARGIS, M.J. CAN, D.R. TALLANT and R.W. LIGHT : Mat. Res.
    Soc. Symp. Proceedings Vol 29, 1984

3. W. HOLBER, G. REKSTEN and R.M. OSGOOD Jr : Appl. Phys. Lett. 46, (2) 201
    (1985).

4. G. REKSTEN, W. HOLBER and R.M. OSGOOD Jr : Appl. Phys. Lett. 48, (8) 551
    (1986)

5. C.I.H. ASHBY : Appl. Phys. Lett. 45, (8) 892 (1984)

6. N. TSUKADA, S. SEMURA, J. Appl. Phys. 55, 3417 (1984)

7. TUNG -PO- LIN : IBM Journal September 1967

8. Y. HORIIKE, M. SHIBAGAKI : Supplement to Japanese Journal of Applied
    Physics Vol 15, 13 (1976)

9. D.L Flamm, V.M. DONNELLY and D.E. IBBOTSON : Semiconductor International
    April 1983, 136

10. J.W. COBURNS and H.F. WINTERS : J. Vac. Sci Technol, 16.2, 391 (1979)

11. C.J. MOGAB, A.C. ADAMS and D.L. FLAMM : J. Appl. Phys. 49, 7 3796 (1978)

12. F.A HOULE and T.J. CHUANG : J. Vac. Sci. Technology 20 (3) 790 (1982)

Nd:YAG LASER PROCESSING FOR CIRCUIT MODIFICATION AND DIRECT
WRITING OF SILICON CONDUCTORS

S.LEPPÄVUORI, J.LENKKERI and J.LEVOSKA
Microelectronics Laboratory, University of Oulu, OULU, Finland

1.INTRODUCTION
The use of focused laser beams makes possible delicate and
flexible fabrication and modification of integrated circuits and
components. This has clear benefits for small-scale production
of circuits and for prototype fabrication. It also makes it
possible to do trouble-shooting and to repair circuits and it
improves the ability to test large circuits. The laser can be
used to cut conductors on integrated circuits using either a
normal air atmosphere [1,2] or an etching gaseous or liquid
medium [3,4]. Using a laser pulse or pulses it is possible to
destroy the thin insulating layer between two conducting layers
and in this way accomplish an electrical contact between the
conducting layers [5]. Other methods of making electrical
connections with lasers have been reported [6-9]. The energy of
the laser beam can also be used to deposit layers of materials
through chemical reactions in laser induced chemical vapour
deposition (LCVD) [10].
   In this paper, work which was carried out on using a single
Nd:YAG laser equipment to cut conductors and to make electrical
connections on integrated circuits using the pulsed action of
the laser is reported. Progress on the use of CW radiation of
the laser to deposit silicon stripes by the LCVD method is also
reported.

2. LASER FACILITY
The facility presently consists of a thick film resistor laser
trimming system which has been modified by reducing the spot
size so as to be able to work on semiconductor ICs. The laser
source is a continuously pumped Nd:YAG laser (λ = 1064 nm)
producing a maximum multimode continuous wave power of 60 W.
The laser is also equipped with an acousto-optic Q-switch which
operates in the pulsed mode from 10 Hz to 5 kHz or as single
shots. The half power pulse duration is 100 ... 120 ns. Using
the frequency doubling crystal the laser runs at its second
harmonic wavelength 532 nm, the pulse length being about 60 ns.
The laser beam is focused through a 57 mm or 16 mm focal length
lens which can be moved over an area of 75x100 mm^2 with a
resolution of 2 μm, either manually or via a microcomputer, by
two perpendicularly mounted translation stages driven by
stepping motors. The monitoring of the work piece is done
through a microscope eyepiece or a CCD camera and monitor.
   The apparatus for laser induced chemical vapour deposition
(LCVD) consists of a gas cabinet with an arrangement for
automatically purging the silane gas, a reaction chamber (0.2 l),

D. J. Ehrlich and V. T. Nguyen (eds.), Emerging Technologies for In Situ Processing, 265–271.

a turbomolecular pump and a decomposition oven for the exhaust gas. The pressure of the process gas can be regulated up to 1 bar.The laser equipment is installed in a clean room of class 10000, the area over the laser being of class 100.

3.LASER CUTTING AND CONNECTING PROCEDURES
3.1. Laser cutting

The Q-switched frequency doubled mode of the Nd:YAG laser was used to cut 5μm wide aluminium and polysilicon conductors through a passivation layer. Each cutting was done with a single 60 ns pulse without damaging the underlying insulation layer. The SEM micrographs (Fig. 1) show that the cuttings are clean with no debris in the gap between the wire ends. Opening of a polysilicon link requires less energy than that required to open an aluminium link and therefore the risk of damaging the substrate is less when polysilicon links are used.

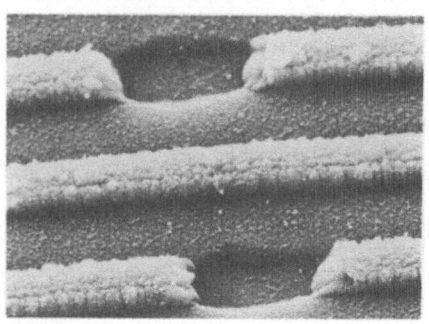
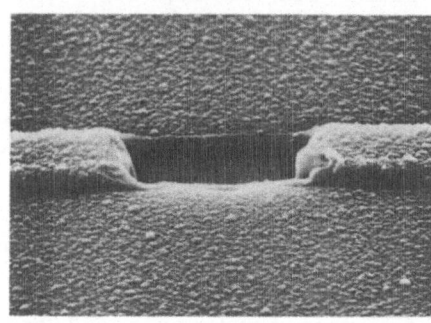

FIGURE 1. SEM micrographs of 5 μm wide laser cut (a) aluminium and (b) polysilicon conductors.

3.2. Polysilicon laser linking

A reliable method of making connections with a laser between two polysilicon conductors without breaking the passivation was developed. The laser link structure consisted of two polysilicon conductor layers separated by a 90 nm thick silicon dioxide insulation layer (Fig. 2). The upper polysilicon layer was covered by a 1 μm thick passivation glass layer. The laser processing was carried out with the pulsed green radiation of the laser. A single 60 ns pulse fused the polysilicon layers together through the 90 nm thick insulator and an electrical connection was established between the conductors. Pictures of the laser links before and after laser programming are shown in Fig. 3. The SEM micrograph in Fig. 3a shows that there was no visible damage of the passivation layer of a laser programmed link. The dark spot on the laser programmed link in the picture taken with the optical microscope (Fig. 3b) shows the hit-point of the laser pulse.

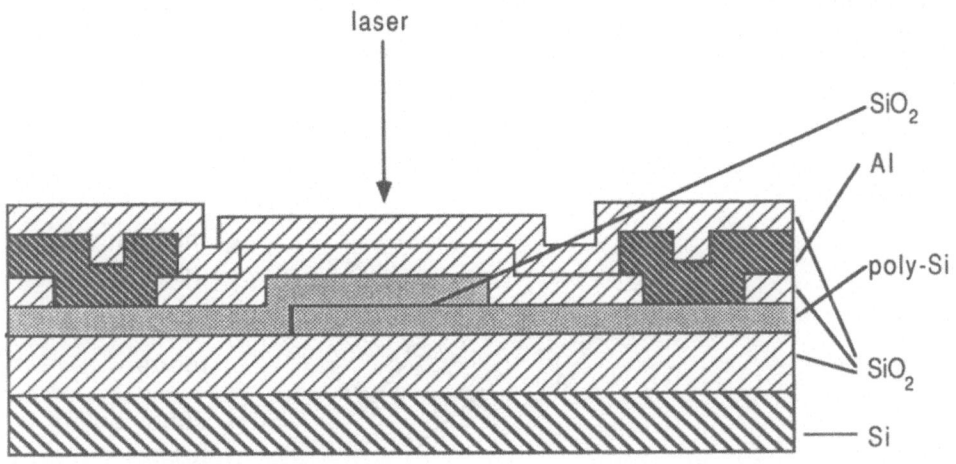

FIGURE 2. Cross-sectional view of the laser link structure.

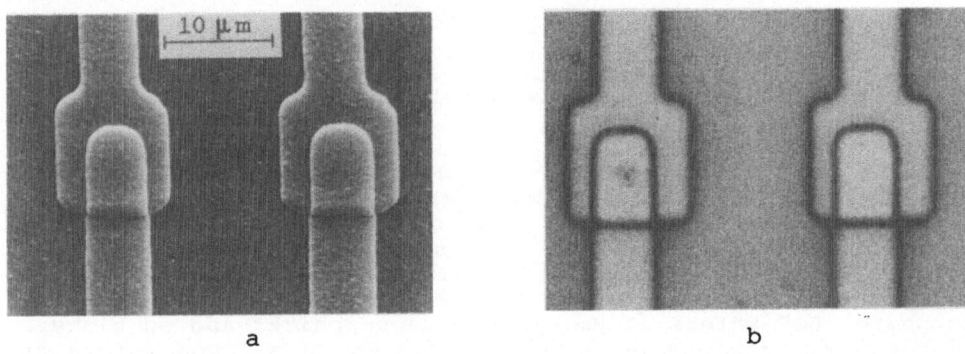

a   b

FIGURE 3. (a) SEM micrograph and (b) optical microscope photograph of two laser programmable links. The left links in both figures were programmed with the laser while the ones on the right were not.

Optimum laser parameters for making connections were determined by varying the spot size and the pulse energy. The spot size was increased by moving the focusing objective upwards and the pulse energy was changed by adjusting the pumping energy of the laser source. The link resistance versus laser pulse energy with three different spot sizes is shown in Fig. 4. It can be seen that with smaller spot sizes the useful energy window is wide and that the link resistance decreases with increasing pulse energy until the passivation layer breaks and the link fails. The minimum link resistance was achieved with a pulse energy just below the threshold energy needed to break the passivation. The lowest link resistance was less than 20 Ω, which is insignificant when compared with the resistance of a polysilicon conductor. A thousand connections were made and the yield was 100%. No changes of the laser programmed links were noted when they were subjected to temperature aging (1000 hours at 125°C).

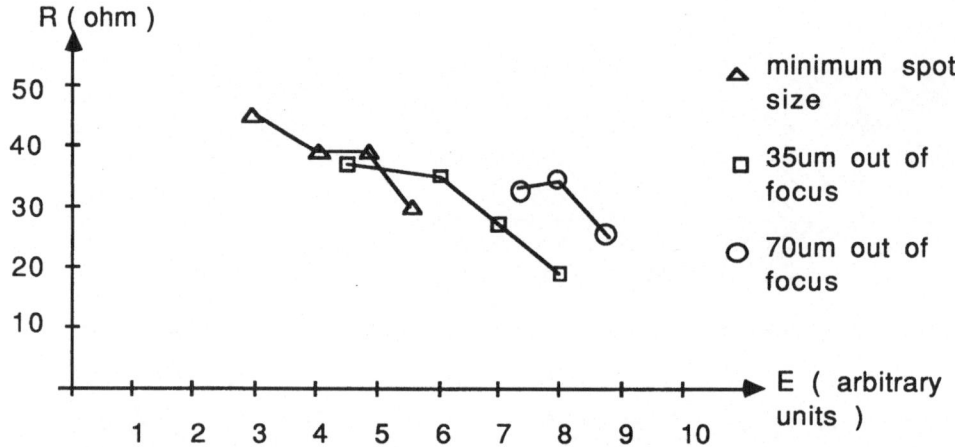

FIGURE 4. Link resistance R versus laser pulse energy E.

4. LASER INDUCED CHEMICAL VAPOUR DEPOSITION OF SILICON STRIPES

The deposition of silicon stripes is based on the pyrolytic decomposition of silane (SiH_4) to silicon and hydrogen. Silicon is deposited on a substrate at a spot heated locally by laser light. The research work started with the deposition of silicon stripes on various Al_2O_3 substrates. Stripes were deposited on sapphire, SOS wafers (1 μm silicon on sapphire) and on sintered alumina. Experiments were also conducted on silicon wafers and on Ag-Pd and Mo stripes which served as conductors on the Al_2O_3 substrates.

The growth process was carried out in a reaction chamber filled with pure silane at 100-200 mbar. The 1064 nm multimode CW beam of the Nd:YAG laser was focused on the substrate by an objective lens with focal length of either 57 or 16 mm. The corresponding minimum widths of the stripes grown were 40 μm and 20 μm; typical thicknesses were 15 μm and 3 μm, respectively.

Pyrolytic decomposition requires a considerable rise in temperature in order to start the reaction on a practical scale. The growth process is therefore strongly dependent on the absorption of light in both the substrate and the deposit. The absorption coefficient at the wavelength of 1064 nm of a Nd:YAG laser is very small for Al_2O_3. This makes the nucleation and the beginning of growth on Al_2O_3 difficult. In practice, the growth can be initiated at places where strongly absorbing material has been deposited by other methods, or at occasional impurity particles on the substrate. The absorption in silicon at room temperature at 1064 nm is also weak but it increases strongly with increasing temperature. When growth has been initiated, silicon stripes can also be deposited on a weakly absorbing substrate, because the absorption of light in the growing deposit can maintain the temperature at an adequate level. The limiting factor for the linear growth rate (scanning speed) is then the deposition rate of silicon on silicon. Controlled deposition is easier on SOS than on sapphire or sintered alumina, due to the absorption of light in the thin silicon layer.

SEM micrographs of the deposited stripes are shown in Fig. 5. According to x-ray diffraction studies, the stripes were found to be polycrystalline with almost random orientation. However, the deposits on sapphire substrates exhibited some preference for epitaxial growth. The minor amount of epitaxy on sapphire can be attributed to the difficulty of keeping the temperature of silicon in the rather narrow temperature region below the melting point where epitaxial growth is the preferred phenomenon. This is partly due to the strong dependence of the absorption of light at 1064 nm in silicon on temperature. According to computer simulations the heating of silicon by a constant intensity light from a Nd:YAG laser (1064 nm) is characterized by a strongly accelerating rise in temperature, compared with the self-controlled behaviour in the case of an argon laser (514 nm). As a matter of fact, undoped silicon subjected to an immovable beam from a Nd:YAG laser is either heated up some tens of degrees K above room temperature, or melted, depending on the power used.

 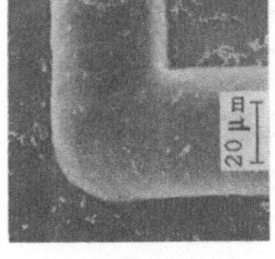 

a   b   c

FIGURE 5. SEM micrographs of silicon stripes deposited on (a) sapphire, (b) SOS and (c) sintered alumina.

The deposited stripes were not intentionally doped but the stripes grown on alumina substrates exhibited a fairly low resistivity of about 4×10^{-3} Ω cm. According to the results from microanalysis, the stripes were found to contain traces of aluminium impurities which must have diffused from the substrate during the growth. The concentration of the charge carriers corresponding to the value of resistivity was about 3×10^{19} cm^{-3}.

270

5. APPLICATIONS

Laser connections and disconnections can be used in a laser programmable logic array (LPLA). A very interesting application for this kind of a logic building block is its usage as a control structure in a large logic design during the prototyping phase [11].

A LPLA was designed using the double polysilicon p-well CMOS-technology. The array consists of two successive NAND-planes which perform PLA's AND-OR function. The basic cell for these NAND-planes is shown in Fig. 6. Basic cells placed in a row compose a multi-input NAND-gate. The input variable could be joined as a part of the resultant NAND-function by making two laser connections which connect the variable to the gates of the transistors. One aluminium connector disconnection was also necessary to remove the wire by-passing the NMOS-transistor in an unused basic cell.

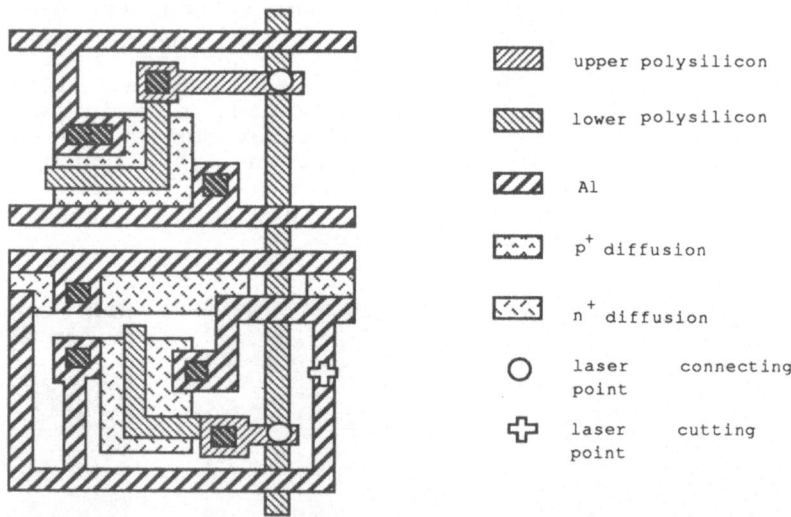

upper polysilicon

lower polysilicon

Al

p^+ diffusion

n^+ diffusion

laser connecting point

laser cutting point

FIGURE 6. Basic cell layout for a laser programmable logic array.

Laser connection between two polysilicon layers is also the basic restructuring method for a gate-array chip now under development. The chip will be composed of a matrix of laser programmable four transistor cells surrounded by programmable wiring channels.

Direct writing of conducting stripes of metals or highly doped silicon can be used to lay down conductor lines on integrated circuits as well as on thick film circuits. In the near future it may be possible to make connections from a silicon chip to the substrate if the problems due to different thermal expansions of the materials can be solved. In the area of sensor technology direct writing methods offer the opportunity of large degree of flexibility for the fabrication of both the sensor elements as well as making connections for these elements.

6. CONCLUSIONS

The versatility and reliability of the Nd:YAG laser has been confirmed. The usefulness in the pulsed mode operation for cutting and linking of conductors has been established. This opens up many possibilities in the area of circuit modification, repair and prototype fabrication. The use of the CW operation for LCVD of silicon allows a fairly high deposition rate. It is possible to clearly view the process of growth with less disturbance compared to when the visible laser radiation is used. However, as discussed earlier, the higher temperature dependence of the absorption coefficient in the case of silicon makes control of the growth fairly difficult.

Future work will be conducted with an argon laser equipped for both visible as well as UV operation. The next stage of the programme is to deposit doped silicon from the gas phase on a variety of substrates.

REFERENCES

1. Yamaguchi H, Hongo M, Miyauchi T, Mitani M, IEEE Journal of Solid-State Circuits, **SC-20** (1985) 1259-1264.
2. Wills S, McPherson J, Proceedings of SPIE, Vol. 385, Los Angeles, California, 1983, 82-86.
3. Osgood RM Jr, Ehrlich DJ, Deutsch TF, Silversmith DJ, Sanchez A, Proceedings of SPIE, Vol. 385, Los Angeles, California, 1983, 112-117.
4. Tsao JY, Ehrlich DJ, Appl.Phys.Lett. **43** (1983) 146-148.
5. Rapeli J, Leppävuori S, Murto T, Proceedings of the 4th International Microelectronics Conference, Kobe, Japan, 1986, 123-126.
6. Calder ID, Naguib HM, IEEE Electron Device Letters **EDL-6** (1985) 557-559.
7. Parker DL, Lin F-Y, Zhang D-K, IEEE Transactions on Components, Hybrids and Manufacturing Technology **CHMT-7** (1984) 438-442.
8. Moulic JR, Kiang YC, Lang RW, Logue JC, Proceedings of SPIE, Vol. 385, Los Angeles, California, 1983, 87-92.
9. Raffel JI, Anderson AH, Chapman GH, Konkle KH, Mathur B, Soares AM, Wyatt PW, IEEE Journal of Solid-State Circuits, **SC-20** (1985) 399-406.
10. Bäuerle D: Laser-Induced Chemical Vapor Deposition, in Laser Processing and Diagnostics, ed. D. Bäuerle, Springer Ser. Chem. Phys., Vol. 39. Berlin, Heidelberg: Springer-Verlag, 1984, 166-182.
11. Logue JC, Kleinfelder WJ, Lowy P, Moulic JR, WU W-W, IBM J. Res. Develop. **25** (1981) 107-115.

STUDY OF EXCIMER LASER ENHANCED ETCHING OF COPPER AND
SILICON WITH (SUB) MONOLAYER COVERAGES OF CHLORINE

T.S. BALLER, G.N.A. van VEEN and J. DIELEMAN
Philips Research Laboratories 5600 JA Eindhoven, The Netherlands

Abstract
The chlorine gas assisted excimer laser etching of copper ans silicon is
studied in the pressure range of 10^{-6} to 10^{-4} mbar. Time of flight
distributions of the etch products are measured as a function of laser
power density and gas pressure. The measured distributions can be
simulated with Maxwell-Boltzmann distributions with temperatures between
800 and 6000K where the temperature increases with increasing chlorine
pressure and laser power density. For Cu the etch rate and the temperature
corresponding to the kinetic energy distributions increases with the
number of laser shots till a steady state situation is reached. In this
steady state the etch rate and temperature are uniquely related to the
value of the laser power density, chlorine pressure and repetition rate. A
RBS measurement shows that in such a steady state situation a chlorinated
layer is formed with a thickness of the order of 1000 A with a chlorine
concentration of the order of 10% in the first few hundred A. For silicon
the chlorinated layer keeps a thickness of at most of the order of one
monolayer. A possible etch mechanism is given.

INTRODUCTION
Laser etching and modification of materials is a relatively new and
interesting field of research especially for the development of innovative
industrial processes. For example the ongoing trend in the electronic
industry towards smaller dimensions has initiated studies on submicron
laser patterning. Many applications are mentioned in the literature using
direct laser writing for mask repair, custom IC's etc. and laser
lithographic systems. However, very little work has been carried out on
the fundamental processes involved in laser etching or deposition. In this
contribution results will be presented on the study of the reactive gas
assisted nanosecond excimer laser etching of copper and silicon. The
velocity distributions of the etch products are measured as a function of
laser power density and gas pressure. From the results a possible etching
mechanism is inferred.

EXPERIMENTAL
The equipment used will be described in detail elsewhere (1) therefore
only a brief description of the apparatus and experimental parameters will
be given here.
    In a UHV apparatus (fig.1) consisting of a reaction chamber and a
differentially pumped detector chamber, both with a base pressure of
1×10^{-9} mbar a target can be irradiated with the 20 ns light pulses of an
excimer laser. The laser power densities used are in the range from 0.07
to 0.55 J/cm^2 at a wavelength of 308nm and a repetition rate of 4 Hz if
not mentioned otherwise. In the reaction chamber, in which the target is
placed, a chlorine pressure between 10^{-6} and 10^{-4} mbar is used during the

273

D. J. Ehrlich and V. T. Nguyen (eds.), Emerging Technologies for In Situ Processing, 273–278.
© *1988 by Martinus Nijhoff Publishers.*

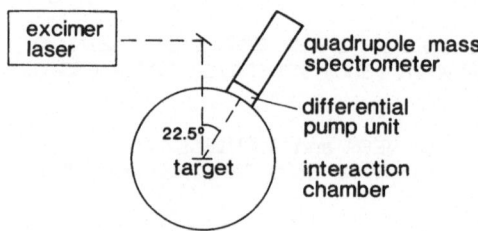

Figure 1 : Schematic drawing of the experimental set-up.

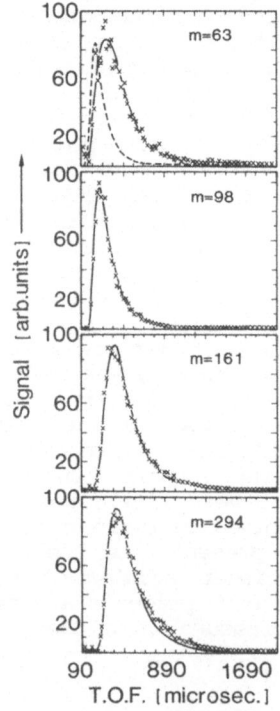

Figure 2 : TOF flux spectra of masses 63,98,161 and 294 obtained by etching of copper at a Cl_2 pressure of 1.2 10^{-5} mbar, a laser power density of 0.2 J/cm² and a repetition rate of 4 Hz. The solid lines indicate Maxwell-Boltzmann (MB) distributions at 4000K. The simulated spectra of m= 63,98 and 161 constitute several contributions as indicated in the mass spectrum in figure 3A. For m= 63 the MB fit at 4000 K for this particular mass is given with a dashed line.

experiment without a detectable influence on the background pressure in the detector chamber. Velocity distributions of the neutral particles, leaving the surface as a result of the laser pulses, are measured by a time of flight (TOF) method. The maximum time resolution is 5 microseconds and the flightpath is 276 mm. As a detector a quadrupole mass spectrometer is used equipped with an electron impact ioniser.

RESULTS AND DISCUSSION

In the laser etching of copper in a chlorine environment all possible masses Cu_xCl_y with x = 1..3 and y = 0..3, except $CuCl_3$, Cu_2, Cu_3 and Cu_3Cl, are detected. For the etching of silicon the main silicon containing masses detected are $SiCl$, Si and a minor amount of $SiCl_2$, $SiCl_3$ and $SiCl_4$. (2). TOF spectra are measured for all detected masses at several laser power densities and gas pressures in the ranges as given above. For the etching of copper a few TOF spectra for different masses measured under identical conditions, indicated in the figure captions, are given in figure 2.

Mass spectra for both of the etch systems, as obtained by integrating the measured TOF spectra after correction for the velocity dependence of the detector, are given, for the parameter values indicated in the figure captions, in figures 3a and 3b. The contribution of $SiCl_4$ is not drawn in figure 3b because it is too small.

In the interpretation of the measured TOF distributions a difficulty arises because various neutral products may contribute to the signal measured at a particular mass. This is due to the fact that during ionisation by electron bombardment in the ionisation chamber of the mass spectrometer, the same ion may be produced from various neutral parents. However it is obvious that the highest mass detected can only originate from one neutral parent.

The TOF distributions can be simulated with Maxwell-Boltzmann (MB) distributions. At equal laser power density and gas pressure the temperatures corresponding to these MB distributions are the same for all neutral products.

For Si all ions observed appear to originate from the corresponding neutral parents. Only a minor part of $SiCl^+$ originates from the neutral parent $SiCl_2$. A higher laser power density or/and a higher gas pressure gives a higher temperature.

For the etching of copper the interpretation of the measured distributions is somewhat complicated, because more masses are observed. The TOF spectra measured for the largest detected particle Cu_3Cl_3 can be simulated very well with MB distributions at mass 294 (Note that only the lowest mass of the isotopic composition is indicated). To obtain good fits at the other detected masses, contributions of higher masses have to be added in the simulations. For instance the TOF spectra measured at masses 231 and 259 are equal to the spectrum measured at 294. The particles measured at these masses result from dissociation of Cu_3Cl_3 during ionisation. The spectrum measured at the mass of copper can be simulated with the sum of two distributions of the same temperature, one from mass 98 and the other from mass 294. Cu particles do not desorb as such. From the simulations it can be concluded that during the etching of copper the major desorbing particles are $CuCl$ and Cu_3Cl_3. In the mass spectra in figure 3 the contributions of the different desorbed particles at the measured masses are indicated.

The temperatures found in the experiments depend both on gas pressure and laser power density and vary between 14000 and 4000K for the etching of silicon and between 800 and 6000K for the copper system. In the etching of copper the temperature and etch rate do not only depend on the above mentioned parameters but also on the number of laser pulses given after the start of the experiment. Just after starting the etch experiment the etch yield and temperature are low. Both increase during processing until a steady state situation is reached. Then the temperature and number of counts per pulse remain constant. The number of pulses needed to achieve a steady state decreases with increasing gas pressure at a constant laser power density (see figure 4). The spectra used to derive the mass spectrum are obtained under these steady state conditions.

Based on the above results a possible mechanism for the laser etching of copper with chlorine can be suggested. A more detailed discussion will be given in a forthcoming publication (1). The fact that it takes a lot of laser pulses to reach a steady state indicates that because of the laser pulses a relatively slow modification of the copper substrate is induced. One possible modification is the indiffusion of chlorine.

To check this a RBS measurement has been made on a single chrystal <111> copper sample irradiated with 50,000 laser pulses of 0.2J/cm^2 at a pressure of 1.2×10^{-5} mbar. From the RBS spectrum a chlorine depth profile is obtained which is given in figure 5. Indeed a chlorinated layer has been formed. The layer has a thickness of about 2000 A and contains about 20% chlorine in the first 300 A.

For copper exposed to chlorine at room temperature it is known that the sticking probability is about 1 for coverages less than a monolayer (3,4), so a monolayer will be formed after an exposure of only 1 L. After the formation of the first monolayer, more chlorine can adsorb on the chlorinated surface, but with a very low sticking probabilité (< 0.002) (3,4). The adsorbed chlorine can diffuse into the metal resulting in the formation of a chlorinated copper layer. Sesselmann and Chuang studied this layer formation extensively (5,6). For the pressures used in our experiments they showed that for very long exposure times (300 s) the ultimate thickness of the chlorinated layer is about 30 A. It therefore can be concluded that with the gas pressures used in our experiments, between two laser pulses the outermost layer can easily be chlorinated because of the high sticking probability. Furthermore the thickness of the "spontaneously" formed chlorinated layer will be small, in contrast with the layer thickness found by the RBS experiment. The difference between a layer formed by exposure to chlorine at room temperature and a layer formed under simultaneous irradiation with laser pulses can be understood as follows. Heating of the solid by a laser pulse will enhance the mobility of the chlorine present in the first few layers. When the chlorine diffuses into the bulk it is possible to produce a thick layer of chlorinated copper if a large number of laser pulses is used. Upon incorporation of Cl into the copper substrate the properties of the mixed layer will alter gradually. It appears that the thermal conductivity of the mixed layer decreases. Because of this decrease in thermal conductivity, the surface will become hotter during the laser pulse if the mixed layer is thicker and/or the concentration of chlorine incorporated in this layer is higher. This is in agreement with the observation that the temperature depends on the number of pulses applied. A steady state situation will be reached when the amount of chlorine adsorbed between two laser pulses is equal to the amount of chlorine desorbed plus the amount of chlorine diffused into the bulk.

In contrast to the Cu + Cl$_2$ case no time dependence is found for the etching of silicon. The chlorine pressure dependence of the heating to

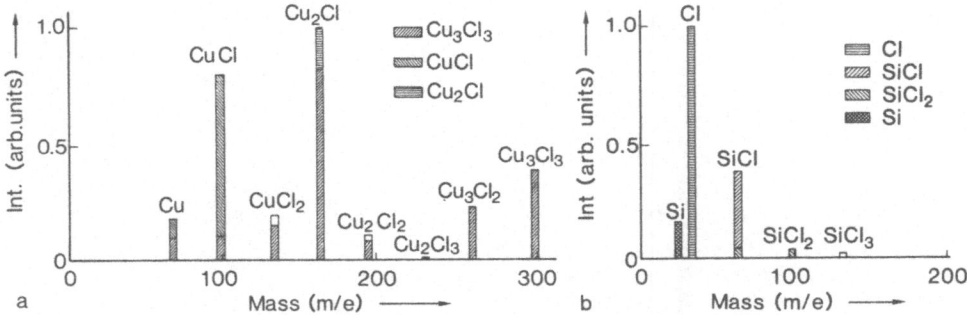

Figure 3A : The mass spectrum derived under the circumstances given in figure 2.

Figure 3B: A typical mass spectrum derived on silicon under irradiation with 308 nm light pulses in a chlorine environment.

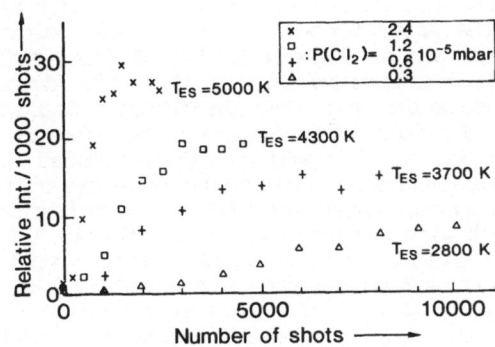

Figure 4 : The evolution of the integral of the TOF spectra of $m = 294$ as a function of the number of laser shots applied for four Cl_2 pressures on copper. The temperature in the steady state (T_{ES}) is also given. The laser power density used is 0.55 J/cm² and the repetition rate is 4 Hz.

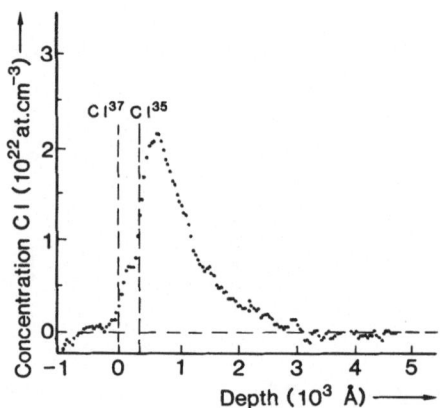

Figure 5 : The Cl depth profile as derived from the RBS spectrum after 50,000 laser pulses (see text).

very high temperatures of copper and particularly for Si is not really understood. More detailed measurements presently being performed, will hopefully shed more light on these phenomena. The observation of TOF distributions which can be simulated with MB distributions for all masses, suggest that, under the present conditions the main desorption process is evaporation. Sesselmann and co-workers concluded from their experiments on the laser etching of chlorinated layers that the desorption is quite thermal (7). An explanation for this difference should be found in a difference between a chlorinated layer formed by exposure of Cu to Cl_2 at room temperature for 300 seconds and a layer produced during additional irradiation with laser pulses, in which the exposure time between pulses is only 0.25 seconds. Preliminary results of measurements at 1.10^{-6} mbar chlorine pressure, but with a repetition rate of only 1 Hz still yield MB distributions, but at higher temperature than for a repetititon rate of 4 Hz. However, measurements with a repetition rate of 1 Hz, but at chlorine pressures of about 10^{-4} mbar result in non-MB distributions, which are drastically shifted to higher kinetic energies, showing that longer exposure times indeed seem to induce kinetic energies in the eV range with a non-MB character.

In conclusion it can be stated that the chemical laser etching of silicon and copper with chlorine under the conditions used in our experiments seems to be a thermal process. The temperatures found are remarkably high. The time behaviour as seen in the etching of obtained can be understood by the build-up of thick chlorinated layers.

REFERENCES
1. G.N.A. van Veen and T.S. Baller, to be published
2. T. Baller, D.J. Oostra, A.E. de Vries and G.N.A. van Veen, J. Appl. Phys. 60(7), 2321 (1986)
3. D. Westphal and A. Goldmann, Surf. Sci., 131, 113 (1983)
4. P.J. Goddard and R.M. Lambert, Surf. Sci., 67, 180 (1977)
5. W. Sesselmann and T.S. Chuang, Surf. Sci., 176, 32 (1986)
6. W. Sesselmann and T.S. Chuang, Surf. Sci., 176, 67 (1986)
7. W. Sesselmann, E.E. Marinero and T.J. Chuang, Appl. Phys. A41, 209 (1986)

SURFACE CHEMICAL PROBES AND THEIR APPLICATION TO THE STUDY
OF IN SITU SEMICONDUCTOR PROCESSING

R.B. JACKMAN AND J.S. FOORD
UNIVERSITY OF OXFORD,
INORGANIC CHEMISTRY LABORATORY,
SOUTH PARKS ROAD, OXFORD, OX1 3QR, ENGLAND.

1. INTRODUCTION
    Reactions at semiconductor-vapour interfaces as
stimulated thermally and by incident photon, electron and
ion radiation permit a wide range of thin film
transformations in the semiconductor surface to be carried
out (1). The processes involved include: (i) dry etching of
semiconductors, metals and dielectrics (ii) epitaxial growth
of elemental and compound layers (iii) growth of dielectrics
(iv) deposition of metals and metal alloys. Since photon,
electron and ion beams can also be spatially defined, either
through the use of physical masks or highly focussed beams,
in theory it should therefore be possible to achieve in situ
the complete fabrication of electronic devices and
integrated circuits or their repair and modification. The
chemistry underlying the processing methods involved is
extremely complex. Even in the most chemically simple
reaction, MBE of semiconducting materials from elemental
beams under ultra-high vacuum, a number of surface processes
must be taken into account in order to achieve a full
description of the phenomena (2). When other processing
techniques are examined, very many more factors must be
looked at. Thus if complex compounds as opposed to the
elements are employed, the chemical conversions demanded
become more sophisticated and the fate of a variety of
species generated in the reaction must be controlled. At
higher pressures (MOVPE etc) the fluid dynamics in the
reaction cell has an important influence. When photon,
electron or ion beams are used elementary steps as excited
electronically and by momentum transfer are brought into
play; such steps can occur in the gaseous, adsorbed or
substrate phases present. Nevertheless, the optimisation of
particular processing methods must rest on achieving a good
description of the underlying reactions. Since a
significant (if not dominating) number of such reactions
actually take place at the substrate-vapour interface,
understanding the surface chemistry must be one of the major
aims. In this article we therefore survey and illustrate
some of the ways which might permit this to be achieved.
2. EXPERIMENTAL STRATEGY
The problem which dominates approaches to surface studies
originates in the very low number of atoms in the surface

D. J. Ehrlich and V. T. Nguyen (eds.), Emerging Technologies for In Situ Processing, 279–288.
© 1988 by Martinus Nijhoff Publishers.

region compared to those in the bulk phases present. This
means that applicable experimental techniques must (a) be
highly sensitive and (b) discriminate strongly between atoms
in the interface and bulk. Highly specialised techniques
have been developed to satisfy these criteria (3) and thus
which solve the problem noted above. Many of such
techniques demand the use of highly ordered single crystal
substrates but this condition is often naturally satisfied
in microelectronics-oriented research. Unfortunately such
methods historically have also tended to place severe
restrictions on the nature of the gas phase in which surface
studies can be carried out. This issue is probably the one
which dominates developments in surface techniques today.
Three distrinct experimental approaches have evolved which
address this problem in different ways (see figure 1)

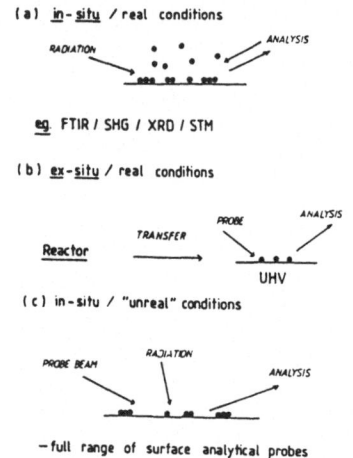

(a) in-situ / real conditions

RADIATION   ANALYSIS

eg. FTIR / SHG / XRD / STM

(b) ex-situ / real conditions

Reactor   TRANSFER   PROBE   ANALYSIS

UHV

(c) in-situ / "unreal" conditions

PROBE BEAM   RADIATION   ANALYSIS

—full range of surface analytical probes

Figure 1.
Three experimental
strategies for
investigating
surface phenomena
involved in semi-
conductor
processing.

2.1 Ex situ Surface Analysis

The simplest way to achieve surface sensitivity is to
find a suitable form of exciting radiation (photons,
electrons, ions, atoms) at the interface and detect a signal
associated with surface scattering in the form of emitted
photons, electrons, ions or atoms. Good sensitivity can be
achieved if the relevant cross-sections are high and a
selection of the wide-range of surface techniques which have
been developed on this basis are noted in Table 1; with such
techniques it is readily possible to obtain an exact
description of the interface under study and thus it might
be thought that understanding the surface chmistry of
semiconductor processing presents no problem. That this is
not the case arises from the fact that virtually all of the
techniques in this category are limited to high vacuum ($<10^{-4}$
mbar) conditions which in the present context are only
satisfied in MBE. Thus such methods cannot be generally

Table 1

Technique	Particle in, particle out	Information provided
Reflection high energy election diffraction	Electron, electron	Surface structure
Surface EXAFS	Photon, electron or ion	Local surface structure
Scanning Auger microprobe	Photon or electron, electron	Surface composition
X-ray photoelectron spectroscopy	Photon, electron	Surface composition and chemical state.
Secondary ion mass spectrometry	Ion, ion	Surface composition
Electron energy loss spectrometry	Electron, electron	Surface vibrational spectrum
Ultra-violet photoelectron spectroscopy	Photon, electron	Surface valence electron structure

applied to the in situ study of chemical reactions during semiconductor processing under realistic experimental conditions. The most established application of these techniques in the field has therefore been for the ex situ analysis of the interface (in ultra-high vacuum) after processing has taken place (under any desired conditions). Many examples of such work exist in the literature. Kowalczyk and Miller (4) have employed XPS to characterise the interfacial structure and composition following photochemical deposition of Sn on GaAs. Ehrlich et al have used Auger depth profiling of photochemically deposited Al (5). Jackman et. al. have used scanning Auger techniques to investigate impurity segregation in ripples formed during LCVD of Fe (6). Such are typical applications.

This approach suffers from two main limitations. Firstly, in its simplest form ex situ analysis simply consists of the withdrawal of the sample from the reactor and its subsequent insertion into a surface analysis facility, during which period exposure to the atmosphere occurs. Since this inevitably results in chemical modification to the surface, reliable information can only be obtained from the near surface region which remains

unperturbed. This problem can be circumvented by attaching
the reaction cell to the surface analysis chamber; direct
transfer can then be achieved without atmospheric
contamination. Nevertheless the surface is still likely to
be perturbed during transfer since weakly bound surface
species are lost during evacuation of the reaction cell. It
is thus still very difficult to infer information on the
reaction mechanisms since a knowledge of the important
intermediates is not obtained.

2.2 In situ Analysis under Real Condition

The ideal solution to the problem identified above would
be to employ surface techniques which are not restricted in
the pressure regime to which they can be applied. This is a
very difficult problem; high (surface) sensitivity results
from a strong interaction of the probe beam with matter but
this also tends to lead to short mean free paths in the gas
phase, hence the pressure limitation. Consequently, no
routinely applicable surface techniques capable of opertion
up to atmospheric pressure as yet exist. However, a number
of experimental approaches have now been demonstrated to be
viable and it is hoped that these can be applied to
semiconductor processing in the near future. The most
promising techniques are breifly reviewed below

(a) Fourrier Transform Infra-Red Spectroscopy (FT-IR)

While transmission IR studies of high surface area
powders is a long established approach to the identification
of surface intermediates, at pressures up to 100 bar, it is
not applicable in semiconductor research where flat surfaces
are employed. For such work, the spectrometer must be
operated in the reflection mode where the absorbance can be
as low as 0.01%, requiring signal-noise ratios of 10^5 or
better. These can now be achieved using good FT-IR
spectrometers and thus the recording of surface IR operation
is now a feasible approach. Since the mean free path of.IR
photons in gases is large, no restrictions in pressure exist
as regards the recording of absorption bands in regions
where no gas phase adsorptions exist; even where the bands
overlap the use of polarisation modulation techniques can be
used to pick out the surface adsorptions. At present the
technique has largely been employed to study simple adsorbed
molecules, but it is clear that the recording of surface
vibrational spectra should provide invaluable information on
the nature of the adsorbed intermediates formed in the much
more complex situations, such as MOVPE, pertaining to
semiconductor processing. Detailed reviews are presented
elsewhere (7).

(b) Grazing incidence X-ray diffraction

Whilst X-ray diffraction is well known as a structural
technique which is capable of operation in wide-ranging
environments it is not normally considered as a surface tool
because of the large penertration depth of X-rays in
solids. However if the angle of incidence is reduced to
less than 1°, total external reflection occurs and the
scattered intensity begins to depend sensitively on the
surface structure present (8-11). Apart from being
insensitive to the gas phase present, the technique also

offers other advantages. Thus by varying the angle of
incidence the analysing depth can be carried from 10 to
10,000A$^{\circ}$ and bound interfaces can be studied. In contrast
to surface electron diffraction techniques, the theoretical
analysis can be straightforward. Glancing incidence XRD
therefore has the potential of being a very useful surface
structural tool for the in situ monitoring of high pressure
epitaxial growth processes.

(c) Second Harmonic Generation

Second order non-linear optical effects give rise to the
generation of a second order harmonic when laser light is
reflected from a conducting surface and the intensity
depends upon the nature of the surface layers present
(12-14). The technique can therefore be used to detect
adsorbed species and again the approach is not restricted to
low pressures. If the second order response of adsorbed
species is first characterised under well-defined
conditions, then the technique can be used under less
defined conditions (e.g. realistic semiconductor processing)
to determine the concentration of such species. The
response time is very rapid and this shall readily permit
the study of surface reaction kinetics.

(d) Scanning Tunnelling microscopy

Application of an electrical potential difference between
a surface and a fine tip separated by 10°A leads to the
flow of an electron tunnelling current, the magnitude of
which depends very sensitively on the surface-tip
separation. Scanning of the tip over the surface at
constant tunnelling current thus provides an elegant way of
measuring the geometric profile of the underlying surface
(15). Importantly, provided the experimental system is
mechanically stable, atomic roughness can be resolved and
hence the technique can provide a real space map of the
surface atomic structure, as revealed in recent publications
(16). While the early work was carried out at ultra-high
vacuum it seems likely that the technique could operate
under a variety of experimental conditions. Again the
technique offers intriguing in situ possibilities as a
structural probe of semiconductor processing.

2.3 In Situ Studies at Low Pressures

As is apparent from the preceding discussion, while in
situ surface physical probes are being developed, such
techniques are far from being routinely applicable in the
complex situations pertaining to semiconductor processing
and it appears at the moment unlikely that detailed surface
characterisation will ever be achievable at high pressures.
The third approach is therefore to attempt to model real
semiconductor processing reactions under idealised low
pressure conditions where all the techniques of surface
physics can be routinely applied. This at present is the
only way in which surface effects can be studied in detail.
We examplify this approach in the following section which
describes results relating to the photochemical deposition
of W, from WF_6 on Si(100).

Refractory metals are attractive materials for use in
device metallisation schemes and the reaction studied has

284

potentially important applications in the field of
semiconductor processing. We have examined the process as
driven by a deuterium lamp under ultra-high vacuum; Auger,
LEED and thermal desorption techniques (17) were employed to
establish the surface photochemical effects involved.

Exposure of a clean Si(100) (2x1) surface to WF_6 vapour
at 77K in the dark results in the formation of an adlayer
which can be characterised by the Auger spectra illustrated
in Figure 2. Heating this surface to 120K results in energy
shifts in the recorded transitions as is also illustrated in
Figure 2. (6).

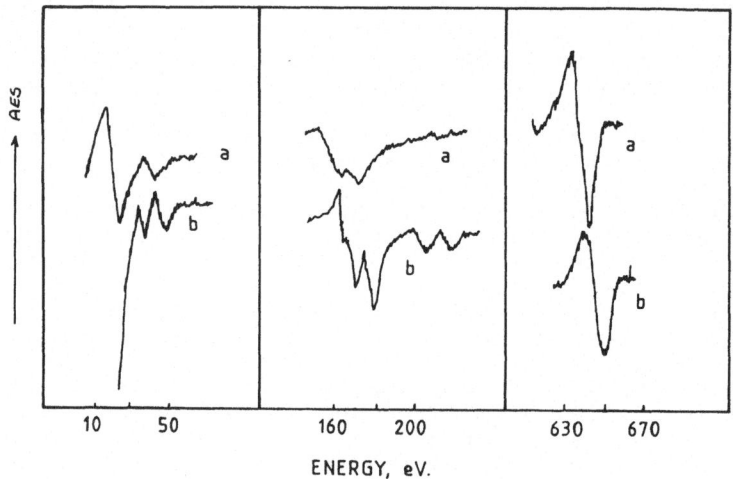

Figure 2   Auger electron spectra for (a) WF_6 adlayer on
            Si(100) and (b) after heating to above 120K.

The peak energies associated with the low temperature phase
remain constant, independent of coverage, from sub-monolayer
up to multilayer concentrations; the energies recorded for
the high temperature phase are characteristic of atomic W
and F species. The data is thus consistent with molecular
adsorption at 77K with facile dissociation as the
temperature is raised.

Although adsorbed WF_6 readily dissociates on Si(100) in
the dark, the adlayer thus generated contains both W and F
and very much more energy needs to be supplied in order to
generate the W adlayer. Thermal programming can be used to
detect the fluorine products (a mixture of SiF_x, x = 1-4
species) as they desorb from the surface in three energy
regimes as show in in Figure 3. The thermal reaction
profile for W deposition from WF_6 is thus as summarised in
Figure 4 (a).

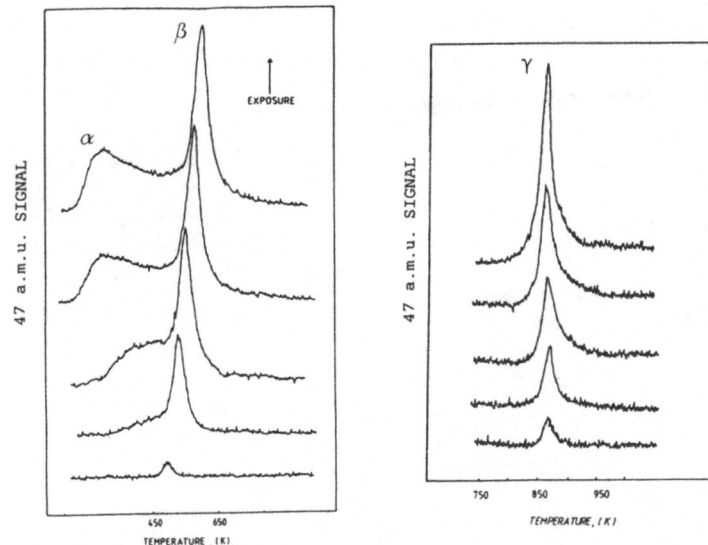

<u>Figure 3</u>   Thermal desorption spectra following increasing
WF$_6$ exposure to Si(100) revealing 3 SiF$_x$ surface
states (α, β and γ).

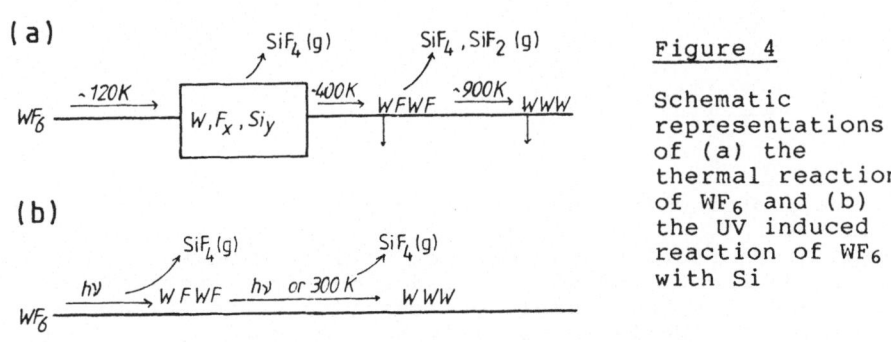

(a)

<u>Figure 4</u>

Schematic
representations
of (a) the
thermal reaction
of WF$_6$ and (b)
the UV induced
reaction of WF$_6$
with Si

(b)

   Three principal effects are observed when ultra-violet
radiation from the deuterium lamp is directed onto the
molecularly adsorbed WF$_6$ surface phase. Mass spectrometric
monitoring of the gap phase during chopping of the light
beam clearly indicates that photon-stimulated desorption of
silicon fluorides takes place (Figure 5). Measurement of
Auger energies (Figure 6) associated with the adsorbed
phases shows the occurence of chemical shifts commensurate
with the photodissociation of WF$_6$. Finally thermal
programming of such adlayers indicates that they very
readily evolve silicon fluorides at low temperatures (Figure
7) in contrast to those produced thermally.

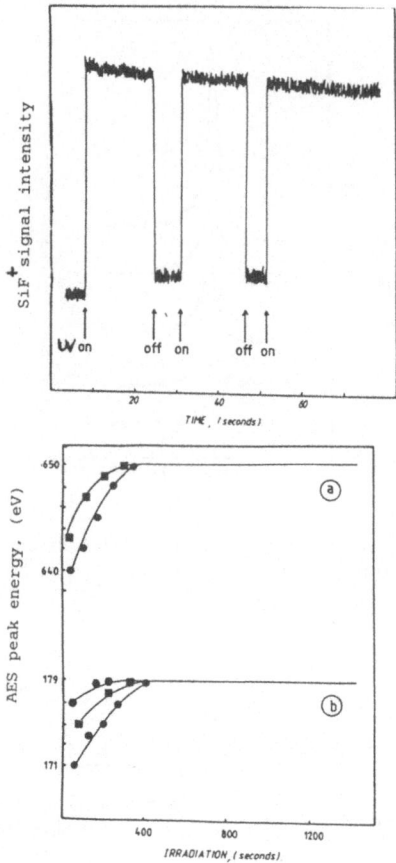

Figure 5 photon stimulated desorption of silicon fluorides during UV irradiation of molecularly adsorbed WF_6

Figure 6 Auger peak energy shift observed during UV irradiation adsorbed WF_6.

The photochemical dissociation pathway is summarised in Figure 4 (b). Clearly photochemistry provides a low temperature route to the energetically demanding step involving removal of silicon fluorides from the surface hence enabling the overall process to operate at low temperature. While the experimental conditions used here are rather different to those conventionally used in LPD, one would expect the same effects to operate in the latter, although clearly new ones could occur. The study nicely illustrates the success and limitations of the experimental strategy employed.

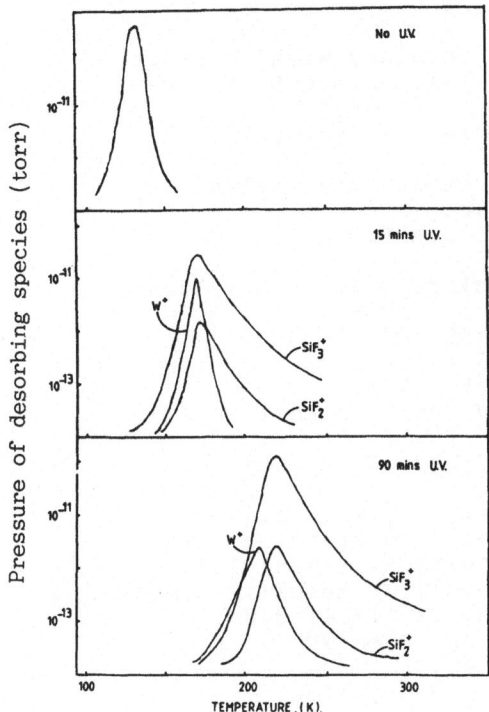

Figure 7 Thermal desorption spectra following molecular adsorption of WF_6 on Si(100) shown as a function of UV irradation prior to desorption.

REFERENCES

1. See, for example, papers contained within "Laser processing and dignastics (11)"Bauere D: Kompa: KL and Laude:L(eds) E-MRS 11, (1986)
2. See, for example, Ploog:K Ann.Rev.Mat.Sci., 12 (1982) 123
3. Czanderna:AW(ed): Methods of Surface Analysis Elesevier, Amersterdam (1985)
4. Kowalczyk:SP and Miller:DL J.APP.Phys. 59 (1986) 287
5. Ehrlich:DJ Tsao:JY and Bozler:CO J.Vac.Sci.Tech. B3 (1985) 1.
6. Jackman:RB Foord:JS Adams:AE and Lloyd:L J.Appl.Phys. 59 (1986) 123
7. Chesters:MA J.Elec.Spec. 38 (1986) 123
8. Marra:WC Eisenberger:P and Cho:AY J.Appl.Phys. 50 (1979) 6927
9. Eisenberger:P and Marra:WC Phys.Rev.Letts 46 (1981) 1081
10. Robinson:IK Phys.Rev.Letts. 50 (1983) 1145
11. Bohr:J Feidenhars:R Nielsen:M Toney:M Johson:RJ and Robinson:IK Phys.Rev.Letts 54 (1985) 1275
12. Tom:HWK Mate:CM Zhu:XD Crowell:JE Heinz:TF Somorzai:GA and Shen:YR Phys.Rev.Letts. 52 (1984) 348
13. Chen:CK Heinz:TF Ricard:D and Shen:YR Phys.Rev.Letts 46 (1981) 1010
14. Tom:HWK Zhu:XD Shen:YR and Somorjai:Ga Surf.Sci. 167 (1986) 167
15. Binning:G Rohrer:H Gerber:Ch and Weibel:E Appl.Phys.Letts 40 (1982) 178
16. Binnig:G Rohrer:H Gerber:Ch and Weibel:E Phys.Rev.Letts 50 (1983) 120
17. Jackman:RB Ebert:H and Foord:JS Surf.Sci. 176 (1986) 183

Index

AlGaAs 55
aluminium 213, 218, 219, 257, 259, 260, 261, 266
annealing 9
applications 270
argon laser processing 171
autocatalytic pyrolysis 121
automated manufacture 45
buried metal layers 55, 57
buried metal structures 56
cadmium mercury telluride 221
cadmium telluride 221
chemical 278
chemical vapor deposition 71 , 265, 268
chemiluminescence spectra 239
chlorine 107 , 109, 110, 111, 213, 214, 215, 216,
    218, 219, 273, 275, 276, 277, 278
chromium oxide 241 , 243, 244, 246
chromyl chloride 241 , 243, 246
combined system 6, 9, 10
compound semiconductor 13
conductors 265
connecting 266
copper 106, 107, 108, 109, 110, 111, 273, 274,
    275, 277, 278
cutting 266
CVD 83, 84, 85, 86, 87, 88, 89, 90
DC plasma 257
deposition 71, 72, 74, 75, 76, 77, 78, 79, 80
deposition kinetics 201
dielectric films 145
direct laser writing 242
direct writing 83, 84, 123, 125, 126, 127, 128
doping 265
dry etching 14, 213
dry process 14
e-beam 9, 131, 132, 134, 135
electron beam annealing 8
electron beam lithography 4
epitaxy 61, 63, 68
etch rate 107, 108, 109, 110, 111, 260, 261, 262,
    263
etching 257, 259, 261, 263, 273, 275, 277
excimer laser 93, 105, 106, 213, 214, 216, 219
FIB 15, 16, 17, 18
focused ion beam 14, 18
focused ion beam implantation 3, 5
fundamental 273
GaAs 56, 57, 58, 59, 61, 62, 63, 68
gaseous environment 111
growth 128
growth on patterned substrates 59

HeNe scattering 235
heterogeneous photolysis 128
heterogeneous pyrolysis 121
IC manufacture 47, 51, 53
III 13
III terials 61
III–V semiconductors 55
image projection 105
impurity induced diffusion 13
in-situ 232
in-situ processing 1, 2, 5, 6, 7, 9, 13, 14, 16, 52, 53,
    58, 59, 221, 222, 223, 225, 229, 230
$In_{0.53}Ga_{0.47}As$ 61, 67, 68
infrared detectors 221, 230, 231
inorganic resists 135
InP 61, 63, 65, 68
integrated circuits 265
integration 13, 14
ion beam 1, 2, 3, 4, 5, 6, 7
ion implantation 14, 61, 65, 68
laser 13, 71, 72, 74, 75, 76, 77, 78, 79, 80, 83, 84,
    85, 86, 88, 89, 90, 137, 139, 140, 143, 265,
    266, 268, 273, 275, 276, 277, 278
laser controlled multipulse etching 105
laser deposition 145, 164, 165
laser direct writing 201, 203
laser driven reactions 240
laser etching 23, 25, 26
laser facets 13, 18
laser induced etching 210
laser processing 173, 213, 249
laser-induced chemical vapor deposition 33
laser-induced etching 33, 39
laser-induced metallization 38
light stimulated process 121
liquid phase 249, 253, 255, 256
lithography 9
mask repair 163, 166, 168
mass spectrometry 275
MBE 1, 2, 7, 9, 10, 55, 56, 58, 59, 60, 61, 67, 68,
    69
mechanisms 71, 77, 78, 79
mercury telluride 221
metal etching 105
metal patterning 23
micro holes 137
microelectronics 83
microconnections 137
MO 61, 69
MOCVD 68
modification 265

modification of interconnection networks 201
molecular beam epitaxy (MBE) 55
nanolithography 1
Nd.YAG 265, 268, 271
OEIC 13, 14, 18, 19
ohmic resistivity 127, 128
optical fiber communication 19
opto-electronic integrated circuits 13
optoelectronics 13
organo-metallic precursors 145
overgrowth 1, 2, 10
oxidation 171, 173, 174, 176, 177, 249, 250, 251,
    252, 255, 256
oxide growth 171, 174, 249
oxygen 171
passivating 13
pattern resolution 107, 110
patterned growth 55
photo-epitaxy 221, 223, 225, 229
photo-MOVPE 223, 224, 225, 226, 227, 228, 229,
    230, 231, 232
photochemistry 71, 76, 77, 78, 80, 241
photodeposition 241, 242
photolysis 241, 243
photolytic deposition 145, 146, 149, 151
photonic 171, 172, 173, 174, 177
photons 171, 174, 175, 176, 177, 250
polycrystalline growth 61, 68
polymer ablation 23
polymer etching 30
pre-nucleation 122
processing 232
prototyping 270
rapid thermal processing 177
reaction enhancement 171
registration 5, 6, 7
repair 265
room temperature processing 121
RTP 177
selected area epitaxy 67, 68
semiconductor etching 23
semiconductor lasers 25
semiconductors 213
silicon 171, 172, 175, 213, 214, 216, 217, 249,
    250, 251, 252, 253, 254, 256, 273, 275, 277
silicon carbide powders 238
silicon dioxide 171, 172, 173, 175, 257, 259, 260,
    261, 263
silicon dioxide growth 171
silicon nitride powders 240
silicon oxidation 171
silicon oxide 171
silicon oxynitride powders 233, 238, 240
silicon powders 235
silicon stripes 265, 268, 269

silicon wafers 137, 139, 140, 142
small-scale production 265
solid phase 249, 250, 251, 252, 255
stripe structures 14
substrate 137
surface 71, 72, 74, 75, 77, 78, 79, 80
surface adsorption 121
surface chemical probes 279
surface degradation 18
surface disorder density 64, 65, 66, 68
temperature 257, 259, 260, 261, 263
thermal oxidation 171
thin films 173, 241, 250
thin metal films 106, 110
time of flight 273, 275
tungsten 55, 56, 57, 58, 164, 168
U.V. Indiced Low Temperature Silicon Layer
    Formation 279
ultra high vacuum 14, 16, 18, 279, 281, 283, 284
ultrafine powder synthesis 240
UV absorption 147, 148
UV lamp 122, 124, 125, 126, 127
UV laser 122, 127, 128
UV light 121
vapour pressure 147, 148, 149, 151
VIA 137, 143
visible laser 122, 124, 125, 126, 127
wavelength dependence 171
X-ray absorption 168
X-ray lithography 163, 164, 165, 166, 168
X-ray masks 131, 135
XeCl 110